# FRICTION AND LUBRICATION IN MECHANICAL DESIGN

# MECHANICAL ENGINEERING

A Series of Textbooks and Reference Books

*Editor*

## L. L. Faulkner

*Columbus Division, Battelle Memorial Institute
and Department of Mechanical Engineering
The Ohio State University
Columbus, Ohio*

# FRICTION AND LUBRICATION IN MECHANICAL DESIGN

## A. A. SEIREG

*University of Wisconsin—Madison*
*Madison, Wisconsin*
*and University of Florida*
*Gainesville, Florida*

CRC Press
Taylor & Francis Group
Boca Raton   London   New York

CRC Press is an imprint of the
Taylor & Francis Group, an **informa** business

CRC Press
Taylor & Francis Group
6000 Broken Sound Parkway NW, Suite 300
Boca Raton, FL 33487-2742

First issued in paperback 2019

ISBN-13: 978-0-8247-9974-8 (hbk)
ISBN-13: 978-0-367-40032-3 (pbk)

**Visit the Taylor & Francis Web site at
http://www.taylorandfrancis.com**

**and the CRC Press Web site at
http://www.crcpress.com**

# Preface

The awareness of friction and attempts to reduce or use it are as old as human history. Scientific study of the friction phenomenon dates back to the eighteenth century and has received special attention in modern times since it is one of the most critical factors in all machinery. The increasing emphasis on material and energy conservation in recent years has added new urgency to the development of practical predictive techniques and information that can be used, in the design stage, for controlling friction and wear. Advances in this field continue to contribute to improved energy efficiency, increased useful life of machines, and reduced maintenance costs.

This book treats friction, lubrication, and wear as empirical phenomena and relies heavily on the experimental studies by the author and his coworkers to develop practical tools for design. Empirical dimensionless relationships are presented, whenever possible, that can be readily applied to a variety of situations confronting the design engineer without the need for extensive theoretical analysis or computation.

The material in the book has been used for many years in an interdisciplinary course on this subject taught at the University of Wisconsin—Madison, and can be used as a text for senior, graduate, or professional development courses. It can also be used as a reference book for practical design engineers because the many empirical equations and design graphs can provide a fundamental parametric understanding to guide their design decisions.

Chapter 1 gives a brief review of the history of this subject and sets the stage for the topics presented in the book. Chapters 2, 3, and 4 summarize the relevant relationships necessary for the analysis of contact mechanics in smooth and rough surfaces, as well as the evaluation of the distribution of the frictional resistance over the contacting surfaces due to the application of tangential loads and twisting moments.

Chapter 5 presents an overview of the mechanism of the transfer of frictional heat between rubbing surfaces and gives equations for estimating the heat partition and the maximum temperature in the contact zone. Chapter 6 deals with the broad aspects of fluid film lubrication with emphasis on the thermal aspects of the problem. It introduces the concept of thermal expansion across the film and provides a method for calculating the pressure that can be generated between parallel surfaces as a result of the thermal gradients in the film. Chapter 7 discusses the problem of friction and lubrication in rolling/sliding contacts and gives empirical equations for calculating the coefficient of friction from the condition of pure rolling to high slide-to-roll ratios. The effect of surface layers is taken into consideration in the analysis. Chapter 8 gives an overview of the different wear mechanisms and includes equations to help the designer avoid unacceptable wear damage under different operating and environmental conditions. Chapter 9 presents selected case illustrations and corresponding empirical equations relating the factors influencing surface durability in important tribological systems such as gears, bearings, brakes, fluid jet cutting, soil cutting, one-dimensional clutches, and animal joints.

Chapter 10 discusses the frictional resistance in micromechanisms. Friction is considered a major factor in their implementation and successful operation. Chapter 11 illustrates the role of friction in the generation of noise in mechanical systems, and Chapter 12 gives an introduction to surface coating technology, an area of growing interest to tribologists.

Finally, Chapter 13 discusses in some detail a number of experimental techniques developed by the author and his coworkers that can be useful in the study of friction, lubrication, wear, surface temperature, and thermally induced surface damage.

I am indebted to my former students, whose thesis research constitutes the bulk of the material in this book. Dr. T. F. Conry for his contribution to Chapter 2; Dr. D. Choi to Chapter 3; Dr. M. Rashid to Chapter 5; Drs. H. Ezzat, S. Dandage, and N. Z. Wang to Chapter 6; Dr. Y. Lin to Chapter 7; Dr. T. F. Conry, Mr. T. Lin, Mr. A. Suzuki, Dr. A. Elbella, Dr. S. Yu, Dr. A. Kotb, Mr. M. Gerath, and Dr. C. T. Chang to Chapter 9; Dr. R. Ghodssi to Chapter 10; Drs. S. A. Aziz and M. Othman to Chapter 11; Dr. K. Stanfill, Mr. M. Unee, and Mr. T. Hartzell to Chapter 12; and Professor E. J. Weiter, Dr. N. Z. Wang, Dr. E. Hsue, and Mr. C. Wang to Chapter 13.

Grateful acknowledgment is also due to Ms. Mary Poupore, who efficiently took charge of typing the text, and to Mr. Joe Lacey, who took upon himself the enormous task of digitizing the numerous illustrations in this book.

*A. A. Seireg*

# Contents

# Unit Conversion Table

1 inch = 25.4 mm = 2.54 cm = 0.0254 meter [m]
1 foot = 0.3048 m
1 mile = 1609 m = 1.609 km
1 lb = 4.48 newton [N] = 0.455 kg
1 lb-ft (moment) = 1.356 N-m
1 lb-ft (work) = 1.356 joule [J]
1 lb-ft/sec = 1/356 watt [W]
1 hp = 0.746 kW
1 lb/in$^2$ (psi) = 6895 pascal [Pa]
1 Btu = 1055 joule [J]
1 centipoise = 0.001 pascal-second [Pa-s]
Degree F = degree C $\times (9/5) + 32$
1 gallon = 3.785 liters = 0.003785 m$^3$
1 quart = 0.946 liter

# FRICTION AND LUBRICATION IN MECHANICAL DESIGN

# 1

# Introduction

## 1.1 HISTORICAL OVERVIEW

The phenomenon of friction has been part of daily life since the beginning of human existence. It is no surprise that some of the earliest human activities involved the reduction of friction when it was wasteful, or the use of friction when it could be beneficial. The first category includes the use of vegetable oils and animal fats as lubricants, as well as the use of rolling motion to take advantage of the resulting low resistance to movement. The second category can be exemplified by the rubbing of twigs to start a fire and the control of motion by braking action.

In most situations, friction is an undesirable phenomenon that should be minimized. It results in hindering movement, wasting effort, generating unwanted heat, and causing wear and damage to the contacting surfaces. It is, however, hard to imagine the world without friction. In a frictionless environment there will be no tractive forces to allow locomotion, gripping, braking, fastening, weaving, and many other situations which are fundamental to human life.

The earliest recorded attempt to reduce friction can be traced back to the 20th century B.C. as illustrated by the temple painting, Figure 1.1, showing a man pouring oil to facilitate the movement of a colossus. Evidence of the use of animal fat on the axles of chariots has also been discovered in Egyptian tombs dating back to the 15th century B.C.

Themistius (390–320 B.C.) observed that rolling friction is much smaller than sliding friction. This is what made the wheel the first major advance in the field of ground transportation.

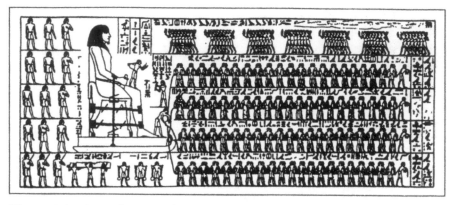

**Figure 1.1** An ancient record dating back to 1900 B.C. showing the use of lubricating oil to reduce friction in moving a colossus.

It was not until the middle ages that Leonardo da Vinci (1452–1519) formulated his basic laws of friction, which provide a predictive rationale for evaluating frictional resistance. He stated that frictional forces are proportional to weight and are independent of the area of contact [1].

Guillaume Amontons (1663–1705) in a paper published in the proceedings of the French Royal Academy of Sciences [2], rediscovered the frictional laws originally proposed by Leonardo da Vinci. The fact that the frictional force was proportional to the normal load was readily accepted by the academy, but the independence of friction on the area of contact was received with skepticism. The senior academician De La Hire (1640–1718) went on to check Amontons' second law and did confirm its validity.

An interesting observation was advanced by John Desaguliers (1683–1744) who pointed out in his book on experimental philosophy, published in 1734, that the frictional resistance between flat metallic bodies may increase as a result of polishing the contacting surfaces. He attributed this to adhesive forces which he called "cohesion." He recognized that such forces exist, but he could not formulate means of accounting for them.

Charles Augustin Coulomb (1736–1806) is generally regarded as the founder of the frictional laws. His understanding of the causes of friction, however, was not completely clear. He recognized the importance of roughness and suggested that friction was due to the work done in dragging one surface up the other. One of the important contributions of Coulomb is his postulation that contact only occurs at the discrete points of asperity contacts. However, he rejected the adhesion theory and reasoned that if adhesion exists, the frictional resistance has to be doubled if the area of contact is doubled. He consequently believed that frictional resistance is due to the work done in moving one surface up the roughness of the other [3].

John Leslie (1766–1832) criticized both the roughness and adhesion theories and believed that friction was due to the work done by deformation of the surface due to roughness. Although these early investigations alluded to the possible mechanism of friction, it took over a century of research to conclude that friction between solids arises from their interaction at the regions where they are in real contact. This is influenced by the geometry of the surfaces, their elastic properties, the adhesive forces at the real contacts and how energy is lost when the surfaces are deformed during sliding.

Friction is generally divided into four regimes: dry, boundary, elastohydrodynamic, and hydrodynamic. In dry friction, surface cleanliness is one of the most important factors influencing the frictional resistance. Even a single molecular layer of grease from the atmosphere or from the fingers may change the coefficient of friction significantly. The influence of surface cleanness is much greater than that of surface roughness. On the other extreme, when the surfaces are separated by a thick film of lubricant, the resistance to movement is determined by the dynamic behavior of the film.

Osborne Reynolds (1842–1912) developed in 1886 the fundamental basis for the hydrodynamic lubrication theory and the frictional resistance [4]. In this case there is no metal-to-metal contact and friction is a result of the shear resistance as influenced by the viscosity of the lubricant and the thickness of the film. Reynolds' analysis was inspired by the experimental findings of Petrov (1836–1920) and Tower (1845–1904). Petrov reported in 1883 [5] that viscosity is the most important fluid property in film lubrication, and not density as previously thought. He also concluded that frictional losses in full film lubrication are the result of viscous shearing of the film.

The experimental studies published by Tower in 1883 and 1885 [6, 7] showed that the load carrying ability of a bearing partially submerged in an oil bath is the result of the high pressures developed in the clearance space between the journal and the sleeve and that the clearance is an essential parameter in achieving full film lubrication and consequently reducing friction in the bearing.

In lubricated concentrated contacts, the pressure in the fluid is usually sufficiently high to deform the solid surfaces. This condition exists in many machine elements such as gears, rolling element bearings, cams and automotive tires on roads covered with water. The analysis of this elastohydrodynamic phenomenon was first investigated by Grubin [8] and Dowson [9, 10] and constitutes an important field of tribology. Both hydrodynamic and elastohydrodynamic friction are highly dependent on speed and the viscosity of the fluid. For low speeds or low viscosity fluids when the lubricating fluid film is not sufficiently thick to separate the asperities on the surface of the contacting solids, the frictional resistance will be much higher than that with full film lubrication but appreciably lower than that for dry surfaces. An early investigation of friction in this regime, which is called boundary lubrication, was undertaken by Sir William Hardy in the early 1920s. His study showed that frictional resistance in the boundary regime is proportional to the normal load. The main advantage of boundary lubrication is to generate a thin fluid film on the surface which reduces the solid-to-solid contacts and consequently reduces friction, wear, and noise [11, 12].

Scientific study of the friction, lubrication, and wear phenomena in all these regimes is now receiving considerable attention in modern engineering. Friction is a primary cause of energy dissipation, and considerable economic savings can be made by better understanding of its mechanism and its control. The operation of most modern engineering systems such as machines, instruments, vehicles and computer hardware, etc. is influenced by the occurrence of friction in some form or another.

Tribology, which is the name currently used to encompass the multitude of activities in this highly interdisciplinary subject, is now attaining a prominent place among the sciences [13]. It continues to present challenges for those who are working in it in response to the ever increasing interest of the mechanical and electronic industries to learn more about the causes of the energy losses due to friction and wear [14]. The enormous energy loss in tribological sinks in the United States is estimated by experts to be $20 billion in 1998.

The emerging technology of micromechanisms is placing new emphasis on tribology on the microscale [15, 16]. Because of their very large surface area to volume ratios, adhesion, friction, surface tension, viscous resistance, and other boundary forces will be the dominant factors which control their design and performance characteristics. Not since the middle ages have tribologists been confronted with new frontiers of such proportions and without precedence in human experience. New challenges are now presenting themselves on how to model, predict, and measure these forces. Understanding friction on the microscale will be the most critical element in the useful utilization of micromechanisms.

## 1.2 THEORIES OF DRY FRICTION

The classical theory of dry friction has been discussed by many workers (e.g., Moore [17] and Rabinowicz [18]). The classical friction laws can be summarized as follows:

1. In any situation where the resultant of tangential forces is smaller than some force parameter specific to that particular situation, the friction force will be equal and opposite to the resultant of the applied forces and no tangential motion will occur.
2. When tangential motion occurs, the friction force always acts in a direction opposite to that of the relative velocity of the surfaces.
3. The friction force is proportional to the normal load.
4. The coefficient of friction is independent of the apparent contact area.
5. The static coefficient is greater than the kinetic coefficient.
6. The coefficient of friction is independent of sliding speed.

Strictly speaking, none of these laws is entirely accurate. Moore indicated that laws (3), (4), (5), and (6) are reasonably valid for dry friction under the following conditions:

For law (3), the normal load is assumed to be low compared to that causing the real area of contact to approach the apparent area.
For law (4), the materials in contact are assumed to have a definite yield point (such as metals). It does not apply to elastic and viscoelastic materials.
Law (5) does not apply for materials with appreciable viscoelastic properties.
Law (6) is not valid for most materials, especially for elastomers where the viscoelastic behavior is very significant.

A number of workers also found some exceptions to the first friction law. Rabinowicz [18] reported that Stevens [19], Rankin [20], and Courtney-Pratt and Eisner [21] had shown that when the tangential force $F$ is first applied, a very small displacement occurs almost instantaneously in the direction of $F$ with a magnitude in the order of $10^{-5}$ or $10^{-6}$ cm.

Seireg and Weiter [22] conducted experiments to investigate the load–displacement and displacement–time characteristics of friction contacts of a ball between two parallel flats under low rates of tangential load application. The tests showed that the frictional joint exhibited "creep" behavior at room temperatures under loads below the gross slip values which could be described by a Boltzmann model of viscoelasticity.

They also investigated the frictional behaviors under dynamic excitation [23, 24]. They found that under sinusoidal tangential forces the "breakaway" coefficient of friction was the same as that determined under static conditions. They also found that the static coefficient of friction in Hertzian contacts was independent of the area of contact, the magnitude of the normal force, the frequency of the oscillatory tangential load, or the ratio of the static and oscillatory components of the tangential force. However, the coefficient of gross slip under impulsive loading was found to be approximately three times higher than that obtained under static or a vibratory load at a frequency of 100 Hz using the same test fixture.

Rabinowicz [25] developed a chart based on a compatibility theory which states that if two metals form miscible liquids and, after solidification, form solid solutions or intermetallic components, the metals are said to be compatible and the friction and wear between them will be high. If, however, they are insoluble in each other, the friction and wear will be low. Accordingly two materials with low compatibility can be selected from the chart to produce low friction and wear.

In the case of lubricated surfaces, Rabinowicz [26] found that the second law of friction was not obeyed. It was found that the direction of the instantaneous frictional force might fluctuate by one to three degrees from the expected direction, changing direction continuously and in a random fashion as sliding proceeded.

The general mechanisms which have been proposed to explain the nature of dry friction are reviewed in numerous publications (e.g., Moore [17]). The following is a summary of the concepts on which dry friction theories are based:

*Mechanical interlocking*. This was proposed by Amontons and de la Hire in 1699 and states that metallic friction can be attributed to the mechanical interlocking of surface roughness elements. This theory gives an explanation for the existence of a static coefficient of friction, and explains dynamic friction as the force required to lift the asperities of the upper surface over those of the lower surface.

*Molecular attraction*. This was proposed by Tomlinson in 1929 and Hardy in 1936 and attributes frictional forces to energy dissipation when the atoms of one material are "plucked" out of the attraction range of their counterparts on the mating surface. Later work attributed adhesional friction to a molecular–kinetic bond rupture process in which energy is dissipated by the stretch, break, and relaxation cycle of surface and subsurface molecules.

*Electrostatic forces*. This mechanism was presented in 1961 and explains the stick–slip phenomena between rubbing metal surfaces by the initiation of a net flow of electrons.

*Welding, shearing and ploughing*. This mechanism was proposed by Bowden in 1950. It suggests that the pressure developed at the discrete contact spots causes local welding. The functions thus formed are subsequently sheared by relative sliding of the surfaces. Ploughing by the asperities of the harder surface through the matrix of the softer material contributes the deformation component of friction.

Dry rolling friction was first studied by Reynolds [27] in 1875. He found that when a metal cylinder rolled over a rubber surface, it moved forward a distance less than its circumference in each revolution of the cylinder. He assumed that a certain amount of slip occurred between the roller and the rubber and concluded that the occurrence of this slip was responsible for the rolling resistance.

Palmgren [28] and Tabor [29] later repeated Reynolds' experiment in more detail and found that the physical mechanism responsible for rolling friction was very different in nature than that suggested by Reynolds. Tabor's experiments showed that interfacial slip between a rolling element and an elastic surface was in reality almost negligible and in any case quite inadequate to account for the observed friction losses. Thus he concluded that rolling resistance arose primarily from elastic–hysteresis losses in the material of the rolling element and the surface.

## 1.3 BOUNDARY LUBRICATION FRICTION

Hardy [30] first used the term "boundary lubrication" to describe the surface frictional behavior of certain organic compounds derived from petroleum products of natural origin such as paraffins, alcohols and fatty acids. Since then, boundary lubrication has been extended to cover other types of lubricants, e.g., surface films and solid mineral lubricants, which do not function hydrodynamically and are extensively used in lubrication.

In the analysis of scoring of gear tooth surfaces, it has been fairly well established that welding occurs at a critical temperature which is reached by frictional heating of the surfaces. The method of calculating such a temperature was published by Blok [31], and his results were adapted to gears in 1952 [32]. Since then, some emphasis has been focused on boundary lubrication. Several studies are available in the literature which deal with the boundary lubrication condition; some of them are briefly reviewed in the following.

Sharma [33] used the Bowden–Leben apparatus to investigate the effects of load and surface roughness on the frictional behavior of various steels over a range of temperature and of additive concentration. The following observations are reported:

Sharp rise in friction can occur but is not necessarily followed by scuffing and surface damage.
Load affects the critical temperature quite strongly.

Neither the smoother surface nor the rougher surface gives the maximum absorption of heat, but there exists an optimum surface roughness.

Nemlekar and Cheng [34] investigated the traction in rough contacts by solving the partial elastohydrodynamic lubrication (EHL) equations. It was found that traction approached dry friction as the ratio of lubricant film thickness to surface roughness approached zero, load had a great influence on friction, and the roller radius had little influence on friction.

Hirst and Stafford [35] examined the factors which influence the failure of the lubricant film in boundary lubrication. It is shown that substantial damage only occurs when a large fraction of the load becomes unsupported by hydrodynamic action. It is also shown that the magnitude of the surface deformation under the applied load is a major factor in breakdown. When the deformation is elastic, the solid surface film (e.g., oxide) remains intact and even a poor liquid lubricant provides sufficient protection against the buildup of the damage. The transition temperature is also much lower. They also discussed the effect of load and surface finish on the transition temperature.

Furey and Appeldoorn [36] conducted an experiment to study the effect of lubricant viscosity on metallic contact and friction in the transition zone between hydrodynamic and boundary lubrication. The system used was one of pure sliding and relatively high contact stress, namely, a fixed steel ball on a rotating steel cylinder. It was found that increasing the viscosity of Newtonian fluids (mineral oils) over the range 2–1100 centipoises caused a decrease in metallic contact. The effect became progressively more pronounced at higher viscosities. The viscosity here was the viscosity at atmospheric pressure and at the test temperature; neither pressure–viscosity nor temperature–viscosity properties appeared to be important factors. On the other hand, non-Newtonian fluids (polymer-thickened oils) gave more contact than their mineral oil counterparts. This suggested that shear–viscosity was important. However, no beneficial effects of viscoelastic properties were observed with these oils. Friction generally decreased with increasing viscosity because the more viscous oils gave less metal-to-metal contact. The coefficient of friction was rather high: 0.13 at low viscosity, dropping to 0.08 at high viscosity. The oils having higher PVIs (pressure–viscosity index $\alpha$) gave somewhat more friction which cannot be solely attributed to differences in metallic contact.

Furey [37] also investigated the surface roughness effects on metallic contact and friction in the transition zone between the hydrodynamic and boundary lubrications. He found that very smooth and very rough surfaces gave less metallic contact than surfaces with intermediate roughness. Friction was low for the highly polished surfaces and rose with increasing surface roughness. The rise in friction continued up to a roughness of about 10 μin, the same general level at which metallic contact stopped increasing. However, whereas further increases in surface roughness caused a reduction in metallic contact, there was no significant effect on friction. Friction was found to be independent of roughness in the range of 10 μin center line average (CLA). He also used four different types of antiwear/antifriction additives (including tricresyl phosphate) and found that they reduced metallic contact and friction but had little effect on reducing surface roughness. The additives merely slowed down the wear-in process of the base oil. Thus he concluded that the "chemical polishing" mechanism proposed for explaining the antiwear behavior of tricresyl phosphate appeared to be incorrect.

Freeman [38] studied several experimental results and summarized them as follows:

An unnecessarily thick layer of boundary lubricant may give rise to excessive frictional resistance because shearing and ploughing of the lubricant film may become factors of importance.

The bulk viscosity of a fluid lubricant appears to have no significance in its boundary frictional behavior.

Coefficients of friction for effective boundary lubricants lie roughly in the range 0.02 to 0.1.

The friction force is almost independent of the sliding velocity, provided the motion is insufficient to cause a rise in bulk temperature. If the motion is intermittent or stick–slip, the frictional behavior may be of a different and unpredictable nature.

In general the variables that influence dry friction also influence boundary friction.

Friction and surface damage depends on the chemical composition of the lubricant and/or the products of reaction between the lubricant and the solid surface.

Lubricant layers only a few molecules thick can provide effective boundary lubrication.

The frictional behavior may be influenced by surface roughness, temperature, presence of moisture, oxygen or other surface contaminants. In general, the coefficient of friction tends to increase with surface roughness.

## 1.4 FRICTION IN FLUID FILM LUBRICATION

Among the early investigations in fluid film lubrication, Tower's experiments in 1883–1884 were a breakthrough which led to the development of lubrication theory [6, 7]. Tower reported the results of a series of experiments intended to determine the best methods to lubricate a railroad journal bearing. Working with a partial journal bearing in an oil bath, he noticed and later measured the pressure generated in the oil film.

Tower pointed out that without sufficient lubrication, the bearing operates in the boundary lubrication regime, whereas with adequate lubrication the two surfaces are completely separated by an oil film. Petrov [5] also conducted experiments to measure the frictional losses in bearings. He concluded that friction in adequately lubricated bearings is due to the viscous shearing of the fluid between the two surfaces and that viscosity is the most important property of the fluid, and not density as previously assumed. He also formulated the relationship for calculating the frictional resistance in the fluid film as the product of viscosity, speed, and area, divided by the thickness of the film.

The observations of Tower and Petrov proved to be the turning point in the history of lubrication. Prior to their work, researchers had concentrated their efforts on conducting friction drag tests on bearings. From Tower's experiments, it was realized that an understanding of the pressure generated during the bearing operation is the key to perceive the mechanism of lubrication. The analysis of his work carried out by Stokes and Reynolds led to a theoretical explanation of Tower's experimental results and to the theory of hydrodynamic lubrication.

In 1886, Osborne Reynolds published a paper on lubrication theory [4] which is derived from the equations of motion, continuity equation, and Newton's shear stress–velocity gradient relation. Realizing that the ratio of the film thickness to the bearing geometry is in the order of $10^{-3}$, Reynolds established the well-known theory using an order-of-magnitude analysis. The assumptions on which the theory is based can be listed as follows.

The pressure is constant across the thickness of the film.
The radius of curvature of bearing surface is large compared with film thickness.
The lubricant behaves as a Newtonian fluid.
Inertia and body forces are small compared with viscous and pressure terms in the equations of motion.
There is no slip at the boundaries.
Both bearing surfaces are rigid and elastic deformations are neglected.

Since then the hydrodynamic theory based on Reynolds' work has attracted considerable attention because of its practical importance. Most initial investigations assumed isoviscous conditions in the film to simplify the analysis. This assumption provided good correlation with pressure distribution under a given load but generally failed to predict the stiffness and damping behavior of the bearing.

A model which predicts bearing performance based on appropriate thermal boundaries on the stationary and moving surfaces and includes a pointwise variation of the film viscosity with temperature is generally referred to as the thermohydrodynamic (THD) model. The THD analyses in the past three decades have drawn considerable attention to the thermal aspects of lubrication. Many experimental and theoretical studies have been undertaken to shed some light on the influence of the energy generated in the film, and the heat transfer within the film and to the surroundings, on the generated pressure.

In 1929, McKee and McKee [39] performed a series of experiments on a journal bearing. They observed that under conditions of high speed, the viscosity diminished to a point where the product of viscosity and rotating speed is a constant. Barber and Davenport [40] investigated friction in several journal bearings. The journal center position with respect to the bearing center was determined by a set of dial indicators. Information on the load-carrying capacity and film pressure was presented.

In 1946, Fogg [41] found that parallel surface thrust bearings, contrary to predictions by hydrodynamic theory, are capable of carrying a load. His experiments demonstrated the ability of thrust bearings with parallel surfaces to carry loads of almost the same order of magnitude as can be sustained by tilting pad thrust bearings with the same bearing area. This observation, known as the Fogg effect, is explained by the concept of the "thermal wedge," where the expansion of the fluid as it heats up produces a distortion of the velocity distribution curves similar to that produced by a converging surface, developing a load-carrying capacity. Fogg also indicated that this load-carrying ability does not depend on a round inlet edge nor the thermal distortion of the bearing pad. Cameron [42], in his experiments with rotating disks, suggested that a hydrodynamic pressure is created in the film between the disks due to the variation of viscosity across the thickness of the film. Viscoelastic lubricants in journal bearing applications were studied by Tao and Phillipoff [43]. The non-Newtonian behavior of viscoelastic liquids causes a flattening in the pressure profile and a shift of the peak film pressure due to the presence of normal stresses in the lubricant. Dubois et al. [44] performed an experimental study of friction and eccentricity ratios in a journal bearing lubricated with a non-Newtonian oil. They found that a non-Newtonian oil shows a lower friction than a corresponding Newtonian fluid under the same operating conditions. However, this phenomenon did not agree with their analytical work and could not be explained.

Maximum bearing temperature is an important parameter which, together with the minimum film thickness, constitutes a failure mechanism in fluid film bearings. Brown and Newman [45] reported that for lightly loaded bearings of diameter 60 in. operating under 6000 rpm, failure due to overheating of the bearing material (babbitt) occurred at about 340°F. Booser et al. [46] observed a babbitt-limiting maximum temperature in the range of 266 to 392°F for large steam turbine journal bearings. They also formulated a one-dimensional analysis for estimating the maximum temperature under both laminar and turbulent conditions.

In a study of heat effects in journal bearings, Dowson et al. [47] in 1966 conducted a major experimental investigation of temperature patterns and heat balance of steadily loaded journal bearings. Their test apparatus was capable of measuring the pressure distribution, load, speed, lubricant flow rate, lubricant inlet and outlet temperatures, and temperature distribution within the stationary bushing and rotating shaft. They found that the heat flow patterns in the bushing are a combination of both radial flows and a significant amount of circumferential flow traveling from the hot region in the vicinity of the minimum film thickness to the cooler region near the oil inlet. The test results showed that the cyclic variation in shaft surface temperature is small and the shaft can be treated as an isothermal component. The experiments also indicated that the axial temperature gradients within the bushing are negligible.

Viscosity is generally considered to be the single most important property of lubricants, therefore, it represents the central parameter in recent lubricant analyses. By far the easiest approach to the question of viscosity variation within a fluid film bearing is to adopt a representative or mean value viscosity. Studies have provided many suggestions for calculations of the effective viscosity in a bearing analysis [48]. When the temperature rise of the lubricant across the bearing is small, bearing performance calculations are customarily based on the classical, isoviscous theory. In other cases, where the temperature rise across the bearing is significant, the classical theory loses its usefulness for performance prediction. One of the early applications of the energy equation to hydrodynamic lubrication was made by Cope [49] in 1948. His model was based on the assumptions of negligible temperature variation across the film and negligible heat conduction within the lubrication film as well as into the neighboring solids. The consequence of the second assumption is that both the bearing and the shaft are isothermal components, and thus all the generated heat is carried out by the lubricant. As indicated in a review paper by Szeri [50], the belief, that the classical theory on one hand and Cope's adiabatic model on the other, bracket bearing performance in lubrication analysis, was widely accepted for a while. A thermohydrodynamic hypothesis was later introduced by Seireg and Ezzat [51] to rationalize their experimental findings.

An empirical procedure for prediction of the thermohydrodynamic behavior of the fluid film was proposed in 1973 by Seireg and Ezzat. This report presented results on the load-carrying capacity of the film from extensive tests. These tests covered eccentricity ratios ranging from 0.6 to 0.90, pressures of up to 750 psi and speeds of up to 1650 ft/min. The empirical procedure applied to bearings submerged in an oil bath as well as to pump-fed bearings where the outer shell is exposed to the atmosphere. No significant difference in the speed–pressure characteristics for these two conditions was observed when the inlet temperature was the same. They showed that the magnitudes of the load-carrying capacity obtained experimentally may differ considerably from those predicted by the insoviscous hydrodynamic theory. The isoviscous theory can either underestimate or overestimate the results depending on the operating conditions. It was observed, however, that the normalized pressure distribution in both the circumferential and axial directions of the journal bearing are almost identical to those predicted by the isoviscous hydrodynamic theory. Under all conditions tested, the magnitude of the peak pressure (or the average pressure) in the film is approximately proportional to the square root of the rotational speed of the journal. The same relationship between the peak pressure and speed was observed by Wang and Seireg [52] in a series of tests on a reciprocating slider bearing with fixed film geometry. A comprehensive review of thermal effects in hydrodynamic bearings is given by Khonsari [53] and deals with both journal and slider bearings.

In 1975, Seireg and Doshi [54] studied nonsteady state behavior of the journal bearing performance. The transient bushing temperature distribution in journal bearing appears to be similar to the steady-state temperature distribution. It was also found that the maximum bushing surface temperature occurs in the vicinity of minimum film thickness. The temperature level as well as the circumferential temperature variation were found to rise with an increase of eccentricity ratio and bearing speed. Later, Seireg and Dandage [55] proposed an empirical thermohydrodynamic procedure to calculate a modified Sommerfeld number which can be utilized in the standard formula (based on the isoviscous theory) to calculate eccentricity ratio, oil flow, frictional loss, and temperature rise, as well as stiffness and damping coefficients for full journal bearings.

In 1980, Barwell and Lingard [56] measured the temperature distribution of plain journal bearings, and found that the maximum bearing temperature, which is encountered at the point of minimum film thickness, is the appropriate value for an estimate of effective viscosity to be used in load capacity calculation. Tonnesen and Hansen [57] performed an experiment on a cylindrical fluid film bearing to study the thermal effects on the bearing performance. Their test bearings were cylindrical and oil was supplied through either one or two holes or through two-axial grooves, 180° apart. Experiments were conducted with three types of turbine oils. Both viscosity and oil inlet geometry were found to have a significant effect on the operating temperatures. The shaft temperature was found to increase with increasing loads when a high-viscosity lubricant was used. At the end of the paper, they concluded that even a simple geometry bearing exhibits over a broad range small but consistent discrepancies when correlated with existing theory. In 1983, Ferron et al. [58] conducted an experiment on a finite-length journal bearing to study the performance of a plain bearing. The pressure and the temperature distributions on the bearing wall were measured, along with the eccentricity ratio and the flow rate, for different speeds and loads. All measurements were performed under steady-state conditions when thermal equilibrium was reached. Good agreement was found with measurements reported for pressure and temperature, but a large discrepancy was noted between the predicted and measured values of eccentricity ratios. In 1986, Boncompain et al. [59] showed good agreement between their theoretical and experimental work on a journal bearing analysis. However, the measured journal locus and calculated values differ. They concluded that the temperature gradient across and along the fluid film is the most important parameter when evaluating the bearing performance.

## 1.5 FRICTIONAL RESISTANCE IN ELASTOHYDRODYNAMIC CONTACTS

In many mechanical systems, load is transmitted through lubricated concentrated contacts where rolling and sliding can occur. For such conditions the pressure is expected to be sufficiently high to cause appreciable deformation of the contacting bodies and consequently the surface geometry in the loaded area is a function of the generated pressure. The study of the behavior of the lubricant film with consideration of the change of film geometry due to the elasticity of the contacting bodies has attracted considerable attention from tribologists over the last half century. Some of the studies related to frictional resistance in this elastohydrodynamic (EHD) regime are briefly reviewed in the following with emphasis on effect of viscosity and temperature in the film.

Dyson [60] interpreted some of the friction results in terms of a model of viscoelastic liquid. He divided the experimental curves of frictional traction versus sliding speed into three regions: the linear region, the nonlinear (ascending) region, and the thermal (descending) region. At low sliding speeds a linear relation exists, the slope of which defines a quasi-Newtonian viscosity, and the behavior is isothermal. At high sliding speeds the frictional force decreases as sliding speed increases, and this can be attributed to some extent to the influence of temperature on viscosity. In the transition region, thermal effects provide a totally inadequate explanation because the observed frictional traction may be several orders of magnitude lower than the calculated values even when temperature effects are considered.

Because of the high variation of pressure and temperature, many parameters such as temperature, load, sliding speed, the ratio of sliding speed to rolling speed, viscosity, and surface roughness have great effects on frictional traction.

Thermal analysis in concentrated contacts by Crook [61, 62], Cheng [63], and Dyson [60] have shown a strong mutual dependence between temperature and friction in EHD contacts. Frictional traction is directly governed by the characteristics of the lubricant film, which, in the case of a sliding contact, depends strongly on the temperature in the contact. The temperature field is in turn governed directly by the heating function.

Crook [61] studied the friction and the temperature in oil films theoretically. He used a Newtonian liquid (shear stress proportional to the velocity gradient in the film) and an exponential relation between viscosity and temperature and pressure. In pure rolling of two disks it has been found that there is no temperature rise within the pressure zone; the temperature rise occurs on the entry side ahead of that zone. When sliding is introduced, it has been found that the temperature on the entry side remains small, but it does have a very marked influence upon the temperatures within the pressure zone, for instance, the introduction of 400 cm/sec sliding causes the effective viscosity to fall in relation to its value in pure rolling by a factor of 50. It has also been shown that at high sliding speeds the effective viscosity is largely independent of the viscosity of oil at entry conditions. This fact carries the important implication that if an oil of higher viscosity is used to give the surfaces greater protection by virtue of a thicker oil film, then there is little penalty to be paid by way of greater frictional heating, and in fact at high sliding speeds the frictional traction may be lower with the thicker film. It has also been found that frictional tractions pass through a maximum as the sliding is increased. This implies that if the disks were used as a friction drive and the slip was allowed to exceed that at which the maximum traction occurs, then a demand for a greater output torque, which would lead to even greater sliding, would reduce the torque the drive can deliver.

Crook [62] conducted an experiment to prove his theory, and found that the effective viscosity of the oil at the rolling point showed that the variation of viscosity, both for changes in pressure and in temperature, decreased as the rolling speed was increased.

Cheng [63] studied the thermal EHD of rolling and sliding cylinders with a more rigorous analysis of temperature by using a two-dimensional numerical method. The effect of the local pressure–temperature-dependent viscosity, the compressibility of the lubricant, and the heat from compression of the lubricant were considered in the analysis. A Newtonian liquid was used. He found that the temperature had major influence on friction force. A slight change in temperature–viscosity exponent could cause great changes in friction data. He also compared his theoretical results with Crook's [62] experimental results and found a high theoretical value at low sliding speed. Thus he concluded that the assumption of a Newtonian fluid in the vicinity of the pressure peak might cease to be valid.

One of the most important experimental studies in EHD was carried out by Johnson and Cameron [64]. In their experiments they found that at high sliding speeds the friction coefficient approached a common ceiling, which was largely independent of contact pressure, rolling speed and disk temperature. At high loads and sliding speeds variations in rolling speed, disk temperature and contact pressure did not appear to affect the friction coefficient. Below the ceiling the friction coefficient increased with pressure and decreased with increasing rolling speed and temperature.

Dowson and Whitaker [65] developed a numerical procedure to solve the EHD problem of rolling and sliding contacts lubricated by a Newtonian fluid. It was found that sliding caused an increase in the film temperatures within the zone, and the temperature rise was roughly proportional to the square of the sliding velocity. Thermal effects restrained the coefficient of friction from reaching the high values which would occur in sliding contacts under isothermal conditions.

Plint [66] proposed a formula for spherical contacts which relates the coefficient of friction with the temperature on the central plane of the contact zone and the radius of the contact zone.

There are other parameters which were investigated for their influence on the frictional resistance in the EHD regime by many tribologists [67–85]. Such parameters include load, rolling speed, shear rate, surface roughness, etc. The results of some of these investigations are utilized in Chapter 7 for developing generalized emperical relationships for predicting the coefficient of friction in this regime of lubrication.

# REFERENCES

1. MacCurdy, E., Leonardo da Vinci Notebooks, Jonathan Cape, London, England, 1938.
2. Amontons, G., Histoire de l'Académie Royale des Sciences avec Les Mémoires de Mathématique et de Physique, Paris, 1699.
3. Coulomb, C. A., Mémoires de Mathématique et de Physique de l'Académie Royale des Sciences, Paris, 1785.
4. Reynolds, O., "On the Theory of Lubrication and Its Application to Mr. Beauchamp Tower's Experiments Including an Experimental Determination of Olive Oil," Phil. Trans., 1886, Vol. 177(i), pp. 157–234.
5. Petrov, N. P. "Friction in Machines and the Effect of the Lubricant," Inzh. Zh., St. Petersburg, Russia, 1883, Vol. 1, pp. 71–140; Vol. 2, pp. 227–279; Vol. 3, pp. 377–436; Vol. 4, pp. 435–464 (in Russian).
6. Tower, B., "First Report on Friction Experiments (Friction of Lubricated Bearings)," Proc. Inst. Mech. Engrs, 1883, pp. 632–659.
7. Tower, B., "Second Report on Friction Experiments (Experiments on Oil Pressure in Bearings)," Proc. Inst. Mech. Engrs, 1885, pp. 58–70.
8. Grubin, A. N., Book No. 30, English Translation DSIR, 1949.
9. Dowson, D., and Higginson, G. R., Elastohydrodynamic Lubrication, Pergamon, New York, NY, 1966.
10. Dowson, D., History of Tribology, Longman, New York, NY, 1979.
11. Bowden, F. P., and Tabor, D., The Friction and Lubrication of Solids, Oxford University Press, New York, NY, 1950.
12. Pinkus, O., "The Reynolds Centennial: A Brief History of the Theory of Hydrodynamic Lubrication," ASME J. Tribol., 1987, Vol. 109, pp. 2–20.
13. Pinkus, O., Thermal Aspects of Fluid Film Tribology, ASME Press, pp. 126–131, 1990.
14. Ling, F. F., Editor, "Wear Life Prediction in Mechanical Components," Industrial Research Institute, New York NY, 1985.
15. Suzuki, S., Matsuura, T., Uchizawa, M., Yura, S., Shibata, H., and Fujita, H., "Friction and Wear Studies on Lubricants and Materials Applicable MEMS," Proc. of the IEEE Workshop on MicroElectro Mechanical Systems (MEMS), Nara, Japan, Feb. 1991.
16. Ghodssi, R., Denton, D. D., Seireg, A. A., and B. Howland, "Rolling Friction in Linear Microactuators," JVSA, Aug. 1993.
17. Moore, A.J., Principles and Applications of Tribology, Pergamon Press, New York, NY, 1975.
18. Rabinowicz, E., Friction and Wear of Materials, John Wiley & Sons, New York, NY, 1965.
19. Stevens, J. S., "Molecular Contact," Phys. Rev., 1899, Vol. 8, pp. 49–56.
20. Rankin, J. S., "The Elastic Range of Friction," Phil. Mag., 7th Ser., 1926, pp. 806–816.
21. Courtney-Pratt, J. S., and Eisner, E., "The Effect of a Tangential Force on the Contact of Metallic Bodies," Proc. Roy. Soc. 1957, Vol. A238, pp. 529–550.

22. Seireg, A., and Weiter, E. J., "Viscoelastic Behavior of Frictional Hertzian Contacts Under Ramp-Type Loads," Proc. Inst. Mech. Engrs, 1966–67, Vol. 181, Pt. 30, pp. 200–206.

23. Seireg, A. and Weiter, E. J., "Frictional Interface Behavior Under Dynamic Excitation," Wear, 1963, Vol. 6, pp. 66–77.

24. Seireg, A. and Weiter, E. J., "Behavior of Frictional Hertzian Contacts Under Impulsive Loading," Wear, 1965, Vol. 8, pp. 208–219.

25. "Designing for Zero Wear – Or a Predictable Minimum," Prod. Eng., August 15, 1966, pp. 41–50.

26. Rabinowicz, E., "Variation of Friction and Wear of Solid Lubricant Films with Film Thickness," ASLE Trans., Vol. 10, n.1, 1967, pp. 1–7.

27. Reynolds, O., Phil. Trans., 1875, p. 166.

28. Palmgren, A., "Ball and Roller Bearing Engineering," S.H. Burbank, Philadelphia, PA, 1945.

29. Tabor, D., "The Mechanism of Rolling Friction," Phil. Mag., 1952, Vol. 43, p. 1066; 1954, Vol. 45, p. 1081.

30. Hardy, W. B., Collected Scientific Papers, Cambridge University Press, London, 1936.

31. Blok, H. "Surface Temperature Under Extreme Pressure Lubrication Conditions," Congr. Mondial Petrole, 2me Cogr., Paris, 1937, Vol. 3(4), pp. 471–486.

32. Kelley, B. W., "A New Look at the Scoring Phenomena of Gears," SAE Trans., 1953, Vol. 61, p. 175.

33. Sharma, J. P., and Cameron, A., "Surface Roughness and Load in Boundary Lubrication," ASLE Trans., Vol. 16(4), pp. 258–266.

34. Nemlekar, P. R., and Cheng, H. S., "Traction in Rough Elastohydrodynamic Contacts," Surface Roughness Effects in Hydrodynamic and Mixed Lubrication", The Winter Annual Meeting of ASME, 1980.

35. Hirst, W., and Stafford, J. V., "Transion Temperatures in Boundary Lubrication," Proc. Instn. Mech. Engrs, 1972, Vol. 186(15/72), 179.

36. Furey, M. J., and Appeldoorn, J. K., "The Effect of Lubricant Viscosity on Metallic Contact and Friction in a Sliding System," ASLE Trans. 1962, Vol. 5, pp. 149–159.

37. Furey, M. J., "Surface Roughness on Metallic Contact and Friction," ASLE Trans., 1963, Vol. 6, pp. 49–59.

38. Eng, B. and Freeman, P., Lubrication and Friction, Pitman, New York, NY, 1962.

39. McKee, S. A. and McKee, T. R., "Friction of Journal Bearing as Influenced by Clearance and Length," ASME Trans., 1929, Vol. 51, pp. 161-171.

40. Barber, E. and Davenport, C., "Investigation of Journal Bearing Performance," Penn. State Coll. Eng. Exp. Stat. Bull., 1933, Vol. 27(42).

41. Fogg, A., "Fluid Film Lubrication of Parallel Thrust Surfaces," Proc. Inst. Mech. Engrs, 1946, Vol. 155, pp. 49–67.

42. Cameron, A., "Hydrodynamic Lubrication of Rotating Disk in Pure Sliding, New Type of Oil Film Formation," J Inst. Petrol., Vol. 37, p. 471.

43. Tao, F. and Phillipoff, W., "Hydrodynamic Behavior of Viscoelastic Liquids in a Simulated Journal Bearing," ASLE Trans., 1967, Vol. 10(3), p. 307.
44. Dubois, G., Ocvrik, F., and Wehe, R., "Study of Effect of a Newtonian Oil on Friction and Eccentricity Ratio of a Plain Journal Bearing," NASA Tech. Note, D-427, 1960.
45. Brown, T., and Newman, A., "High-Speed Highly Loaded Bearings and Their Development," Proc. Conf. on Lub. and Wear, Inst. Mech. Engrs., 1957.
46. Booser et al. "Performance of Large Steam Turbine Journal Bearings," ASLE Trans., Vol. 13, n.4, Oct. 1970, pp. 262–268. Also, "Maximum Temperature for Hydrodynamic Bearings Under Steady Load," Lubric. Eng., Vol. 26, n.7, July 1970, pp. 226–235.
47. Dowson, D., Hudson, J., Hunter, B., and March, C., "An Experimental Investigation of the Thermal Equilibrium of Steadily Loaded Journal Bearings," Proc. Inst. Mech. Engrs, 1966–67, Vol. 101, 3B.
48. Cameron, A., The Principles of Lubrication, Longmans Green & Co., London, England, 1966.
49. Cope, W., "The Hydrodynamic Theory of Film Lubrication," Proc. Roy. Soc., 1948, Vol. A197, pp. 201–216.
50. Szeri, A. Z., "Some Extensions of the Lubrication Theory of Osborne Reynolds," J. of Tribol., 1987, pp. 21–36.
51. Seireg, A. and Ezzat, H., "Thermohydrodynamic Phenomena in Fluid Film Lubrication," J. Lubr. Technol., 1973, pp. 187–194.
52. Wang, N. Z. and Seireg, A., Experimental Investigation in the Performance of the Thermohydrodynamic Lubrication of Reciprocating Slider Bearing, ASLE paper No. 87-AM-3A-3, 1987.
53. Khonsari, M. M., "A Review of Thermal Effects in Hydrodynamic Bearings, Part I: Slider and Thrust Bearings," ASLE Trans., 1986, Vol. 30, pp. 19–25.
54. Seireg, A., and Doshi, R. C., "Temperature Distribution in the Bush of Journal Bearings During Natural Heating and Cooling," Proceedings of the JSLE-ASLE International Lubrication Conference, Tokyo, 1975, pp. 194–201.
55. Seireg, A., and Dandage S., "Empirical Design Procedure for the Thermohydrodynamic Behavior of Journal Bearings," ASME J. Lubr. Technol., 1982, pp. 135–148.
56. Barwell, F.T., and Lingard, S., "The Thermal Equilibrium of Plain Journal Bearings," Proceedings of the 6th Leeds–Lyon Symposium on Tribology, Dowson, D. et al., Editors, 1980, pp. 24–33.
57. Tonnesen, J., and Hansen, P. K., "Some Experiments on the Steady State Characteristics of a Cylinderical Fluid-Film Bearing Considering Thermal Effects," ASME J. Lubr. Technol., 1981, Vol. 103, pp. 107–114.
58. Ferron, J., Frene, J., and Boncompain, R., "A Study of the Thermohydrodynamic Performance of a Plain Journal Bearing, Comparison Between Theory and Experiments," ASME J. Lubr. Technol., 1983, Vol. 105, pp. 422–428.
59. Boncompain, R., Fillon, M., and Frene, J., "Analysis of Thermal Effects in Hydrodynamic Bearings," J. Tribol., 1986, Vol. 108, pp. 219–224.

60. Dyson, A., "Frictional Traction and Lubricant Rheology in Elastohydrodynamic Lubrication," Phil. Trans. Roy. Soc., Lond., 1970, Vol. 266(1170), pp. 1–33.

61. Crook, A. W., "The Lubrication of Rollers," Phil. Trans. Roy. Soc., Lond., 1961, Vol. A254, p. 237.

62. Crook, A. W., "The Lubrication of Rollers," Phil. Trans. Roy. Soc., Lond., 1963, Vol. A255, p. 281.

63. Cheng, H.S., "A Refined Solution to the Thermal Elastohydrodynamic Lubrication of Rolling and Sliding Cylinders," ASLE Trans., 1965, Vol. 8, pp. 397–410.

64. Johnson, K. L., and Cameron, R., "Shear Behavior of Elastohydrodynamic Oil Films at High Rolling Contact Pressures," Proc. Inst. Mech. Engrs, 1967–68, Vol. 182, Pt. 1, No. 14.

65. Dowson, D., and Whitaker, A. V., "A Numerical Procedure for the Solution of the Elastohydrodynamic Problem of Rolling and Sliding Contacts Lubricated by a Newtonian Fluid," Proc. Inst. Mech. Engrs, 1965–66, Vol. 180, Pt. 3B, p. 57.

66. Plint, M. A., "Traction in Elastohydrodynamic Contacts," Proc. Inst. Mech. Engrs, 1967–68, Vol. 182, Pt. 1, No. 14, p. 300.

67. O'Donoghue, J. P., and Cameron, A., "Friction and Temperature in Rolling Sliding Contacts," ASLE Trans., 1966, Vol. 9, pp. 186–194.

68. Benedict, G. H., and Kelley, B. W., "Instaneous Coefficients of Gear Tooth Friction," ASLE Trans., 1961, Vol. 4, pp. 59–70.

69. Misharin, J. A., "Influence of the Friction Conditions on the Magnitude of the Friction Coefficient in the Case of Rolling with Sliding," International Conference on Gearing, Proceedings, Sept. 1958.

70. Hirst, W., and Moore, A. J., "Non-Newtonian Behavior in Elasto-hydrodynamic Lubrication," Proc. Roy. Soc., 1974, Vol. A337, pp. 101–121.

71. Johnson, K. L., and Tevaarwerk, J. L., "Shear Behavior of Elastohydrodynamic Oil Films," Proc. Roy. Soc., 1977, Vol. A356, pp. 215–236.

72. Conry, T. F., Johnson, K. L., and Owen, S., "Viscosity in the Thermal Regime of Elastohydrodynamic Traction," 6th Lubrication Symposium, Lyon, Sept., 1979.

73. Trachman, E. G., and Cheng, H. S., "Thermal and Non-Newtonian Effects on Traction in Elastohydrodynamic Contacts," Elastohydrodynamic Lubrication, 1972 Symposium, p. 142.

74. Trachman, E. G., and Cheng, H. S., "Traction in Elastohydrodynamic Line Contacts for Two Synthesized Hydrocarbon Fluids," ASLE Trans., 19??, Vol. 17(4), pp. 271–279.

75. Winer, W. O., "Regimes of Traction in Concentrated Contact Lubrication," Trans. ASME, 1982, Vol. 104, p. 382.

76. Sasaki, T., Okamura, K., and Isogal, R., "Fundamental Research on Gear Lubrication," Bull. JSME, 1961, Vol. 4(14), p. 382.

77. Sasaki, T. Okamura, K. Konishi, T., and Nishizawa, Y. "Fundamental Research on Gear Lubrication," Bull. JSME, 1962, Vol. 5(19), p. 561.
78. Drozdov, Y. N., and Gavrikov, Y. A., "Friction and Scoring Under the Conditions of Simultaneous Rolling and Sliding of Bodies," Wear, 1968, Vol. 11, p. 291.
79. Kelley, B. W., and Lemaski, A.J., "Lubrication of Involute Gearing," Proc. Inst. Mech. Engrs, 1967–68, Vol. 182, Pt. 3A, p. 173.
80. Dowson, D., "Elastohydrodynamic Lubrication," Interdisciplinary Approach to the Lubrication of Concentrated Contacts, Special Publication No. NASA-SP-237, National Aeronautics and Space Administration, Washington, D.C., 1970, p. 34.
81. Wilson, W. R. D., and Sheu, S., "Effect of Inlet Shear Heating Due to Sliding on EHD Film Thickness," ASME J. Lubr. Technol., Apr. 1983, Vol. 105, p. 187.
82. Greenwood, J. A., and Tripp, J. H., "The Elastic Contact of Rough Spheres," J. Appl. Mech., March 1967, p. 153.
83. Lindberg, R. A., "Processes and Materials of Manufacture," Allyn and Bacon, 1977, pp. 628–637.
84. "Wear Control Handbook", ASME, 1980.
85. Szeri, A. Z., Tribology: Friction, Lubrication and Wear, Hemisphere, New York, NY, 1980.

# 2

# The Contact Between Smooth Surfaces

## 2.1 INTRODUCTION

It is well known that no surface, natural or manufactured, is perfectly smooth. Nonetheless the idealized case of elastic bodies with smooth surfaces is considered in this chapter as the theoretical reference for the contact between rough surfaces. The latter will be discussed in Chapter 4 and used as the basis for evaluating the frictional resistance.

The equations governing the pressure distribution due to normal loads are given without detailed derivations. Readers interested in detailed derivations can find them in some of the books and publications given in the references at the end of the chapter [1–39].

## 2.2 DESIGN RELATIONSHIPS FOR ELASTIC BODIES IN CONTACT

### Case 1: Concentrated Normal Load on the Boundary of a Semi-Infinite Solid

The fundamental problem in the field of surface mechanics is that of a concentrated, normal force $P$ acting on the boundary of a semi-infinite body as shown in Fig. 2.1. The solution of the problem was given by Boussinesq [1] as:

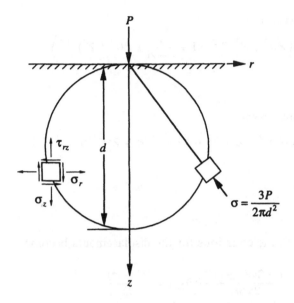

**Figure 2.1** Concentrated load on a semi-infinite elastic solid.

$\sigma_r$ = horizontal stress at any point

$$= \frac{P}{2\pi}\left\{(1-2v)\left[\frac{1}{r^2} - \frac{Z}{r^2}\left(\frac{3P}{2\pi}Z^3(r^2Z^2)^{-5/2}\right)^{-1/2}\right] - 3r^2Z(r^2+Z^2)^{-5/2}\right\}$$

$\sigma_Z$ = vertical stress at any point

$$= \frac{3P}{2\pi}Z^3(r^2Z^2)^{-5/2}$$

$\tau_{rZ}$ = shear stress at any point

$$= \frac{3P}{2\pi}rZ^2(r^2+Z^2)^{-5/2}$$

where

$v$ = Poisson's ratio

The resultant principal stress passes through the origin and has a magnitude:

$$\sigma = \sqrt{\sigma_Z^2 + \tau_{rZ}^2} = \frac{3P}{2\pi(r^2+Z^2)} = \frac{3P}{2\pi d^2}$$

The displacements produced in the semi-infinite solid can be calculated from:

$u$ = horizontal displacement

$$= \frac{(1-2v)(1+v)}{2\pi Er} P\left(Z(r^2+Z^2)^{-1/2} - 1 + \frac{1}{1-2v} r^2 Z(r^2+Z^2)^{-3/2}\right)$$

and

$w$ = vertical displacement

$$= \frac{P}{2\pi E}[(1+v)Z^2(r^2+Z^2)^{-3/2} + 2(1-v^2)(r^2+Z^2)^{-1/2}]$$

where

$E$ = elastic modulus

At the surface where $Z = 0$, the equations for the displacements become:

$$(u)_{Z=0} = -\frac{(1-2v)(1+v)P}{2\pi Er} \quad (w)_{Z=0} = \frac{P(1-v^2)}{\pi Er}$$

which increase without limit as $r$ approaches zero. Finite values, however, can be obtained by replacing the concentrated load by a statically equivalent distributed load over a small hemispherical surface at the origin.

### Case 2: Uniform Pressure over a Circular Area on the Surface of a Semi-Infinite Solid

The solution for this case can be obtained from the solution for the concentrated load by superposition. When a uniform pressure $q$ is distributed over a circular area of radius $a$ (as shown in Fig. 2.2) the stresses and deflections are found to be:

$(w)_{r=a}$ = deflection at the boundary of the loaded circle

$$= \frac{4(1-v^2)qa}{\pi E}$$

$(w)_{r=0}$ = deflection at the center of the loaded circle

$$= \frac{2(1-v^2)qa}{E}$$

$(\sigma_Z)_{r=0}$ = vertical stress at any point on the $Z$-axis

$$= q\left(-1 + \frac{Z^3}{(a^2+Z^2)^{3/2}}\right)$$

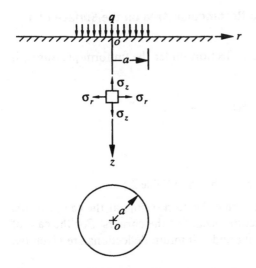

**Figure 2.2**  Uniform pressure on a circular area.

$(\sigma_r)_{r=0}$ = horizontal stress at any point on the Z-axis

$$= \frac{q}{2}\left[ -(1+2v) + \frac{2(1+v)Z}{\sqrt{a^2 Z^2}} - \left( \frac{Z}{\sqrt{a^2 + Z^2}} \right)^3 \right]$$

$(\tau)_{r=0} = \frac{1}{2}(\sigma_r - \sigma_Z)_{r=0}$

= maximum shear stress at any point on the Z-axis

$$= \frac{q}{2}\left[ \frac{1-2v}{2} + (1+v)\frac{Z}{\sqrt{a^2 + Z^2}} - \frac{3}{2}\left( \frac{Z}{\sqrt{a^2 + Z^2}} \right)^3 \right]$$

From the above equation it can be shown that the maximum combined shear stress occurs at a point given by:

$$Z = a\sqrt{\frac{2(1+v)}{7-2v}}$$

and its value is:

$$\tau_{max} = \frac{q}{2}\left( \frac{1-2v}{2} + \frac{2}{9}(1+v)\sqrt{2(1+v)} \right)$$

### Case 3: Uniform Pressure over a Rectangular Area on the Surface of a Semi-Infinite Solid

In this case (Fig. 2.3) the average deflection under the uniform pressure $q$ is calculated from:

$$w_{\text{ave}} = k(1 - v^2)\frac{q}{E}\sqrt{A}$$

where

$A$ = area of rectangle

$k$ = factor dependent on the ratio $b/a$ as shown in Table 2.1

It should be noted that the maximum deflection occurs at the center of the rectangle and the minimum deflection occurs at the corners. For the case of a square area $(b = a)$ the maximum and minimum deflections are given by:

$$(w_{\text{max}})_{\frac{b}{a}=1} = 1.12\,\frac{q\sqrt{A}(1 - v^2)}{E}$$

$$(w_{\text{min}})_{\frac{b}{a}=1} = 0.56\,\frac{q\sqrt{A}(1 - v^2)}{E}$$

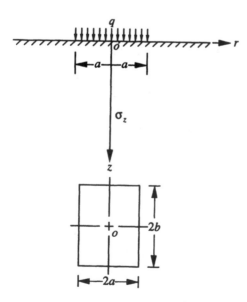

**Figure 2.3**  Uniform pressure over a rectangular area.

**Table 2.1** Values of Factor $k$

| $b/a$ | $k$ |
|---|---|
| 1 | 0.95 |
| 1.5 | 0.94 |
| 2 | 0.92 |
| 3 | 0.88 |
| 5 | 0.82 |
| 10 | 0.71 |
| 100 | 0.37 |

## Case 4: A Rigid Circular Cylinder Pressed Against a Semi-Infinite Solid

In this case, which is shown in Fig. 2.4, the displacement of the rigid cylinder is calculated from:

$$w = \frac{P(1 - \nu^2)}{2aE}$$

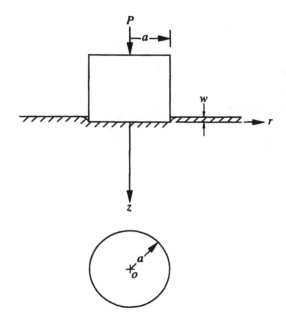

**Figure 2.4**  Rigid cylinder over a semi-infinite elastic solid.

where

$p$ = total load on the cylinder
$a$ = radius of the cylinder

The pressure distribution under the cylinder is given by

$$q = \frac{P}{2\pi a\sqrt{a^2 - r^2}}$$

which indicates that the maximum pressure occurs at the boundary ($r = a$) where localized yielding is expected. The minimum pressure occurs at the center of the contact area ($r = 0$) and has half the value of the average pressure.

### Case 5: Two Spherical Bodies in Contact

In this case (Fig. 2.5) the area of contact is circular with radius $a$ given by

$$a = 0.88 \sqrt[3]{\frac{P}{E_e}}\, R_e$$

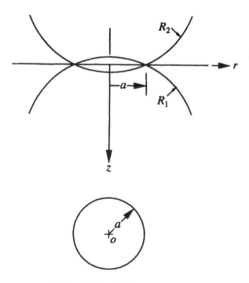

**Figure 2.5**  Spherical bodies in contact.

assuming a Poisson's ratio $\nu = 0.3$ and the pressure distribution over this area is:

$$q = 0.616 \sqrt[3]{\frac{P}{R_e^2} E_e^2} \left[ 1 - \left(\frac{r}{a}\right)^2 \right]^{1/2}$$

The radial tensile stress and maximum combined shear stress at the boundary of the contact area can be calculated as:

$$(\sigma_r)_{r=a} = (\tau_{\max})_{r=a} = 0.082 \sqrt[3]{\frac{P}{R_e^2} E_e^2}$$

where

$P$ = total load

$$\frac{1}{R_e} = \left(\frac{1}{R_1} + \frac{1}{R_2}\right)$$

$$\frac{1}{E_e} = \frac{1}{E_1} + \frac{1}{E_2}$$

$E_1, E_2$ = modulus of elasticity for the two materials

## Case 6: Two Cylindrical Bodies in Contact

The area of contact in this case (Fig. 2.6) is a rectangle with width $b$ and length equal to the length of the cylinders. The design relationships in this case are:

$$b = 2.15 \sqrt{\frac{P'}{E_e} R_e}$$

$q$ = pressure on the area of contact

$$= 0.591 \sqrt{\frac{P'}{R_e} E_e} \left[ 1 - \left(\frac{2r}{b}\right)^2 \right]^{1/2}$$

where

$P'$ = load per unit length of the cylinders

$$\frac{1}{R_e} = \frac{1}{R_1} + \frac{1}{R_2}$$

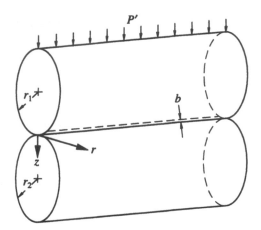

**Figure 2.6** Two cylindrical bodies in contact.

$R_1, R_2$ = radii of cylinders (positive when convex and negative when concave)

$$\frac{1}{E_e} = \frac{1}{E_1} + \frac{1}{E_2}$$

$E_1, E_2$ = modulus of elasticity for the two materials

### Case 7: General Case of Contact Between Elastic Bodies with Continuous and Smooth Surfaces at the Contact Zone

Analysis of this case by Hertz can be found in Refs 1 and 2. A diagrammatic representation of this problem is shown in Fig. 2.7 and the contact area is expected to assume an ellipitcal shape. Assuming that $(R_1, R_1')$ and $(R_2, R_2')$ are the principal radii of curvature at the point of contact for the two bodies respectively, and $\psi$ is the angle between the planes of principle curvature for the two surfaces containing the curvatures $1/R_1$ and $1/R_2$, the curvature consants $A$ and $B$ can be calculated from:

$$A + B = \frac{1}{2}\left(\frac{1}{R_1} + \frac{1}{R_1'} + \frac{1}{R_2} + \frac{1}{R_2'}\right)$$

$$A - B = \frac{1}{2}\left[\left(\frac{1}{R_1} - \frac{1}{R_1'}\right)^2 + \left(\frac{1}{R_2} - \frac{1}{R_2'}\right)^2 + 2\left(\frac{1}{R_1} - \frac{1}{R_1'}\right)\left(\frac{1}{R_2} - \frac{1}{R_2'}\right)\cos 2\psi\right]^{1/2}$$

These expressions can be used to calculate the contact parameter $P$ from the relationship:

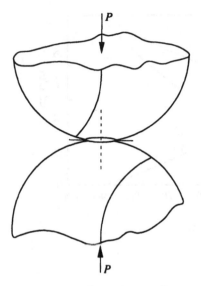

**Figure 2.7** General case of contact.

$$\cos\theta = \frac{B - A}{A + B}$$

The semi-axes of the elliptical area are:

$$a = m\sqrt[3]{\frac{3\pi}{4}\frac{P(k_1 + k_2)}{(A + B)}} \qquad b = n\sqrt[3]{\frac{3\pi}{4}\frac{P(k_1 + k_2)}{(A + B)}}$$

where

$m, n$ = functions of the parameter $\theta$ as given in Fig. 2.8
$P$ = total load
$$k_1 = \frac{1 - v_1^2}{\pi E_1} \qquad k_2 = \frac{1 - v^2}{\pi E^2}$$
$v_1, v_2$ = Poisson's ratios for the two materials
$E_1, E_2$ = corresponding modulii of elasticity

## Case 8: Beams on Elastic Foundation

The general equation describing the elastic curve of the beam is:

$$EI\,\frac{dy^4}{dx^4} = -ky$$

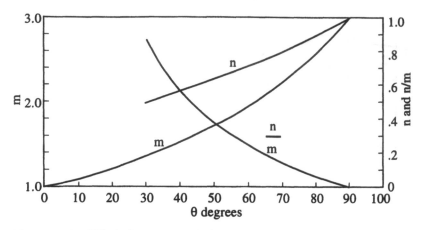

**Figure 2.8**   Elliptical contact coefficients.

where

$k$ = foundation stiffness per unit length
$E$ = modulus of elasticity of beam material
$I$ = moment of inertia of the beam

With the notation:

$$\beta = \sqrt[4]{\frac{k}{4EI}}$$

the general solution for beam deflection can be represented by:

$$y = e^{\beta x}(A \cos \beta x + B \sin \beta x) + e^{-\beta x}(C \cos \beta x + D \sin \beta x)$$

where $A$, $B$, $C$ and $D$ are integration constants which must be determined from boundary conditions.

For relatively short beams with length smaller than $(0.6/\beta)$, the beam can be considered rigid because the deflection from bending is negligible compared to the deflection of the foundation. In this case the deflection will be constant and is:

$$\delta = \frac{P}{kL}$$

and the maximum bending moment = $PL/4$.

For relatively long beams with length greater than $(5/\beta)$ the deflection will have a wave form with gradually diminishing amplitudes. The general solution can be found in texts on advanced strength of materials.

Table 2.2 lists expressions for deflection $y$, slope $\theta$, bending moment $M$ and shearing force $V$ for long beams loaded at the center.

### Case 9: Pressure Distribution Between Rectangular Elastic Bars in Contact

The determination of the pressure distribution between two bars subjected to concentrated transverse loads on their free boundaries is a common problem in the design of mechanical assemblies. This section presents an approximate solution with an empirical linear model for the local surface contact deformation. The solution is based on an analytical and a photo-elastic study [18]. A diagrammatic representation of this problem is shown

**Table 2.2**  Beam on Flexible Supports

| Condition | Governing equations[a] |
|---|---|
| | $y = \dfrac{P\beta}{2k}\,\phi(\beta x)$ <br><br> $\theta = -\dfrac{P\beta^2}{k}\,\zeta(\beta x)$ <br><br> $M = \dfrac{P}{4\beta}\,\psi(\beta x)$ <br><br> $V = -\dfrac{P}{2}\,\gamma(\beta x)$ |
| | $y = \dfrac{M_o\beta^2}{k}\,\zeta(\beta x)$ <br><br> $\theta = \dfrac{M_o\beta^2}{k}\,\psi(\beta x)$ <br><br> $V = -\dfrac{M_o\beta}{k}\,\phi(\beta x)$ |

[a]Where:
$(\beta x) = e^{-\beta x}(\cos\beta x + \sin\beta x)$
$\zeta(\beta x) = e^{-\beta x}(\sin\beta x)$
$\psi(\beta x) = e^{-\beta x}(\cos\beta x - \sin\beta x)$
$\gamma(\beta x) = e^{-\beta x}(\cos\beta x)$

in Fig. 2.9a. The problem is approximately treated as two beams on an elastic foundation, as shown in Fig. 2.9b. The equations describing the system are:

$$E_1 I_1 \frac{d^4 y_1}{dx^4} = k_1 z_1 = q_x \tag{2.1}$$

$$E_2 I_2 \frac{d^4 y_2}{dx^4} = k_2 z_2 = q_x \tag{2.2}$$

where

$q_x$ = load intensity distribution at the interface (lb/in.)

$I_1, I_2$ = moments of inertia of beam cross-sections

$E_1, E_2$ = modulii of elasticity

$z_1, z_2$ = local surface deformations

$y_1, y_2$ = beam deflections

$k_1, k_2$ = empirical linear contact stiffness for the two bars respectively calculated as:

$$k_1 = \frac{Et}{0.544} \qquad k_2 = \frac{Et}{0.544}$$

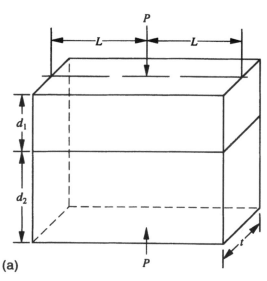

(a)

**Figure 2.9a**  Two rectangular bars in contact.

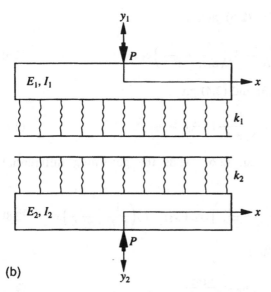

**Figure 2.9b** Simplified model for two beams in contact.

The criterion for contact requires, in the absence of initial separations, the total elastic deflection to be equal to the rigid body approach at all points of contact, therefore:

$$y_1 + y_2 + z_1 + z_2 = \alpha \tag{2.3}$$

where $\alpha$ is the rigid body approach defining the compliance of the entire joint between the points where the loads are applied:

$$z_1, z_2 \geq 0$$

The continuity of force at the interface yields:

$$k_1 z_1 = k_2 z_2 \tag{2.4}$$

Equations (2.1)–(2.4) are a system of four equations in four unknowns. This system is now reduced to a single differential equation as follows. Adding Eqs (2.1) and (2.2) gives:

$$\frac{d^4 y_1}{dx^4} + \frac{d^4 y_2}{dx^4} - \frac{k}{E_1 I_1} z_1 - \frac{k_2}{E_1 I_1} z_2 = 0 \tag{2.5}$$

Substituting Eq. (2.4) into Eq. (2.5) yields:

$$\frac{d^4y_1}{dx^4} + \frac{d^4y_2}{dx^4} - k_1\left(\frac{1}{E_1I_1} + \frac{1}{E_2I_2}\right)z_1 = 0 \tag{2.6}$$

The combination of Eqs. (2.3) and (2.4) gives:

$$z_1 = \frac{k_2}{k_1 + k_2}(\alpha - y_1 - y_2) \tag{2.7}$$

Substituting Eq. (2.7) into Eq. (2.6) yields the governing differential equation:

$$\frac{d^4}{dx^4}(y_1 + y_2) + K_e\left(\frac{1}{E_1I_1} + \frac{1}{E_1I_2}\right)(y_1 + y_2) = K_e\left(\frac{1}{E_1I_1} + \frac{1}{E_2I_2}\right)\alpha \tag{2.8}$$

where $K_e$ is an effective stiffness:

$$K_e = \frac{k_1k_2}{k_1 + k_2} \tag{2.9}$$

With the following notation:

$$\beta \equiv \sqrt[4]{\frac{K_e}{4}\left(\frac{1}{E_1I_1} + \frac{1}{E_2I_2}\right)} \tag{2.10}$$

$$\nu \equiv y_1 + y_2$$

Eq. (2.10) may now be rewritten as:

$$\frac{d^4\nu}{dx^4} + 4\beta^4\nu = 4\beta^4\alpha \tag{2.11}$$

The following are the boundary conditions which the solution of (2.11) must satisfy, provided $L \geq \ell$, where $\ell$ is half the length of the pressure zone:

The beam deflections are zero at the center location.
The slope is zero at the center.
The summation of the interface pressure equals the applied load.
The pressure at the end of the pressure zone is zero.
The moment at the end of the pressure zone is zero.
The shear force at the end of the pressure zone is zero.

The six unknowns to be determined by the above boundary conditions are the four arbitrary constants of the complementary solution, the rigid body approach $\alpha$, and the effective half-length of contact $\ell$. The four constants and the rigid body approach are determined as a function of the parameter $\lambda = \beta \ell$. A plot of the rigid body approach versus $\lambda$ is shown in Fig. 2.10. At $\lambda = \pi/2$, the slope of the curve is zero. For values of $\lambda$ greater than $\pi/2$, the values of $z_1$ and $z_2$ become negative, which is not permitted. The maximum permitted values of $\lambda$ is then $\pi/2$ and the effective half-length of contact is $\ell = \pi/(2\beta)$. If $\pi/(2\beta)$ is greater than $L$, the effective length is then $2L$.

The expression for the load intensity at the interface is:

$$q = k_1 z_1 = \frac{P\beta}{2}\left[\frac{(\cosh 2\lambda + \cos 2\lambda + 2)}{\sinh 2\lambda + \sin 2\lambda}\right]\cosh \beta x \cos \beta x$$
$$- \frac{(\cosh 2\lambda - \cos 2\lambda)}{(\sinh 2\lambda + \sin 2\lambda)}\sinh \beta x \sin \beta x \qquad (2.12)$$
$$- \sinh \beta x \cos \beta x + \cosh \beta x \sin \beta x$$

where

$\lambda = \beta \ell$
$q =$ the load intensity (lb/in.)
$x =$ restricted to be $0 \leq x \leq \ell$

The normalized load intensity versus position is shown in Fig. 2.11 for $\beta \ell = \pi/2$. This figure represents a generalized dimensionless pressure distribution for cases where $\ell \ll L$.

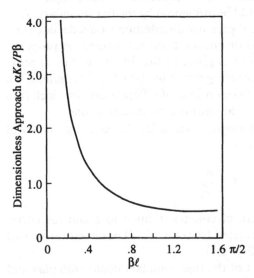

**Figure 2.10**   Dimensionless approach of the two beams versus the parameter $\beta \ell$.

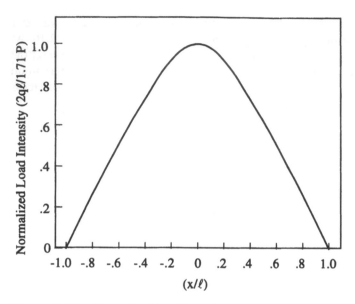

**Figure 2.11** Normalized load intensity over contact region.

The assumption of a constant contact stiffness can be considered adequate as long as the theoretical contact length is far from the ends of the beams. For cases where $\ell$ approaches $L$, it is expected that the compliance as well as the stress distribution would be influenced by the free boundary. As a result, it is expected that the actual pressure distribution would deviate from the theoretical distribution based on constant contact stiffness. A proposed model for treating such conditions is given in the following. The approximate model gives a relatively simple general method for determining the contact pressure distributions between beams of different depths which is in general agreement with experimental photoelastic investigations.

In the model the contact half-length $\ell$ is calculated from the geometry of the beam according to the formula:

$$\ell = \frac{\pi}{2\beta}$$

When $\ell \ll L$, the true half-length of contact is equal to $\ell$ and the corresponding pressure is directly calculated from Eq. (2.12) or directly evaluated from the dimensionless plot of Fig. 2.11.

As $\ell$ approaches $L$, the effect of the free boundary comes into play and the constant stiffness model can no longer be justified. An empirical method to deal with the boundary effect for such cases is explained in the following. The method can be extended for the cases where $\ell \geq L$.

Because of the increase in compliance at the boundaries of a finite beam as the stressed zone approaches it, a fictitious theoretical contact length $\ell'(\ell' > \ell)$ can be assumed to describe a hypothetical contact condition for equivalent beams with $L' \gg \ell$ (according to the empirical relationship given in Fig. 2.12. The pressure distribution for this hypothetical contact condition is then calculated. Because the actual half-length of the beam is $L$, it would be expected that the pressure between $\ell'$ and $L$ would have to be carried over the actual length $L$ for equilibrium. The redistribution of the pressure outside the physical boundaries of the beam is assumed to follow a mirror image, as shown in Fig. 2.13.

The superposition of this reflected pressure on the pressure within the boundaries of the beam gives the total pressure distribution.

The general procedure cam be summarized as follows:

1. Calculate $\beta$ from geometry and the material of the contacting bars according to Eq. (2.10).
2. Calculate $\ell$ from the equation $\ell = \pi/(2\beta)$.
3. Using $\ell$ and $L$, find $\ell'$ from Fig. 2.12. Notice that for $\ell \ll L$, $\ell' = \ell$.

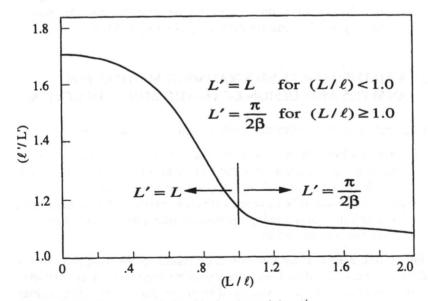

**Figure 2.12** Empirical relationship for determining $\ell'$.

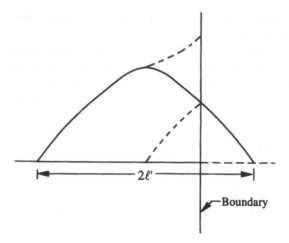

**Figure 2.13** The "mirror image" procedure.

4.  Evaluate the pressure distribution over $\ell'$ from the normalized graph, Fig. 2.11.
5.  For $\ell \ll L$, the pressure distribution as calculated in step 4 is the true contact pressure.
6.  When $\ell'$ is greater than $L$, the distribution calculated by step 4 is modified by reflection (as a mirror image of the pressure outside the physical boundaries defined by the length $L$).

## 2.3  A MATHEMATICAL PROGRAMMING METHOD FOR ANALYSIS AND DESIGN OF ELASTIC BODIES IN CONTACT

The general contact problem can be divided into two categories:

Situations where the interest is the evaluation of the contact area, the pressure distribution, and rigid body approach when the system configuration, materials and applied loads are known;

Systems which are to be designed with appropriate surface geometry for the objective of obtaining the best possible distribution of pressure over the contact region.

In this section a general formulation is discussed for treating this class of problems using a modified linear programming approach. A simplex-type algorithm is utilized for the solution of both the analysis and design situations. A detailed treatment of this problem can be found in Refs 18 and 19.

### 2.3.1   The Formulation of the Contact Problem

The contact problems which are analyzed here are restricted to normal surface loading conditions. Discrete forces are used to represent distributed pressures over finite areas. The following assumptions are made:

1.   The deformations are small.
2.   The two bodies obey the laws of linear elasticity.
3.   The surfaces are smooth and have continuous first derivatives.

Problem formulation and geometric approximations can therefore be made within the limits of elasticity theory.

### 2.3.2   Condition of Geometric Compatibility

At any point $k$ in the proposed zone of contact (Fig. 2.14), the sum of the elastic deformations and any initial separations must be greater than or equal to the rigid body approach. This condition is represented as:

$$w_{k(1)} + w_{k(2)} + \varepsilon_k - \alpha \geq 0 \qquad (2.13)$$

where

$$\varepsilon_k = \text{initial separation at point } k$$
$$w_{k(1)}, w_{k(2)} = \text{elastic deformations of the two bodies respectively at point } k$$
$$\alpha = \text{rigid body approach}$$

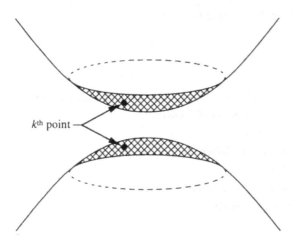

$k^{\text{th}}$ point

**Figure 2.14**   Zone of contact.

### 2.3.3 Condition of Equilibrium

The sum of all the forces $F_k$ acting at the discrete points ($k = 1, \ldots, N$ where $N$ is the number of candidate points for contact) must balance the applied load ($P$) normal to the surface. The equilibrium condition can therefore be written as:

$$\sum_{k=1}^{N} F_k = P \tag{2.14}$$

### 2.3.4 The Criterion for Contact

At any point $k$, the left-hand side of the inequality constraint in Eq. (2.13) may be strictly positive or identically zero. Defining a slack variable $Y_k$ representing a final separation, the contact problem can be formulated as follows.

Find a solution $(F, \alpha, Y)$ which satisfies the following constraints:

$$-SF + \alpha e + IY = \varepsilon$$
$$e^T F = P \tag{2.15}$$

Either

$$F_k = 0 \text{ or } Y_k = 0$$
$$F_k \geq 0, \qquad Y_k \geq 0, \qquad \alpha \geq 0$$

where

$$S_{kj} = a_{kj,(1)} + a_{kj(2)}$$

$a_{kj(1)}, a_{kj(2)}$ = influence coefficients for the deflection of the two bodies respectively

$S_{kj} = N \times N$ matrix of influence coefficients

$F = N \times 1$ vector of forces

$Y = N \times 1$ vector of slack variables (or final separation)

$e = N \times 1$ vector of 1's

$\varepsilon = N \times 1$ vector of initial separations

$\alpha$ = rigid body approach, a scalar

## 2.4  A GENERAL METHOD OF SOLUTION BY A SIMPLEX-TYPE ALGORITHM

The problem as formulated in Eq. (2.15) can be solved using a modification of the simplex algorithm used in linear programming. The changes required for the modification are minor and are similar to those given by Wolfe [8]. When Eq. (2.15) is represented in a tableau form in Table 2.3, the condition for the solution can be stated as:

> Find the set of column vectors corresponding to $(F, \alpha, Y)$ subject to the conditions, either $F_k = 0$ or $Y_k = 0$, such that the right-hand side is a nonnegative linear combination of these column vectors. These column vectors are called a basis.

For a problem with $N$ discrete points, the number of possible combinations of these column vectors taken $(N + 1)$ at a time is:

$$C_N = 2^N - 1$$

Because of the very large number of combinations, an efficient method is required for finding the unique, feasible solution. The following algorithm proved to be effective for the problem under investigation.

The original problem as formulated in Eq. (2.15) can be rewritten as:

$$\text{Minimize} \sum_{j=1}^{N+1} Z_j$$

such that

$$-SF + \alpha e + IY + I\overline{Z} = \varepsilon$$
$$e^T F + Z_{N+1} = P$$

(2.16)

**Table 2.3**  Representation of Eq. (2.15)

| $F_1$ | $F_2$ | ... | $F_N$ | $\alpha$ | $Y_1$ | $Y_2$ | ... | $Y_N$ | |
|---|---|---|---|---|---|---|---|---|---|
| $-S_{11}$ | $-S_{12}$ | $-\dots$ | $-S_{1N}$ | 1 | 1 | | | | $= \varepsilon_1$ |
| $-S_{12}$ | $-S_{22}$ | $-\dots$ | $-S_{2N}$ | 1 | | 1 | | | $= \varepsilon_2$ |
| $\vdots$ | | | $\vdots$ | $\vdots$ | | | $\ddots$ | $\vdots$ | $\vdots$ |
| $-S_{N1}$ | $-S_{N2}$ | $-\dots$ | $-S_{NN}$ | 1 | | 1 | | 1 | $= \varepsilon_N$ |
| $+1$ | $+1$ | $+\dots$ | $+1$ | | | | | | $= P$ |

Subject to the conditions that

Either

$$F_k = 0 \text{ or } Y_k = 0$$
$$F_k \geq 0, Y_k \geq 0, \alpha \geq 0, Z_j \geq 0$$

where

$Z_j$ = artificial variables which are required to be nonnegative $(j = 1, \ldots, N + 1)$

$\overline{Z}$ = an $N \times 1$ vector of artificial variables with components $Z_1, \ldots, Z_N$

The above problem can be classified as a linear programming problem [13] if it were not for the condition that either $F_k = 0$ or $Y_k = 0$. The simplex algorithm for linear programming can, however, be utilized to solve by making a modification of the entry rules.

The conditions of Eq. (2.15) require some restrictions on the entering variables. Suppose the entering variable is chosen as $F_s$. A check must be made to see if the $Y_s$ is not in the basis, $F_s$ is free to enter the basis.

The actual replacement of variables is accomplished by an operation called pivoting. This pivot operation consists of $N + 1$ elementary operations which replace a system by an equivalent system in which a specified variable has a coefficient of unity in one equation and zero elsewhere [13].

A flow diagram of the modified simplex algorithm is shown in Fig. 2.15.

Computational experience has shown the simplex-type algorithm to converge to the unique feasible point in at most $(3/2)(N + 1)$ cycles, the majrity of cases converge in $N + 1$ cycles.

The simplex-type algorithm for the solution of the contact problem requires less computer storage space when compared to available solution algorithms such as Rosen's gradient projection method [14] or the Frank–Wolfe algorithm [15]. Only minor modifications of the well-known simplex algorithm are required. This algorithm is also readily adaptable to the design problem which is discussed later in this section.

**EXAMPLE 1.** The classical problem of two spheres in contact is considered as an example. In this case the influence coefficient matrix $S$ in Eq. (2.15) is calculated according to a Boussinesq model as discussed earlier in this chapter:

$$a_{ki} = \frac{(1 - v^2)}{\pi E} \frac{1}{d_{kj}}$$

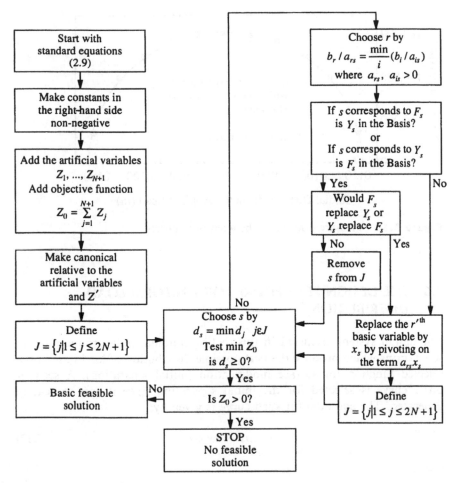

**Figure 2.15** Flow diagram for the simplex-type algorithm.

where

$v$ = Poisson's ratio

$d_{kj}$ = distance from point $k$ to point $j$ in the contact zone

Figure 2.16 shows a comparison between the classical Hertzian pressure distribution and that obtained by the described technique. The spheres considered are steel with radii of 1 in. and 10 in., respectively and the applied load is 100 lb. The algorithm solution gave a value of 0.000281 in. for the rigid body approach which compares favorably with 0.000283 in. for the classical Hertz solution.

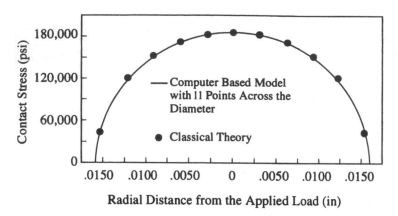

**Figure 2.16** Pressure distribution between two spheres.


## 2.5 THE DESIGN PROCEDURE FOR UNIFORM LOAD DISTRIBUTION

The design system discussed in this section automatically produces initial separations which produce the best possible distribution of load based on a selected function for surface modification (initial separation). A second-order curve is selected for the initial separation since it can be readily generated. The equation for such curve is given by:

$$y = ax^2 + bx + c \tag{2.17}$$

where

$y =$ initial separation profile and is required to be $\geq 0$
$x =$ axial position along the face

The correction profile can be attained by modifying one or both of the contacting surfaces. The objective of the design system is to evaluate the constants $(a, b, c)$ for the optimal corrections corresponding to the distribution giving the minimum possible value for the maximum load intensity.

In the formulation of the design system the compatibility condition given in Eq. (2.15) is used with $\varepsilon$ being replaced by Eq. (2.17). Accordingly:

$$-SF + \alpha e + IY - a\overline{X}^2 - b\overline{X} - c = 0 \tag{2.18}$$

where

$\overline{X}^2 = N \times 1$ vector whose $k$th element is $x_k^2$

$\overline{X} = N \times 1$ vector whose $k$th element is $x_k$

$x_k$ = position of the $k$th point

The condition of equilibrium and the criterion for contact are the same as in Eqs (2.14) and (2.15).

The initial separations are required to be nonnegative, therefore:

$$a\overline{X}^2 + b\overline{X} + c \geq 0$$

where $a$ governs the sign of the second derivative.

If we define $\Delta k$ as the length of the line segment at the $k$th point, the average load intensity over that segment is $F_k/\Delta k$. The value of $p_{\max}$ must be greater than the average load intensities at all the candidate points. This constraint is written as follows:

$$DF \leq p_{\max}\, e$$

where $D$ is a diagonal matrix whose $k$th element is $1/\Delta k$.

The design system is now stated in a concise form as:

$$\text{Minimize } p_{\max}$$

such that

$$
\begin{aligned}
-SF + IY + \alpha e - a\overline{X}^2 - b\overline{X} - c &= 0 \\
DF - p_{\max}\, e &\leq 0 \\
-a\overline{X}^2 - b\overline{X} - c &\leq 0 \\
e^T F &= P \\
F, Y, \alpha, c &\geq 0
\end{aligned}
\qquad (2.19)
$$

Subject to the condition that either

$$Y_k = 0 \text{ or } F_k = 0$$

It should be noted that an upper bound must be given to $c$ to keep the values of $c$ and $\alpha$ finite in Eq. (2.19).

The algorithm for solving the design problem (Fig. 2.17) is divided into two parts. The first part finds a feasible solution for the load distribution while the initial separations are constrained to be zero. The second part minimizes the maximum load intensity using the parameters $(a, b, c)$ as design variables. The simplex-type algorithm is used in both parts.

The minimization of the maximum load intensity is a nonlinear programming problem, the objective function is linear but the constraints are nonlinear [17]. Since all the constraints are linear except for the criterion for contact, the basic simplex algorithm can again be used with the modified entry rules as discussed previously.

**EXAMPLE 2.** The case of a steel beam on an elastic foundation is considered here as an illustration of the design system. It is required in this case to calculate the necessary initial separations which produce, as closely as possible, a uniform pressure. Given in this example are:

$$L, t, \text{ and } d = \text{length, width, and depth of beam}$$
$$= 8.9 \text{ in., } 1.0 \text{ in., and } 4.0 \text{ in., respectively}$$
$$k = \text{foundation modulus} = 10^7 \text{ lb/in./in.}$$

The results from the solution algorithm with a quadratic modification are given in Figs 2.18 and 2.19 and the pressure distribution without initial separation is shown in Fig. 2.18 for comparison. The initial separation as calculated from the analysis program for a uniform pressure distribution is also shown in Fig. 2.19. It can be seen that the quadratic modification, although it does not provide an exactly uniform pressure distribution, represents the best practical initial separation for the stated objective.

An approximation for evaluation of the surface modification can be obtained by assuming a uniform load distribution, computing the necessary initial separations and then fitting these data to a curve with the stated form of surface modification. In the process of curve fitting, the main objective is to approximate the computed initial separations without regard to the resulting load distribution.

The same approach can be readily applied to the surface modification of bolted joints to produce uniform pressure in the joint and consequently minimize the tendency for leakage or fretting depending on the application.

**EXAMPLE 3.** In this example, the same approach is applied for determining the initial separation necessary to produce uniform pressure at the interface between multiple-layered beams. The case considered for illustration is shown in Fig. 2.20 where three cantilever beams are subjected to

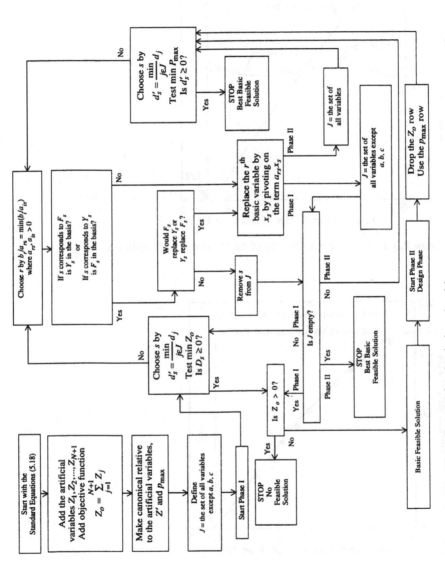

**Figure 2.17** Flow diagram for the design algorithm.

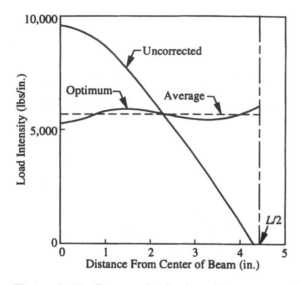

**Figure 2.18**  Pressure distribution of beam on elastic foundation.

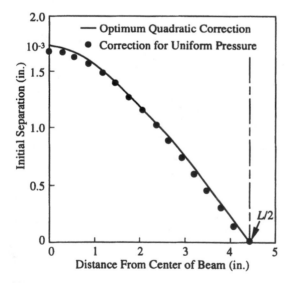

**Figure 2.19**  Initial separation for uniform pressure distribution and optimal quadratic correction.

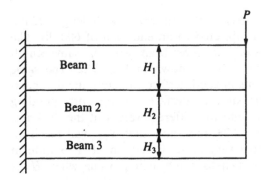

**Figure 2.20** Multiple cantilever beams.

an end load. The applied load $P$ is assumed to be 6000 lb, the length of the beam is 12 in. and width is 1 in. and the thicknesses are 5 in., 2 in. and 5 in., respectively. The beams are made of steel with modulus of elasticity equal to $30 \times 10^6$ psi. The interface areas are divided into 24 segments and the force on each segment is found to be 70.62 lb for both interfaces which is equivalent to 141.24 psi. The calculated initial separations are given in Fig. 2.21.

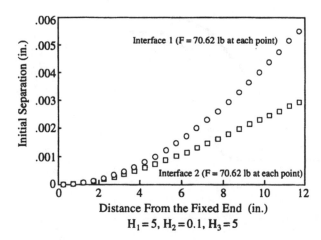

**Figure 2.21** Initial separations.

**EXAMPLE 4.** In this case a steel cantilever beam with length 12 in., width 1 in. and thickness 3 in. is subjected to an end load of 6000 lb. The beam is supported by another steel cantilever beam with the same length and width and different thickness $H_2$ as shown in Fig. 2.22. The same algorithm is used with 24 segments at the interface to determine the maximum attainable uniform pressure at the interface and the corresponding initial separation $S_{max}$ at the free end for different values of the thickness $H_2$. The results are given in Fig. 2.23 and show that $S_{max}$ will reach an asymptotic limit when $H_2$ is either very large or very small. The uniform load on each segment $F_{max}$ is shown to reach a limit value when $H_2$ is very large.

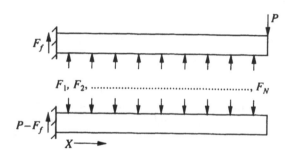

**Figure 2.22**  Discrete forces.

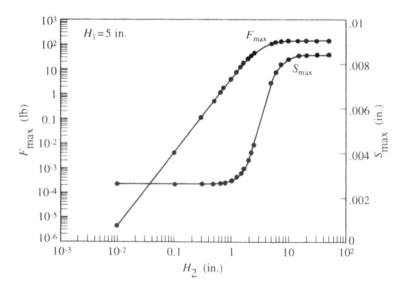

**Figure 2.23**  Limits for uniform pressure distribution and initial separation.

Numerous illustrative examples for simulated bolted joints, multiple layer beams and elastic solids with finite dimensions are given in Ref. 20.

A list of some of the publications dealing with different aspects of the contact problem is given in Refs 21–39.

## REFERENCES

1. Timoshenko, S.P., Theory of Elasticity, McGraw Hill Book Company, New York, 1951.
2. Love, A. E. H., A Treatise on the Mathematical Theory of Elasticity, Dover, New York, 1944.
3. Timoshenko, S. P., Strength of Materials, Part II, D. Van Nostrand, New York, 1950.
4. Galin, L. A., Contact Problems in the Theory of Elasticity, Translation by H. Moss, North Carolina State College, 1961.
5. Keer, L. M., "The Contact Stress Problem for an Elastic Sphere Indenting an Elastic Layer," Trans. ASME, Journal of Applied Mechanics, 1964, Vol. 86, pp. 143–145.
6. Tu, Y., "A Numerical Solution for an Axially Symmetric Contact Problem," Trans. ASME, Journal of Applied Mechanics, 1967, Vol. 34, pp. 283.
7. Tsai, N., and Westmann, R. A., "Beam on Tensionless Foundation," Proc. ASCE, J. Struct. Div., April 1966, Vol. 93, pp. 1–12.
8. Wolfe, P., "The Simplex Method for Quadratic Programming," Econometrica, 1959, Vol. 27, pp. 328–398.
9. Dorn, W. S., "Self-Dual Quadratic Programs," SIAM J. Appl. Math., 1961, Vol. 9, pp. 51–54.
10. Cottle, R. W., "Nonlinear Programs with Positively Bounded Jacobians," JSIAM Appl. Math., 1966, Vol. 14(1), pp. 147–158.
11. Kortanek, K., and Jeroslow, R., "A Note on Some Classical Methods in Constrained Optimization and Positively Bounded Jacobians," Operat. Res., 1967, Vol. 15(5), pp. 964–969.
12. Cottle, R. W., "Comments on the Note by Kortanek and Jeroslow," Operat. Res., 1967, Vol. 15(5), pp. 964–969.
13. Dantzig, G. W., Linear Programming and Extensions, Princeton University Press, Princeton, NJ, 1963.
14. Rosen, J. B., "The Gradient Projection Method for Non-Linear Programming," SIAM J. Appl. Math., 1960, Vol. 8, pp. 181–217; 1961, Vol. 9, pp. 514–553.
15. Frank, M., and Wolfe, P., "An Algorithm for Quadratic Programming," Naval Research Logist. Q., March–June 1956, Vol. 3(1 & 2), pp. 95–110.
16. Kerr, A. D., "Elastic and Viscoelastic Foundation Models," Trans. ASME, J. Appl. Mech., 1964, Vol. 86, pp. 491–498.
17. Mangasarian, O. L., Nonlinear Programming, McGraw Hill Book Company, New York, NY, 1969.

18. Conry, T. F., "The Use of Mathematical Programming in Design for Uniform Load Distribution in Nonlinear Elastic Systems," Ph.D. Thesis, The University of Wisconsin, 1970.

19. Conry, T. F., and Seireg, A., "A Mathematical Programming Method for Design of Elastic Bodies in Contact," Trans. ASME, J. Appl. Mech., 1971, Vol. 38, pp. 387–392.

20. Ni, Yen-Yih, "Analysis of Pressure Distribution Between Elastic Bodies with Discrete Geometry," M.Sc. Thesis, University of Florida, Gainesville, 1993.

21. Johnson, K. L., Contact Mechanics, Cambridge University Press, New York, NY, 1985.

22. Ahmadi, N., Keer, L. M., and Mura, T., "Non-Hertzian Contact Stress Analysis – Normal and Sliding Contact," Int. J. Solids Struct., 1983, Vol. 19, p. 357.

23. Alblas, J. B., and Kuipers, M., "On the Two-Dimensional Problem of a Cylindrical Stamp Pressed into a Thin Elastic Layer," Acta Mech., 1970, Vol. 9, p. 292.

24. Aleksandrov, V. M., "Asymptotic Methods in Contact Problems," PMM, 1968, Vol. 32, pp. 691.

25. Andersson, T., Fredriksson, B., and Persson, B. G. A., "The Boundary Element Method Applied to 2-Dimensional Contact Problems," New Developments in Boundary Element Methods. CML Publishers, Southampton, England, 1980.

26. Barovich, D., Kingsley, S. C., and Ku, T. C., "Stresses on a Thin Strip or Slab with Different Elastic Properties from that of the Substrate," Int. J. Eng. Sci., 1964, Vol. 2, p. 253.

27. Beale, E. M. L., "On Quadratic Programming," Naval Res. Logist. Q., 1959, Vol. 6, p. 74.

28. Bentall, R. H., and Johnson, K. L., "An Elastic Strip in Plane Rolling Contact," Int. J. Mech. Sci., 1968, Vol. 10, p. 637.

29. Calladine, C. R., and Greenwood, J. A., "Line and Point Loads on a Non-Homogeneous Incompressible Elastic Half-Space," Quarterly Journal of Mechanics and Applied Mathematics, 1978, Vol. 31, p. 507.

30. Comniou, M., "Stress Singularities at a Sharp Edge in Contact Problems with Friction," ZAMP, 1976, Vol. 27, p. 493.

31. Dundurs, J., Properties of Elastic Bodies in Contact, Mechanics of Contact between Deformable Bodies, University Press, Delft, Netherlands, 1975.

32. Greenwood, J. A., and Johnson, K. L., "The Mechanics of Adhesion of Viscoelastic Solids," Philosphical Magazine, 1981, Vol. 43, p. 697.

33. Matthewson, M. J., "Axi-Symmetric Contact on Thin Compliant Coatings," Journal of Mechanics and Physics of Solids, 1981, Vol. 29, p. 89.

34. Maugis, D., and Barquins, M., "Fracture Mechanics and the Adherence of Viscoelastic Bodies," Journal of Physics D (Applied Physics), 1978, Vol. 11.

35. Meijers, P., "The Contact Problems of a Rigid Cylinder on an Elastic Layer," Applied Sciences Research, 1968, Vol. 18, p. 353.

36. Mossakovski, V. I., "Compression of Elastic Bodies Under Conditions of Adhesion," PMM, 1963, Vol. 27, p. 418.
37. Pao, Y. C., Wu, T. S. and Chiu, Y. P., "Bounds on the Maximum Contact Stress of an Indented Layer," Trans. ASME Series E, Journal of Applied Mechanics, 1971, Vol. 38, p. 608.
38. Sneddon, I. N., "Boussinesq's Problem for a Rigid Cone," Proc. Cambridge Philosphical Society, 1948, Vol. 44, p. 492.
39. Vorovich, I. I., and Ustinov, I. A., "Pressure of a Die on an Elastic Layer of Finite Thickness," Applied Mathematics and Mechanics, 1959, Vol. 23, p. 637.

# 3

# Traction Distribution and Microslip in Frictional Contacts Between Smooth Elastic Bodies

## 3.1 INTRODUCTION

Frictional joints attained by bolting, riveting, press fitting, etc., are widely used for fastening structural elements. This chapter presents design formulae and methods for predicting the distribution of frictional forces and microslip over continuous or discrete contact areas between elastic bodies subjected to any combination of applied tangential forces and moments. The potential areas for fretting due to fluctuation of load without gross slip are discussed.

The analysis of the contact between elastic bodies has long been of considerable interest in the design of mechanical systems. The evaluation of the stress distribution in the contact region and the localized microslip, which exists before the applied tangential force exceeds the frictional resistance, are important factors in determining the safe operation of many structural systems.

Hertz [1] established the theory for elastic bodies in contact under normal loads. In his theory, the contact area, normal stress distribution and rigid body approach in the direction of the common normal can be found under the assumption that the dimensions of the contacting bodies are significantly larger than the contact areas.

Various extensions of Hertz theory can be found in the literature [2–15], and the previous chapter gives an overview of procedures for evaluating the area of contact and the pressure distribution between elastic bodies of arbitrary smooth surface geometry resulting from the application of loading.

An important class of contact problems is that of two elastic bodies which are subjected to a combination of normal and tangential forces.

The evaluations of the traction distribution and the localized microslip on the contact area due to tangential loads are important factors in determining the safe operation of many structural systems. Several contributions are available in the literature which deal with the analytical aspects of this problem [16–19]. The contact areas considered in all these studies are, however, limited to either a circle or an ellipse, and a brief summary of the results of both cases is given in the following section.

This chapter also presents algorithmic solutions which can be utilized for the analysis of the general case of frictional contacts. Three types of interface loads are to be expected: tangential forces, twisting moments, and different combinations of them. When the loads are lower than those necessary to cause gross slip, the microslip corresponding to these loads may cause fretting and surface cracks. The prediction of the areas of microslip and the energy generated in the process are therefore of considerable interest to the designer of frictional joints.

## 3.2 TRACTION DISTRIBUTION, COMPLIANCE, AND ENERGY DISSIPATION IN HERTZIAN CONTACTS

### 3.2.1 Circular Contacts

As shown in Fig. 3.1, when two spherical bodies are loaded along the common normal by a force $P$, they will come into contact over an area with radius $a$. When the system is then subjected to a tangential force $T < fP$, Mindlin's theory [16] for circular contacts defines the traction distribution over the contact area and can be summarized as follows:

$$a^* = a\left(1 - \frac{T}{fP}\right)^{1/3}$$

$$u = \frac{3fP(2-v)}{16Ga}\left[1 - \left(1 - \frac{T}{fP}\right)^{2/3}\right] \qquad \rho \le a^*$$

$$v = 0 \qquad \rho \le a^*$$

$$F_x = \frac{3fP}{2\pi a^2}\left(1 - \frac{\rho^2}{a^2}\right)^{1/2} \qquad a^* \le \rho \le a \tag{3.1}$$

$$F_x = \frac{3fP}{2\pi a^2}\left(1 - \frac{\rho^2}{a^2}\right)^{1/2} - \frac{3fPa^*}{2\pi a^3}\left(1 - \frac{\rho^2}{a^{*2}}\right)^{1/2} \qquad \rho \le a^* \tag{3.2}$$

$$F_y = 0 \text{ over the entire surface}$$

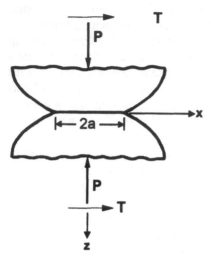

**Figure 3.1**   The contact of spherical bodies.

where

$F_x, F_y$ = traction stress components at any radius $\rho$

$a$ = radius of a circular contact area

$a^*$ = radius defining the boundary between the slip and no-slip regions

$\rho = (x^2 + y^2)^{1/2}$ = polar coordinate of any point within the contact area

$T$ = tangential force

$P$ = normal load

$f$ = coefficient of friction

$G$ = shear modulus of the material

$v$ = Poisson's ratio

Figure 3.2 illustrates the traction distribution as defined by Eqs (3.1) and (3.2). It can be seen that

$$a^* = a \text{ for } T = 0$$

and no microslip occurs;

$$a^* = 0 \text{ for } T = fP$$

and the entire contact area is in a state of microslip and impending gross slip.

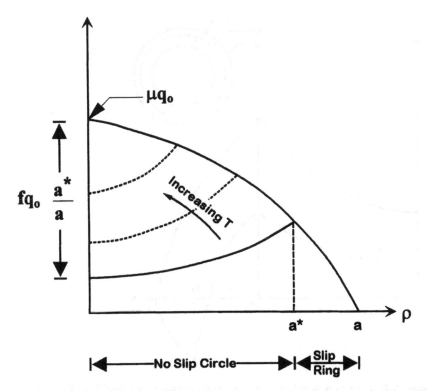

**Figure 3.2** Traction distribution for circular contact ($q_0$ =maximum contact pressure $= 3P/(2\pi a^2)$).

The deflection $\delta$ (rigid body tangential movement) due to the any load $T \geq fP$ can be calculated from:

$$\delta_T = \frac{3(2-v)fP}{16Ga}\left[1-\left(1-\frac{T}{fP}\right)^{2/3}\right] \qquad (3.3)$$

Consequently, at the condition of impending gross slip, $T = fP$:

$$(\delta_T)_0 = \frac{3(2-v)fP}{16Ga} \qquad (3.4)$$

The traction distribution and compliance for a tangential load fluctuating between $\pm T^*$ (where $T^* < fP$) can be calculated as follows (see Figs 3.3 and 3.4):

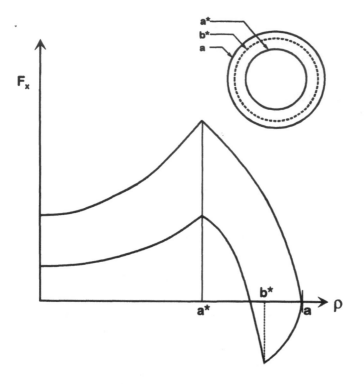

**Figure 3.3**   Traction distribution for decreasing tangential load $T < T^*$.

$b^*$ = inner radius of slip region

$$= a\left(1 - \frac{(T^* - T)}{2fP}\right)^{1/3}$$

$a^*$ = inner radius of slip region at the peak tangential load $T^*$

$$= a\left(1 - \frac{T^*}{fP}\right)^{1/3}$$

$$F_x = -fq_0\left[1 - \left(\frac{\rho}{a}\right)^2\right]^{1/2} \qquad b^* \le \rho \le a$$

$$F_x = -fq_0\left[1 - \left(\frac{\rho}{a}\right)^2\right]^{1/2} + 2fq_0\left(\frac{b^*}{a}\right)\left[1 - \left(\frac{\rho}{b}\right)^2\right] \qquad a^* \le \rho \le b^*$$

$$F_x = -fq_0\left[1 - \left(\frac{\rho}{a}\right)^2\right]^{1/2} + 2fq_0\left(\frac{b^*}{a}\right)\left[1 - \left(\frac{\rho}{b^*}\right)^2\right]^{1/2} - fq_0\left(\frac{a^*}{a}\right)\left[1 - \left(\frac{\rho}{a^*}\right)^2\right]^{1/2} \qquad \rho < a^*$$

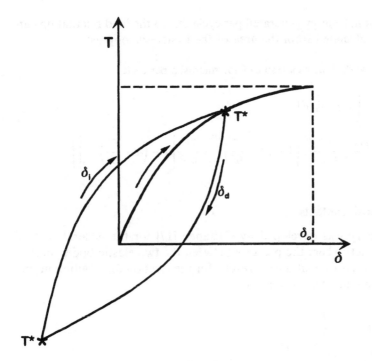

**Figure 3.4** Hysteresis loop.

where

$$q_0 = \text{maximum contact pressure} = \frac{3P}{2\pi a^2}$$

The deflection can be calculated from:

$$\delta_d = \text{deflection for decreasing tangential load}$$

$$= \frac{3(2-v)fP}{16Ga}\left[2\left(1-\frac{T^*-T}{2fP}\right)^{2/3}-\left(1-\frac{T^*}{fP}\right)^{2/3}\right]$$

for $T$ decreasing from $T^*$ to $-T^*$;

$$\delta_i = -\delta_d(-T)$$

$$= -\frac{3(2-\delta)fP}{16Ga}\left[2\left(1-\frac{T^*+T}{2fP}\right)^{2/3}-\left(1-\frac{T^*}{fP}\right)^{2/3}-1\right]$$

for $T$ increasing from $-T^*$ to $T^*$.

The frictional energy generated per cycle due to the load fluctuation can therefore be calculated from the area of the hysteresis loop as:

$$w = \text{work done as a result of the microslip per cycle}$$

$$= \int_{-T^*}^{T^*} (\delta_d - \delta_i)\, dT$$

$$= \frac{g(2-v)f^2 P^2}{10Ga}\left\{ 1 - \left(1 - \frac{T^*}{fP}\right)^{5/3} - \frac{5T^*}{6fP}\left[ 1 - \left(1 - \frac{T^*}{fP}\right)^{2/3}\right]\right\}$$

(3.5)

### 3.2.2 Elliptical Contacts

A similar theory was developed by Cattaneo [17] for the general case of Hertzian contacts where the pressure between the two elastic bodies occurs over an elliptical area. Cattaneo's results for the traction distribution in this case can be summarized as follows:

$$\left(\frac{a^*}{a}\right)^3 = \left(\frac{b^*}{b}\right)^3 = 1 - \frac{T}{fP}$$

$$F_x = \frac{3fP}{2\pi ab}\left(1 - \frac{x^2}{a^2} - \frac{y^2}{b^2}\right)^{1/2} \qquad \text{on slip region}$$

$$F_x = \frac{3fP}{2\pi ab}\left(1 - \frac{x^2}{a^2} - \frac{y^2}{b^2}\right)^{1/2} - \frac{3(fP-T)}{2\pi a^* b^*}\left(1 - \frac{x^2}{a^{*2}} - \frac{y^2}{b^{*2}}\right)^{1/2} \qquad \text{on no-slip region}$$

$$F_y = 0 \qquad \text{for the entire surface}$$

where

    $a, b$ = major and minor axes of an elliptical contact area

    $a^*, b^*$ = inner major and minor axes of the ellipse defining the boundary between the slip and no-slip regions

## 3.3 ALGORITHMIC SOLUTION FOR TRACTION DISTRIBUTION OVER CONTACT AREA WITH ARBITRARY GEOMETRY SUBJECTED TO TANGENTIAL LOADING BELOW GROSS SLIP

This section presents a computer-based algorithm for the analysis of the traction distribution and microslip in the contact areas between elastic bodies subjected to normal and tangential loads. The algorithm utilizes a modified linear programming technique similar to that discussed in the previous chapter. It is applicable to arbitrary geometries, disconnected contact areas, and different elastic properties for the contacting bodies. The analysis assumes that the contact areas are smooth and the pressure distribution on them for the considered bodies due to the normal load is known beforehand or can be calculated using the procedures discussed in the previous chapter.

### 3.3.1  Problem Formulation

The following nomenclature will be used:

$x, y$ = rectangular coordinates of position

$u, v$ = rectangular coordinates of displacement in the $x$- and $y$-directions respectively

$F_x, F_y$ = rectangular components of traction on a contact area

$E$ = Young's modulus

$v$ = Poisson's ratio

$G$ = modulus of rigidity

$P$ = applied normal force

$T$ = applied tangential force

$f$ = coefficient of friction

$N$ = number of discrete elements in the contact grid

$F_k$ = discretized traction force in the direction of the tangential force at any point $k$

$u_k$ = discretized displacement force in the direction of the tangential force at any point $k$

$y_{1k}$ = displacement slack variables in the direction of the tangential force at point $k$

$y_{2k}$ = force slack variables in the direction of the tangential force at point $k$

The contact area is first discretized into a finite number of rectangular grid elements. Discrete forces can be assumed to represent the distributed shear traction over the finite areas of the mesh. Since the two bodies in contact obey the laws of linear elasticity, the condition for compatibility of deformation can therefore be stated as follows:

$$u_k = \beta \qquad \text{in the no-slip region}$$
$$u_k < \beta \qquad \text{in the slip region} \tag{3.6}$$

where the difference between the rigid body movement $\beta$ and the elastic deformation $u_k$ at any element in the slip region is the amount of slip. The constraints on the traction values can also be stated as:

$$F_k < fP_k \qquad \text{in the no-slip region}$$
$$F_k = fP_k \qquad \text{in the slip region} \tag{3.7}$$

where

$F_k = $ the discretized traction force in the $x$ direction at any point $k$
$P_k = $ the discretized normal force at any point $k$
$\quad f = $ the coefficient of friction

The condition for equilibrium can therefore be expressed as:

$$\sum_{k=1}^{N} F_k = T \tag{3.8}$$

Introducing a set of nonnegative slack variables $Y_{1k}$ and $Y_{2k}$, Eqs (3.6) and (3.7) can be rewritten as follows:

$$u_k + Y_{1k} = \beta \tag{3.9}$$

where

$y_{1k} = 0 \qquad$ in the no-slip region
$y_{1k} > 0 \qquad$ in the slip region

$$F_k + Y_{2k} = fP_k \tag{3.10}$$

where

$Y_{2k} > 0 \qquad$ in the no-slip region
$Y_{2k} = 0 \qquad$ in the slip region

Since a point $k$ must be either in the no-slip region or in the slip region, therefore:

$$\text{either } Y_{1k} = 0 \text{ or } Y_{2k} = 0 \tag{3.11}$$

## 3.3.2 General Model for Elastic Deformation

Since both bodies are assumed to obey the laws of elasticity, the elastic deformation $u_k$ at a point $k$ is a linear superposition of the influences of all the forces $F_j$ acting on a contact area. Accordingly:

$$u_k = \sum_{j=1}^{N} a_{kj} F_j \qquad (3.12)$$

where

$a_{kj}$ = the deformation in the $x$-direction at point $k$ due to a unit force at point $j$

The discrete contact problem can now be formulated in a form similar to that given in Chapter 2 as:

Find $(\overline{F}, \overline{Y}_1, \overline{Y}_2, \beta)$ which satisfies the following constraints:

$$A\overline{F} + I\overline{Y}_1 = \beta e$$
$$I\overline{F} + I\overline{Y}_2 = f\overline{P}$$
$$e^T\overline{F} = T$$

$$
\begin{array}{llll}
Y_{1k} = 0 & \text{or} & Y_{2k} = 0 & \text{for } k = 1, \dots, N \\
F_k \geq 0, & Y_{1k} \geq 0, & Y_{2k} \geq 0, \quad \beta \geq 0 & \text{for } k = 1, \dots, N
\end{array}
$$

where

$A = N \times N$ matrix of influence coefficients
$I = N \times N$ identity matrix
$\overline{F}$ = discretized tangential force vector
$\overline{Y}_1, \overline{Y}_2$ = slack variable vectors
$\overline{P}$ = discretized normal force vector
$e$ = vector of 1's

The problem can be restated in a form suitable for solution by a modified linear program as follows:

$$\text{Minimize} \sum_{i=1}^{2N+1} z_i$$

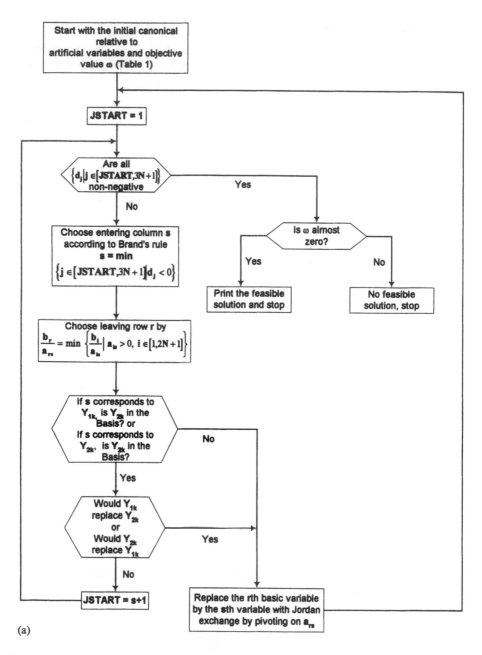

(a)

**Figure 3.5** (a) Flow chart for the analysis algorithm. (b) Initial table.

| $\overline{\mathbf{F}}$ | $\overline{\mathbf{Y}}_1$ | $\overline{\mathbf{Y}}_2$ | $\beta$ | |
|---|---|---|---|---|
| A | I | 0 | E | 0 |
| I | 0 | I | 0 | $f\overline{P}$ |
| $\mathbf{e}^T$ | 0 | 0 | 0 | T |
| $\overline{\mathbf{c}}$ | $-\mathbf{e}^T$ | $-\mathbf{e}^T$ | $-N$ | $\omega$ |

where    $\overline{\mathbf{c}}$ : a cost coefficient vector of length N

$$c_j = -\sum_{i=1}^{N} a_{ij} - 2$$

$\omega$ : initial merit value

(b)         $\omega = -fP - T$

subject to

$$A\overline{F} + I\overline{Y}_1 - \beta e + I\overline{Z}_1 = 0$$
$$I\overline{F} + I\overline{Y}_2 + I\overline{Z}_2 = f\overline{P}$$
$$e^T\overline{F} + z_{2N+1} = T$$

$Y_{1k} = 0$     or     $Y_{2k} = 0$     for $k = 1, \ldots, N$

$F_k \geq 0,$     $Y_{1k} \geq 0,$    $Y_{2k} \geq 0,$    $\beta \geq 0$      for $k = 1, \ldots, N$

$z_i \geq 0$     for $i = 1, \ldots, 2N + 1$

where

$\overline{Z}_1 =$ first $N$ artificial variable vector

$\overline{Z}_2 =$ next $N$ artificial variable vector

$z_{2N+1} =$ artificial variable for the equilibrium equation

The above problem can be solved as a linear programming problem [20] with a modification of the entry rule. Suppose the entering variable is chosen as $Y_{1s}$. A check must be made to see if the $Y_{2s}$ corresponding $Y_{1s}$ is in the basis and if the $Y_{2s}$ is not in the basis, $Y_{1s}$ is free to enter the basis. If $Y_{2s}$ is in the basis, then it must be in the leaving row, $r$, for $Y_{1s}$ to enter the basis. If $Y_{2s}$ is not in the leaving row, $r$, $Y_{1s}$ cannot enter the basis and a new entering variable must be chosen. The same logic can be applied when the chosen entering variable is $Y_{2s}$.

The flow chart for the algorithm utilizing linear programming with modified entry rule is shown in Fig. 3.5.

It is assumed for the circular and elliptical contacts that the surface of contact is very small compared to the radii of curvature of the bodies; therefore, the solution obtained for semi-infinite bodies subjected to point loads can be employed. Accordingly, the influence coefficients, $a_{kj}$, can be expressed as follows [21, 22]:

If $k \neq j$, then

$$A_{kj} = \frac{1}{\pi r_{kj}} \frac{1 - v^2}{E} + \frac{x_{kj}^2}{\pi r_{kj}^3} \frac{v(1 + v)}{E}$$

If $k = j$, then

$$A_{kk} = \frac{1}{\sqrt{\pi A_k}} \frac{(1 + v)(2 - v)}{E}$$

where

$x_{kj} = x_k - x_j$

$y_{kj} = y_k - y_j$

$r_{kj} = (x_{kj}^2 + y_{kj}^2)^{1/2}$

$A_k$ = area of the grid element $k$

### 3.3.3 Illustrative Examples

**EXAMPLE 1: Circular Hertzian Contact with Similar Materials.** The first application of the developed algorithm is finding the traction distribution over the contact area of a steel sphere of 1 in. radius on a steel half space. The normal load is taken as 2160 lbf, the tangential load is 144 lbf and the coefficient of friction is 0.1. A grid with 80 elements is used in this case to approximate the circular contact. A comparison between Mindlin's theory, which is discussed in Section 3.2 (solid line) and the numerical results (symbol s) obtained by the modified linear program is shown in Fig. 3.6 and good agreement can be seen. The rigid body movement ($0.66196 \times 10^{-4}$ in.) was also found to compare favorably with Mindlin's prediction ($0.67139 \times 10^{-4}$ in.) with a deviation of 1.41%.

**EXAMPLE 2: Circular Hertzian Contact with Different Materials.** The contact of steel sphere of 1 in. radius on a rubber half space is considered. The material constants used are as follows:

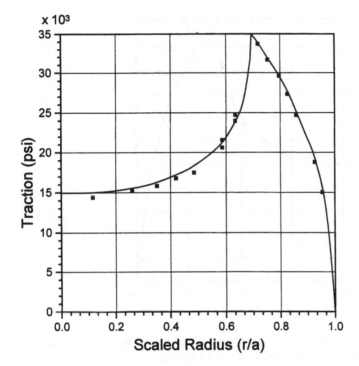

**Figure 3.6** Traction distribution on the circular contact between two bodies of the same material as compared with Mindlin's theory.

$$E_{steel} = 30 \times 10^6 \, psi$$
$$E_{rubber} = 1 \times 10^6 \, psi$$
$$\nu_{steel} = \nu_{rubber} = 0.3$$

A normal load of 270 lbf, a tangential load of 36 lbf, and a coefficient of friction of 0.2 are used in this case.

As shown in Fig. 3.7, the traction distribution (symbol s) shows good agreement with Mindlin's theory (solid line). The rigid body movement of the rubber half space ($0.39469 \times 10^{-3}$ in.) was found to be 30 times that of the steel sphere ($0.13156 \times 10^{-4}$ in.) and both agree well with Mindlin's prediction with a 2.29% deviation when a grid with 80 elements was used.

**EXAMPLE 3: Elliptical Hertzian Contact.** Four cases were investigated in this example with different ratios between the major and minor axes

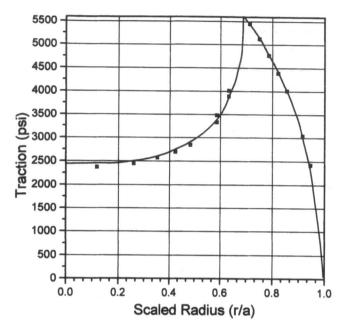

**Figure 3.7** Traction distribution on the circular contact between two bodies of differrent materials as compared with Mindlin's theory.

using different rectangular grid elements, as shown in Table 3.1. A Hertzian-type pressure distribution was assumed in all cases.

The results, which are plotted in Figs 3.8 to 3.11, respectively, show good agreement with Cattaneo's theory [17]. The rectangular grid elements were used in order to save conveniently in computer storage. If a square grid element had been used, better correlation would have been obtained.

The tangential force is applied in the direction of the $a$-axis in all cases.

**Table 3.1** The Four Elliptical Hertzian Contact Cases

| Contact area | Number of grids | Resulting figure |
|---|---|---|
| $a/b = 2.0$ | 112 | Fig. 3.8 |
| $a/b = 0.5$ | 112 | Fig. 3.9 |
| $a/b = 8.0$ | 116 | Fig. 3.10 |
| $a/b = 0.125$ | 116 | Fig. 3.11 |

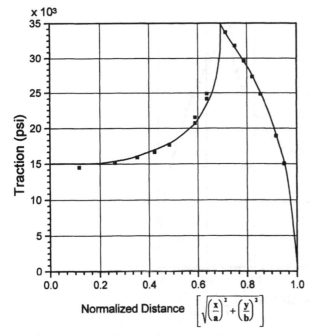

**Figure 3.8** Traction distribution on the elliptical contact as compared with Cattaneo's theory ($a/b = 2$).

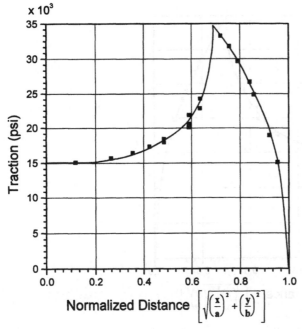

**Figure 3.9** Traction distribution on the elliptical contact as compared with Cattaneo's theory ($a/b = 0.5$).

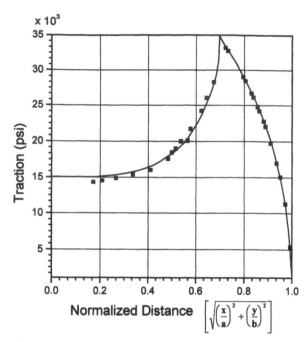

**Figure 3.10** Traction distribution on the elliptical contact as compared with Cattaneo's theory ($a/b = 8$).

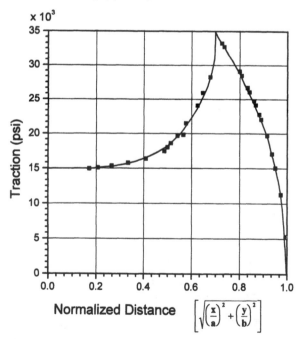

**Figure 3.11** Traction distribution on the elliptical contact as compared with Cattaneo's theory ($a/b = 0.125$).

**EXAMPLE 4: Square Contact Area on Semi-Infinite Bodies with Uniform Pressure Distribution.** A hypothetical square contact area between two steel bodies with uniform contact pressure of 10,000 psi and a coefficient of friction, $f = 0.12$, is discretized with 100 square grid elements. The equal traction contours are shown in Figs (3.12) and (3.13) for a tangential force, $T = 800$ lbf and 1000 lbf, respectively. The development of the slip region with increasing tangential load and the rigid body movement is shown in Figs. 3.14 and 3.15, respectively.

**EXAMPLE 5: Discrete Contact Area on Semi-Infinite Bodies.** Two disconnected square areas of the same size (0.6 in. × 0.6 in) on semi-infinite steel bodies are in contact with uniform pressures assumed on each contact region. The centroids of the two squares are placed 1.0 in. apart.

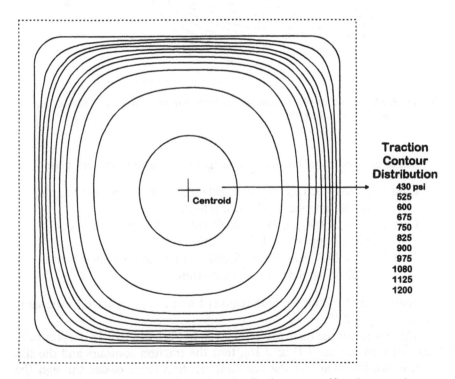

**Figure 3.12** Contour plot of traction distribution on a uniformly pressed square contact area with $T = 800$ lbf.

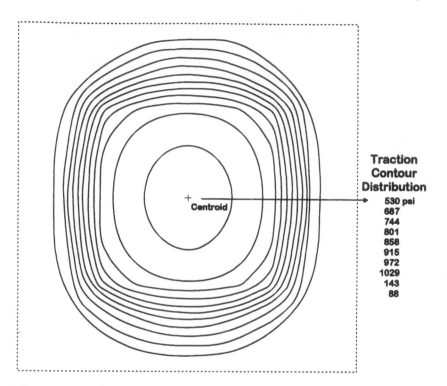

**Figure 3.13** Contour plot of traction distribution on a uniformly pressed square contact area with $T = 1000\,\text{lbf}$.

Three conditions of normal and tangential loading are used here:

Case 1. $P_1 = 10,000\,\text{psi}$, $P_2 = 10,000\,\text{psi}$ and the tangential load is 800 lbf applied at 45° inclination.

Case 2. $P_1 = 20,000\,\text{psi}$, $P_2 = 10,000\,\text{psi}$ and the tangential load is 800 lbf applied at 45° inclination.

Case 3. $P_1 = 20,000\,\text{psi}$, $P_2 = 10,000\,\text{psi}$ and the tangential load is 1200 lbf applied at 45° inclination.

The coefficient of friction on both regions 1 and 2 is assumed to be the same where $f_1 = f_2 = 0.12$.

The results for the three cases are given in Figs 3.16 to 3.18, respectively. It can be seen in Case 1 (Fig. 3.16), that the traction contours and the slip patterns are identical and the resultant traction force passes through the centroid. As would be expected, the other two cases show different traction distributions in the two disconnected contact areas and the resultant trac-

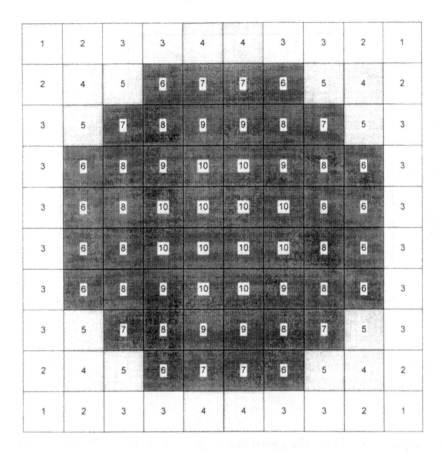

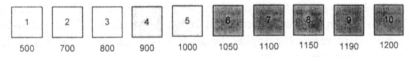

**Figure 3.14** Development of slip regions on a uniformly pressed square contact area with increasing tangential load.

tion force is consequently found to be displaced from the centroid. The effect of the tangential load on the change in distribution of load between the two areas, for the case with region 1 and 2 subjected to normal pressures of 20,000 psi and 10,000 psi respectively, is shown in Fig. 3.19a, and the location of the resultant force for each area, as well as for the entire contact area from the whole area centroid, is given in Fig. 3.19b. An illustration of

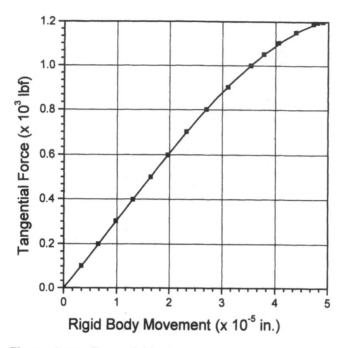

**Figure 3.15**   Tangential load versus rigid body movement curve for a uniformly pressed square contact area.

the sequence of slip for the above case is shown in Fig. 3.20. It can be seen that, in this case, region 2 reached the condition of full slip at a load of 1000 lbf, whereas gross slip occurred for the total contact at 1296 lbf. Some slip is also shown to occur in region 1 below 1000 lbf. A plot of the rigid body movement versus the applied tangential load can be seen in Fig. 3.21 for the three considered cases.

## 3.4   FRICTIONAL CONTACTS SUBJECTED TO A TWISTING MOMENT

### 3.4.1   Preprocessor

One of the boundary conditions, in this case, is that the direction of the displacement in the no-slip region should be circumferential with respect to the center of rotation on the contact surface [23]. A preprocessor determines the direction of the traction at each grid point by satisfying the above

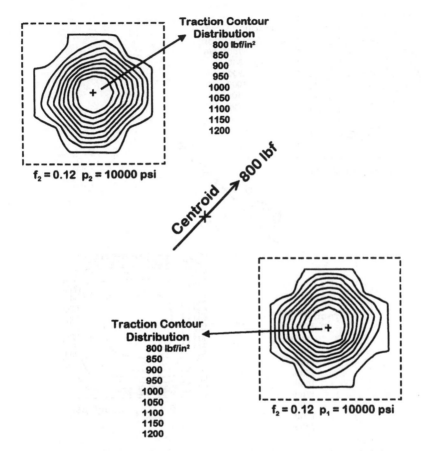

**Figure 3.16** Traction distribution for contact on two discrete square areas under an 800 lbf tangential load applied at a 45° inclination. $P_1 = 10,000$ psi, $P_2 = 10,000$ psi (Case 1).

boundary condition under the assumption of no slip on the entire contact area to linearize the problem [24]. This assumption implicitly implies that the directions of discretized traction forces will not deviate significantly with slip from those with no slip.

### 3.4.2  Problem Formulation

For compatibility of deformation, the circumferential deformation should be equal to the product of the angle of rigid rotation and the radial distance

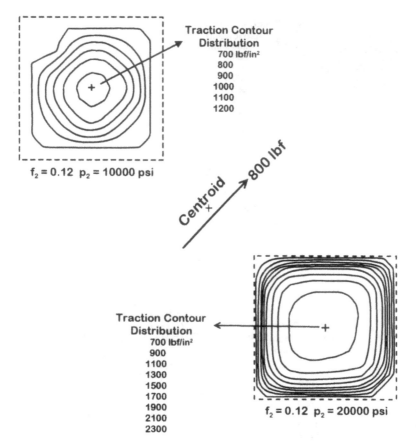

**Figure 3.17** Traction distribution for contact on two discrete square areas under an 800 lbf tangential load applied at a 45° inclination. $P_1 = 20,000$ psi, $P_2 = 10,000$ psi (Case 2).

from the center of rotation in the no-slip region and less than that in the slip region. Because the bodies in contact are assumed to obey the law of linear elasticity, the deformation at a grid point can be expressed as a linear superposition of the effect of all the discretized traction forces acting on a contact surface.

The traction force value should be less than the frictional resistance (the product of the coefficient of friction and the normal force) in the no-slip region and equal to the frictional resistance in the slip region.

For the equilibrium condition, the sum of the moment produced by the discretized traction forces should be equal to the applied twisting moment.

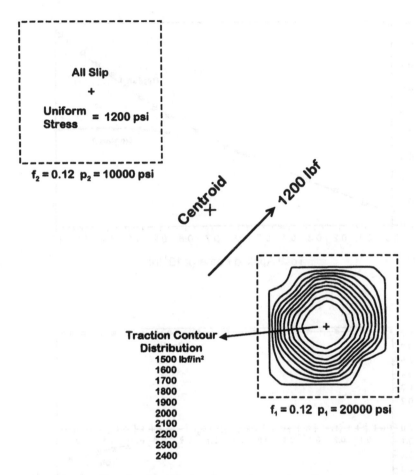

**Figure 3.18** Traction distribution for contact on two discrete square areas under 1200 lb tangential load applied at a 45° inclination. $P_1 = 20,000$ psi, $P_2 = 10,000$ psi (Case 3).

Because the slip region is not known before hand, the complementary condition (a grid point must be either in the no-slip region or in the slip region) should be observed.

The problem is to find a set of the discretized traction forces which satisfies all the above conditions with the assumed center of rotation. The modified linear programming technique offers a readily suitable formulation and is used to obtain the solution.

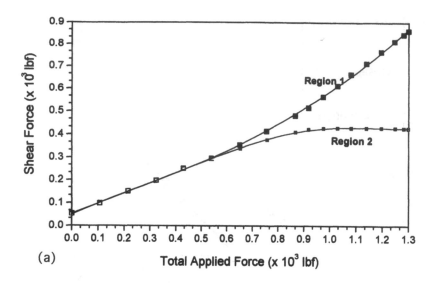

(a)

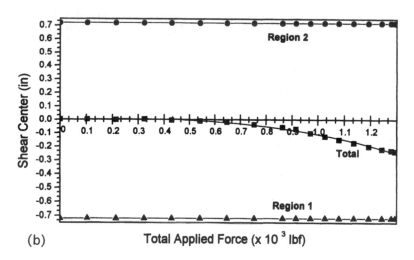

(b)

**Figure 3.19** (a) Tangential load sharing between the two contact zones at different applied loads. (b) Distance between the line of action of the resultant frictional resistance and centroid at different applied loads.

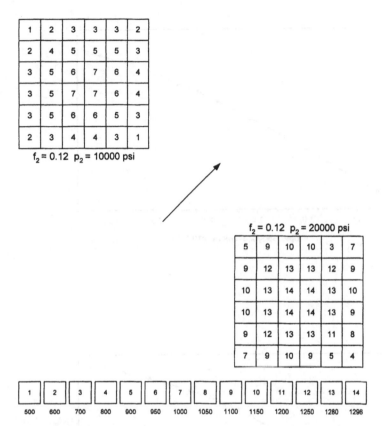

**Figure 3.20** Development of slip regions on a discrete contact area with increasing tangential load.

### 3.4.3 Iterative Procedure

The modified linear programming formulation is first implemented with an initial guess for the center of rotation in order to find the discretized traction forces whose directions are predetermined by the preprocessor. Now the residual forces (the rectangular components of the sum of the traction forces) can be calculated. These residual forces must be equal to zero when the real center of rotation is found, since no tangential forces are applied. The residual forces are then used to modify the center of rotation and the process is repeated until the residual forces vanish. The real center of rotation, the traction force distribution, the microslip region, and the angle of rigid body rotation are determined by this iterative procedure, as depicted in the flow chart (Fig. 3.22).

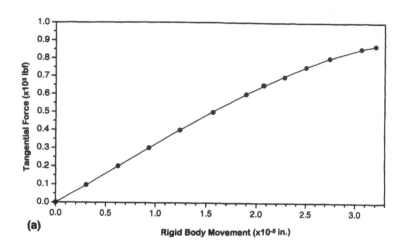

**(a)**

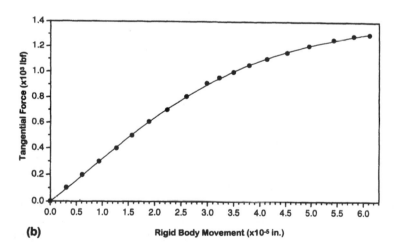

**(b)**

**Figure 3.21**  (a) Joint compliance under same tangential loads before gross slip (Case 1). (b) Joint compliance under different tangential loads before gross slip (Cases 2 and 3).

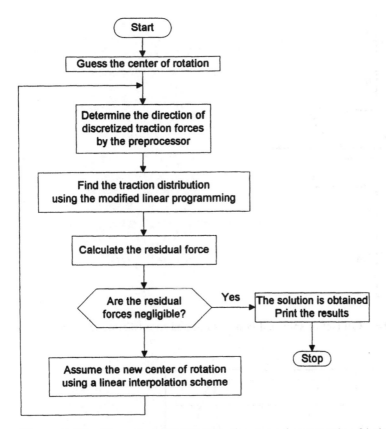

**Figure 3.22** Flow chart for the iterative procedure to solve frictional contact problem subjected to a twisting moment.

### 3.4.4 Illustrative Examples

**EXAMPLE 1: Circular Hertzian Contact.** The contact between two steel spheres of 1 in. radius (Fig. 3.23) is first considered in order to compare the result from the developed procedure with the analytical solution by Lukin [23]. The normal load is taken as 2160 lbf, the twisting moment is 2.45 in. -lbf, and the coefficient of friction is 0.1. A grid with 80 square elements is used to discretize the circular contact area of $0.36628 \times 10^{-1}$ in. radius.

A comparison between Lubkin's theory (solid line) and the numerical results (symbol s) is plotted in Fig. 3.24 and very good agreement can be seen. The angle of rigid rotation ($0.10641 \times 10^{-2}$ rad) is also found to compare favorably with Lubkin's theory ($0.11119 \times 10^{-2}$ rad) with a deviation of 4.30%. The center of rotation is at the centroid.

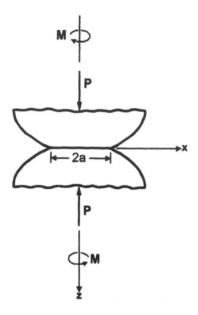

**Figure 3.23** Contact of spherical bodies subjected to a twisting moment.

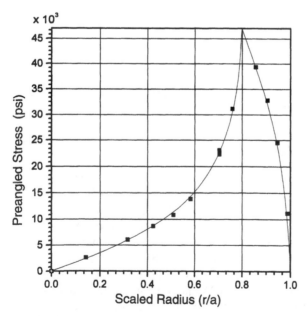

**Figure 3.24** Traction distribution on the circular contact as compared with Lubkin's theory.

**EXAMPLE 2: Elliptical Hertzian Contact.** The elliptical Hertzian contact area with an aspect ratio of 2 is assumed to occur when a normal load of 2160 lbf is applied on two steel bodies. The pressure distribution is assumed to be Hertzian in this case. The coefficient of friction is taken to be equal to 0.1 and a twisting moment of 3.8 in.-lbf is applied on the interface. A grid with 80 rectangular elements of the side ratio of 2 is used to discretize the elliptical contact area.

The contours of the magnitude and the direction of the tractions are plotted in Figs. 3.25 and 3.26. The border line between the no-slip region and the slip region is also shown as a broken line. The centroid in this case is the center of rotation.

## Interpreted Magnitude of Stress

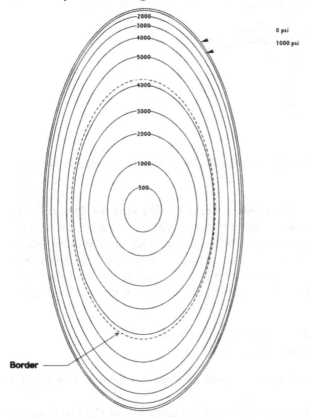

**Figure 3.25** Contour plot for the magnitude of traction on the elliptical contact area.

**Direction of Predirected Stress (N=80)**

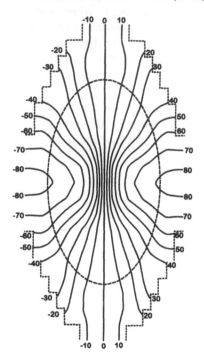

**Figure 3.26**   Contour plot for the direction of traction on the elliptical contact.

**EXAMPLE 3: Disconnected Contact Areas on Semi-Infinite Bodies.**   Two disconnected square areas of the same size (0.6 in. × 0.6 in.) on a semi-infinite steel body are assumed to be in contact with another semi-infinite steel body. The centroids of the two squares are located 1 in. apart (Fig. 3.27). Uniform pressure is assumed on each contact region and the coefficient of friction is 0.12 for both regions. Two cases of normal loading are considered here.

Case 1. $P_1 = 10,000$ psi and $P_2 = 10,000$ psi;
Case 2. $P_1 = 20,000$ psi and $P_2 = 10,000$ psi.

Each region is discretized with 36 square elements to have 72 elements for the entire contact area.

The contours of the magnitude and the direction of the tractions are plotted in Figs 3.27 and 3.28 for Case 1 with a twisting moment of 500 in.-lbf and in Figs 3.29 and 3.30 for Case 2 with a twisting moment of 700 in.-lbf. It

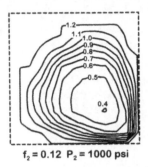

f$_2$ = 0.12 P$_2$ = 1000 psi

500 in-lbf ⊕

f$_1$ = 0.12 P$_1$ = 1000 psi

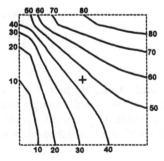

**Figure 3.27** Contour plot for the magnitude of traction on the contact area of disconnected squares with $M = 500$ in.-lbf and normal loading of Case 1.

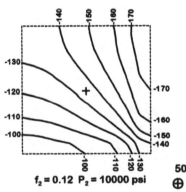

f$_2$ = 0.12 P$_2$ = 10000 psi

500 in-lbf ⊕

f$_1$ = 0.12 P$_1$ = 10000 psi

**Figure 3.28** Contour plot of the direction of traction on the contact area of disconnected squares with $M = 500$ in.-lbf and normal loading of Case 1.

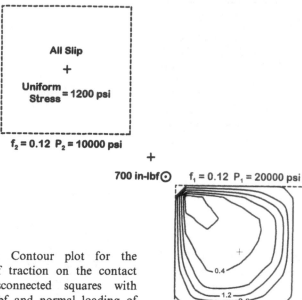

**Figure 3.29** Contour plot for the magnitude of traction on the contact area of disconnected squares with $M = 700$ in.-lbf and normal loading of Case 2.

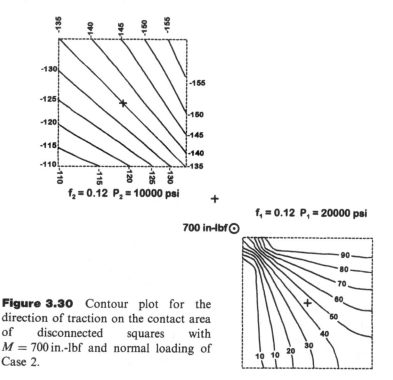

**Figure 3.30** Contour plot for the direction of traction on the contact area of disconnected squares with $M = 700$ in.-lbf and normal loading of Case 2.

*88*

can be seen that the traction contours and the slip patterns for both regions 1 and 2 are identical and the center of rotation is the centroid for the symmetric normal loading. As would be expected, the case of asymmetric normal loading shows different traction distributions in the two disconnected contact areas and the center of rotation is consequently found to be displaced from the centroid. Also notice that region 2 reaches the state of total slip for Case 2, with a twisting moment of 700 in.-lbf, and circumferential tractions are assumed for region 2.

The center of rotation always occurred on the line connecting the centroids of two disconnected squares. The x-distance between the center of rotation and the centroid for Case 2 versus the applied twisting moment is plotted in Fig. 3.31.

The development of the slip region with the increasing twisting moment is shown in Fig. 3.32 for Case 2. It can be seen that region 2 reaches a state of total slip at a twisting moment of 700 in.-lbf, and that gross slip occurs at 770 in.-lbf. Some slip is also shown to occur in region 1 below 700 in.-lbf.

The compliance curve relating the angle of rigid rotation and the twisting moment is plotted in Fig. 3.33a for Case 1 and in Fig. 3.33b for Case 2.

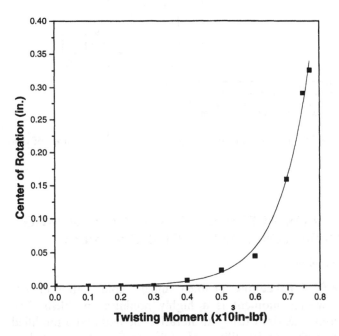

**Figure 3.31** Locations of the center of rotation from the centroid versus applied twisting moments on the contact area of disconnected squares for Case 2.

| 1 | 2 | 2 | 3 | 3 | 2 |
|---|---|---|---|---|---|
| 2 | 3 | 4 | 4 | 4 | 3 |
| 2 | 4 | 5 | 5 | 5 | 4 |
| 3 | 4 | 5 | 5 | 5 | 5 |
| 3 | 4 | 5 | 5 | 5 | 5 |
| 2 | 3 | 4 | 5 | 5 | 5 |

$f_2 = 0.12 \quad p_2 = 10000$ psi

⊙

$f_2 = 0.12 \quad p_2 = 20000$ psi

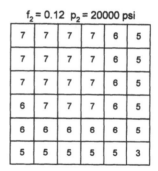

| 7 | 7 | 7 | 7 | 6 | 5 |
|---|---|---|---|---|---|
| 7 | 7 | 7 | 7 | 6 | 5 |
| 7 | 7 | 7 | 7 | 6 | 5 |
| 6 | 7 | 7 | 7 | 6 | 5 |
| 6 | 6 | 6 | 6 | 6 | 5 |
| 5 | 5 | 5 | 5 | 5 | 3 |

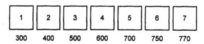

| 1 | 2 | 3 | 4 | 5 | 6 | 7 |
|---|---|---|---|---|---|---|
| 300 | 400 | 500 | 600 | 700 | 750 | 770 |

**Figure 3.32** Progression of slip with increasing twisting for contact area of disconnected squares.

## 3.5 FRICTIONAL CONTACTS SUBJECTED TO A COMBINATION OF TANGENTIAL FORCE AND TWISTING MOMENT

### 3.5.1 Iterative Procedure

The analysis of the frictional contact problem under a combination of tangential force and twisting moment is a highly nonlinear problem. The problem is piecewisely linearized using an iterative method and a modified linear programming technique is utilized at each iteration. The procedure followed in the iterative method is shown in Fig. 3.34.

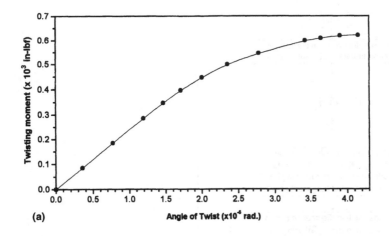

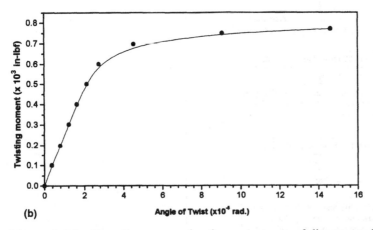

**Figure 3.33** Compliance curve for the contact area of disconnected squares: (a) Case 1; (b) Case 2.

## 3.5.2 Illustrative Examples

**EXAMPLE 1: Circular Hertzian Contact.** The first example is an analysis of the contact between two steel spheres of 2 in. radius (Fig. 3.35). The circular contact area of $5.15 \times 10^{-2}$ in. radius results from a normal load of 3000 lbf and the coefficient of friction is taken as 0.1. The tangential force of 146 lbf and the twisting moment of 4.8 in.-lbf are applied on the contact surface. A grid with 80 square elements is used to discretize the circular contact area.

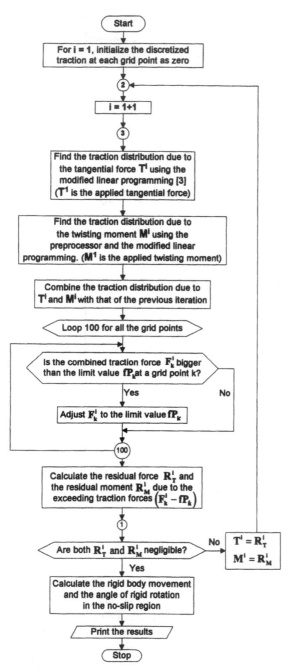

**Figure 3.34** Flow chart for the iterative procedure.

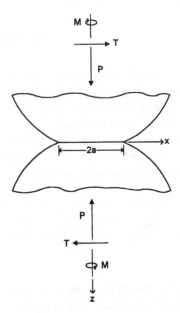

**Figure 3.35** Contact of spherical bodies subjected to a combination of tangential force and twisting moment.

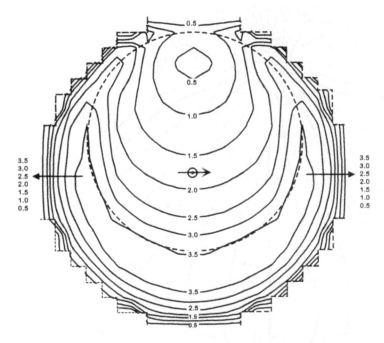

**Figure 3.36** Contour plot for the magnitude of traction on the circular contact subjected to a combined load (using iterative procedure).

The contours of the magnitude and the direction of the traction distribution using the iterative procedure are plotted in Figs 3.36 and 3.37. The border-line between the no-slip region and the slip region is also shown as a broken line. The center of rotation is found to be located at the centroid.

The rigid body movement and the angle of rigid rotation obtained by the iterative procedure ($0.68224 \times 10^{-4}$ in. and $0.10670 \times 10^{-2}$ rad) agree well with those obtained by using a nonlinear programming formulation [24].

The elapse CPU time on a Harris 800 to obtain the above results using the iterative procedure is 14 min, whereas that using the nonlinear programming technique is 31 min when the solution obtained by the iterative procedure is used as an initial guess.

**EXAMPLE 2: Disconnected Contact Area on Semi-Infinite Bodies.** Consider two disconnected square areas of the same size (0.6 in. × 0.6 in.) on a

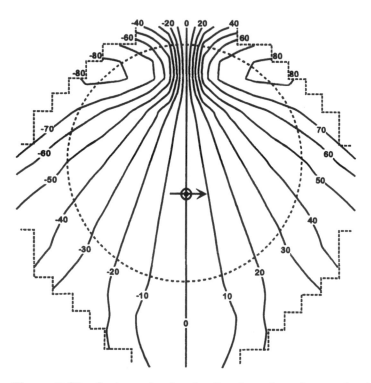

**Figure 3.37** Contour plot for the direction of traction on the circular contact subjected to a combined load (using iterative procedure).

semi-infinite steel body. The centroids of the two squares are located 1 in. apart (Fig. 3.38). Uniform pressure is assumed on each contact region and the coefficient of friction is 0.12 for both regions. Each region is discretized with 36 square elements (72 elements for the entire contact area). Two cases of loading are considered here:

Case 1. $P_1 = 20,000\,\text{psi}$, $P_2 = 10,!000\,\text{psi}$, $T = 500\,\text{lbf}$, $M = 300\,\text{in.-lbf}$.
Case 2. $P_1 = 10,000\,\text{psi}$, $P_2 = 20,000\,\text{psi}$, $T = 600\,\text{lbf}$, $M = 400\,\text{in.-lbf}$.

For Case 1, the contours of the magnitude and the direction of the traction distribution obtained by the iterative procedure are plotted in Figs 3.38 and 3.39 and found to compare favorably with those obtained by applying the nonlinear programming technique [24].

The corresponding results for Case 2 are shown in Figs 3.40 and 3.41. In this case, region 1 is found to be in a state of total slip.

The rigid body motions from the iterative procedure ($0.16436 \times 10^{-4}$ in. and $0.12323 \times 10^{-4}$ rad for Case 1 and $0.30509 \times 10^{-4}$ in. and $0.32721 \times 10^{-4}$ rad for Case 2) compare favorably with those from the nonlinear

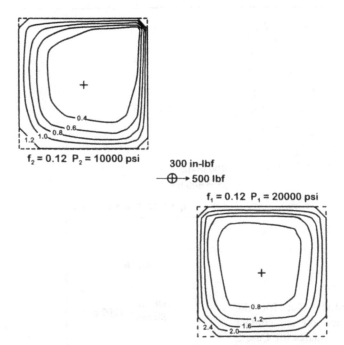

**Figure 3.38** Contour plot for the magnitude of traction on the contact area of disconnected squares for Case 1 (using iterative procedure).

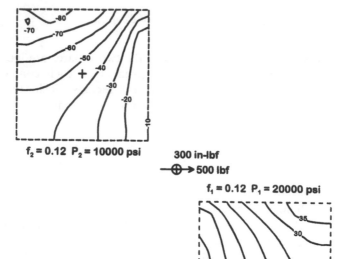

$f_2 = 0.12$  $P_2 = 10000$ psi

300 in-lbf
⊕→ 500 lbf

$f_1 = 0.12$  $P_1 = 20000$ psi

**Figure 3.39** Contour plot for the direction of traction on the contact area of disconnected squares for Case 1 (using iterative procedure).

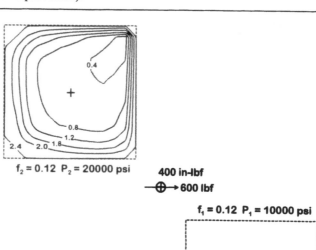

$f_2 = 0.12$  $P_2 = 20000$ psi

400 in-lbf
⊕→ 600 lbf

$f_1 = 0.12$  $P_1 = 10000$ psi

All Slip

+

Uniform Stress = 1200 psi

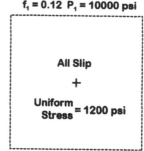

**Figure 3.40** Contour plot for the magnitude of traction on the contact area of disconnected squares for Case 2 (using iterative procedure).

96

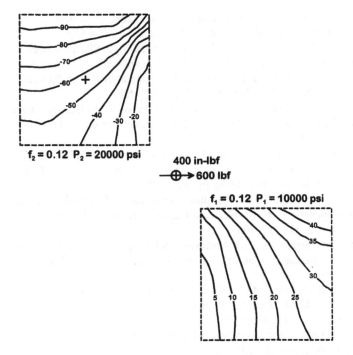

**Figure 3.41** Contour plot for the direction of traction on the contact area of disconnected squares for Case 2 (using iterative procedure).

programming technique $(0.16502 \times 10^{-4}$ in. and $0.12263 \times 10^{-4}$ rad for Case 1 and $0.31257 \times 10^{-4}$ in. and $0.30892 \times 10^{-4}$ rad for Case 2), with deviations of 2.39% and 0.49% for Case 1, and 2.31% and 5.92% for Case 2, respectively.

The elapse CPU times on a Harris 800 to obtain the above results by the iterative procedure are 2 min for Case 1, and 8 min for Case 2, whereas those necessary to obtain the results from the nonlinear programming technique are 18 min for Case 1, and 46 min for Case 2, respectively, when the solutions obtained by the iterative procedure are used as initial guesses.

## REFERENCES

1. Hertz, H., "Miscellaneous Papers" translated by Jones, D. E., and Schott, G. A., Macmillan, New York, NY, 1896, pp. 146–162, 163–183.
2. Lundberg, G., "Elastische Beruhrung Zweier Halbraume,"Forsch. Ingenieurw., 1939, Vol. 10, pp. 201–211.

3. Cattaneo, C., "Teoria del contatto elasiico in seconda approssimazione," University of Rome, Rend., Mat. Appl., 1947, Vol. 6, pp. 504–512.

4. Conway, H. D., "The Pressure Distribution between Two Elastic Bodies in Contact," Z. Angew. Math. Phys., 1956, Vol. 7, pp. 460–465.

5. Greenwood, J. A., and Tripp, J. H., "The Elastic Contact of Rough Spheres," J. Appl. Mech., Trans. ASME, March 1967, pp. 153–159.

6. Schwartz, J., and Harper, E. Y., "On the Relative Approach of Two Dimensional Elastic Bodies in Contact," Int. J. Solids Struct., Dec. 1971, Vol. 7(12), pp. 1613–1626.

7. Tsai, K. C., Dundurs, J., and Keer, L. M., "Contact between an Elastic Layer with a Slightly Curved Bottom and a Substrate," J. Appl. Mech., Trans. ASME, Sept. 1972, Ser. E., Vol. 39(3), pp. 821–823.

8. Kalker, J. J., and Van Randen, Y., "Minimum Principle for Frictionless Elastic Contact with Application to Non-Hertzian Contact Problems,"J. Eng. Math., April 1972, Vol. 6(2), pp. 193–206.

9. Conry, T. F., and Seireg, A., "A Mathematical Programming Method for Design of Elastic Bodies in Contact," J. Appl. Mech., Trans. ASME, June 1971, pp. 387–392.

10. Erdogan, F., and Ratwani, M., "Contact Problem for an Elastic Layer Supported by Two Elastic Quarter Planes," J. Appl. Mech., Trans. ASME, Sept. 1974, Ser. E, Vol. 41(3), pp. 673–678.

11. Nuri, K. A., "Normal Approach between Curved Surfaces in Contact," Wear, Dec. 1974, Vol. 30(3), pp. 321–335.

12. Francavilla, A., and Zienkiewicz, O. C., "Note on Numerical Computation of Elastic Contact Problems," Int. J. Numer. Meth. Eng., 1975, Vol. 9(4), pp. 913–924.

13. Haug, E., Chand, R., and Pan, K., "Multibody Elastic Contact Analysis by Quadratic Programming," J. Optim. Theory Appl., Feb. 1977, Vol. 21(2), pp. 189–198.

14. Kravchuk, A. S., "On the Hertz Problem for Linearly and Non-Linearly Elastic Bodies of Finite Dimensions," Appl. Math. Mech., 1977, Vol. 41(2), pp. 320–328.

15. Goriacheva, I. G., "Plane and Axisymmetric Contact Problems for Rough Elastic Bodies," Appl. Math. Mech., 1979, Vol. 43(1), pp. 104–111.

16. Mindlin, R. D., "Compliance of Elastic Bodies in Contact," J. Appl. Mech., Trans. ASME, 1949, Vol. 16, pp. 259–268.

17. Cattaneo, C., "Sul Contatto di due Corpi Elastici: Distribuzione Locale Degli Sforzi," Accad. Lincei, Rendic., 1938, Ser. 6, Vol. 27, pp. 342–348, 434–436, 474–478.

18. Johnson, K. L., "Surface Interaction Between Elastically Loaded Bodies Under Tangential Forces," Proc. Roy. Soc. (Lond.), 1955, A, Vol. 230, pp. 531–548.

19. Deresiewicz, H., "Oblique Contact of Non-Spherical Elastic Bodies," J. Appl. Mech., Trans. ASME, 1967, Vol. 24, pp. 623–624.

20. Danzig, G. W., Linear Programming and Extensions, Princeton University Press, Princeton, NJ, 1963.
21. Love, A. E. H., A Treatise on the Mathematical Theory of Elasticity, 4th ed., Dover Book Company, New York, NY, 1944.
22. Timoshenko, S. P., and Goodies, J. N., Theory of Elasticity, 3rd ed., McGraw Hill Book Company, New York, NY, 1970.
23. Lubkin, J. L., "The Torsion of Elastic Spheres in Contact," J. Appl. Mech., Trans. ASME, 1951, Vol. 73, pp. 183–187.
24. Choi, D., "An Algorithmic Solution for Traction Distribution in Frictional Contacts," Ph.D. Thesis, The University of Wisconsin–Madison, 1986.

# 4

# The Contact Between Rough Surfaces

## 4.1  SURFACE ROUGHNESS

All surfaces, natural or manufactured, are not perfectly smooth. The smoothest surface in natural bodies is that of the mica cleavage. The mica cleavage has a roughness of approximately 0.08 μin. The roughness of manufactured surfaces vary from a few microinches to 1000 μin. depending on the cutting process and surface treatment. Representative examples of some of these are given in Table 4.1.

Roughness represents the deviation from a nominal surface and is a composite of waviness and asperities. Both of these are shallow curved surfaces with the latter having wavelengths orders of magnitude smaller than the former. Asperities can be also considered as wavy surfaces on a microscale with their height being in the order of 2–5% of the wavelength, as illustrated in Fig. 4.1.

## 4.2  SURFACE ROUGHNESS GENERATION

Surface roughness plays an important role in machine design. During the metal cutting operation, a machined surface is created as a result of the movement of the tool edge relative to the workpiece. The quality of the surface is a factor of great importance in the evaluation of machine tool productivity. The results from a large number of theoretical and experimental studies on surface roughness during turning are available in the

*100*

**Table 4.1**    Average Surface Roughness for Some Common Manufacturing Processes

| Manufacturing process | Average surface roughness (μin.) |
|---|---|
| Super finishing | 2–8 |
| Lapping | 2–16 |
| Polishing | 4–16 |
| Honing | 4–32 |
| Grinding | 4–63 |
| Electrolytic grinding | 8–16 |
| Barrel finishing | 8–32 |
| Boring, turning | 16–250 |
| Die casting | 32–63 |
| Cold rolling, drawing | 32–125 |
| Extruding | 32–125 |
| Reaming | 32–125 |
| Milling | 32–250 |
| Mold casting | 63–125 |
| Drilling | 63–250 |
| Chemical milling | 63–250 |
| Elect. discharge machining | 63–250 |
| Planing, shaping | 63–500 |
| Sawing | 63–1000 |
| Forging | 125–500 |
| Snagging | 250–1000 |
| Hot rolling | 500–1000 |
| Flame cutting | 500–1000 |
| Sand casting | 500–1000 |

literature [1–18]. Although various factors affect the surface condition of a machined part, it is generally accepted that the cutting parameters such as speed, feed, rate, depth of cut, and tool nose radius have significant influence on the surface geometry for a given machine tool and workpiece setup. There is also general agreement that surface roughness improves with increasing machine tool stiffness, cutting speed, and tool nose radius, and decreasing feed rate [1–3]. It has also been reported [2, 9] that at speeds less than a certain value, discontinuous or semidiscontinuous chips and built-up-edge formation may occur, which can give rise to poor surface finish. At speeds above that specific value, the built-up-edge size decreases and the surface finish improves. This specific speed limit depends on many factors such as workpiece, tool conditions, and the state of the machine tool. Sata [9] reported 22.86 m/min as the speed limit in his experiment. Of the factors influencing the surface roughness, the depth of cut was found to have the

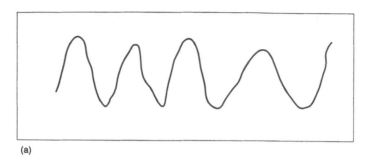

(a)

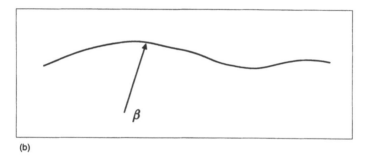

(b)

**Figure 4.1**  (a) Profile of surface roughness. (b) Asperity curvature without height modification.

least effect. Statistical techniques, developed by Box and Wilson, have been applied to establish a predictive equation for the relationship between tool life or surface roughness and cutting conditions [6, 7, 19].

Other studies indicated that tool wear causes the surface finish to deteriorate rapidly and has a direct effect on the maximum roughness [4]. Also, the principal cutting edge and hardness of the workpiece material itself are found to affect the surface roughness [5, 20]. Some studies presented stochastic models to characterize the indeterministic components of surface texture [16, 20].

For a tool with a finite radius in an idealized cutting condition with a rigid machine tool, the peak-to-valley roughness, $R_{max}$, which is known as kinematic roughness, in the case of small values of feed can be shown to be [14]:

$$R_{max} = \frac{f^2}{8r} \text{ for } f \leq 2r\sin D_\theta \qquad (4.1)$$

where

$R_{max}$ = peak-to-valley surface roughness
  $f$ = feed rate
  $r$ = tool radius
  $D_\theta$ = end relief angle

Based on an extensive experimental study using one lathe, Hasegawa et al. [8] developed a statistical relationship of peak-to-valley surface roughness, $R_{max}$ in terms of the cutting speed, feed rate, depth of cut, and tool nose radius using a response surface method. The statistical relationship is given approximately by the following:

$$R_{max} = 625 V^{-0.433} f^{0.813} d^{0.034} r^{-0.47} \qquad (4.2)$$

where

$v$ = cutting speed
$d$ = depth of cut

It can be seen that considerable variations in the calculated surface roughness may result from the use of Eqs (4.1) and (4.2), as well as any of the numerous empirical equations available in the literature [1–8]. One of the main reasons for the discrepancies is that the vibratory behavior is defined by the relative movement of the tool with respect to the workpiece in the machining process and has deterministic and stochastic components. As reviewed above, the stochastic component is assumed to be the results of the random excitation during the cutting process and the deterministic component depends on the dynamic characteristics of the machine tool. In order to select or modify a machine tool, which can be used to generate a particular surface quality, it is necessary to quanfity the influence of its dynamic parameters on surface roughness [21–28].

A piecewise dynamic simulation of the interaction between the tool and the workpiece system in turning is reported by Jang and Seireg [29]. A generalized computer-based model is developed for predicting surface roughness for any given condition which takes into consideration all the important parameters influencing the deterministic vibratory behavior of the machine tool–workpiece system. The parameters considered in the simulation are: the feed rate, cutting speed, depth of cut, radius of cutting edge, the dimensions of the workpiece, and the mass, stiffness, and damping of the machine structure as well as the cutting tool assembly.

Extensive numerical results from the simulation suggest that the uncoupled natural frequency of the tool assembly is the fundamental parameter controlling the generated surface roughness. The parametric equation, which is developed from the simulation results, is found to be in good agreement with the data obtained from an extensive series of tests using mild steel specimens. The simulation results also show that the well-known kinematic equation for predicting surface roughness based on the geometry of the cutting process gives essentially the same results as the simulation when the tool natural frequency is greater than 150 Hz.

Figure 4.2 gives a sample result from the simulation. It shows the average generated surface roughness along the generatrix of the workpiece. The roughness values were obtained by displacing the tool edge an amount equal to the sum of the average relative vibration at each axial location and the kinematic roughness.

The simulation was utilized to develop generalized equations for surface roughness based on the output from the vibratory model. A generalized equation of the following form is assumed:

$$R_{\max} = e^c V^{k1} f^{k2} d^{k3} r^{k4} \tag{4.3}$$

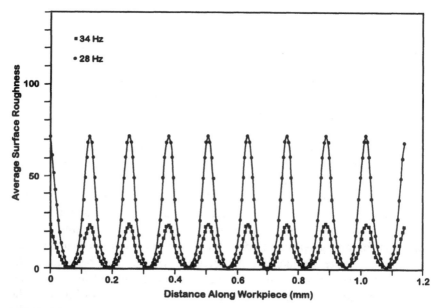

**Figure 4.2** Average surface roughness along the axis of workpiece ($V = 61\,\mathrm{m/min}$, $f = 0.127\,\mathrm{mm/rev}$, $d = 0.305\,\mathrm{mm}$, $r = 0.794\,\mathrm{mm}$). (■) $\bar{\omega} = 34\,\mathrm{Hz}$, (●) $\bar{\omega} = 28\,\mathrm{Hz}$.

The results as generated from the simulation for cutting conditions covering the practical range of applications are used in a regression analysis to obtain the best fit for the equation parameters $C, k_1, k_2, k_3,$ and $k_4$.

The values of the system parameters of Eq. (4.3) for the case of a chuck mass $(M_s)$ of 34 kg and machine structures with stiffness $K_s$ greater than $10^8$ N/m were found to be independent of $K_s$ and are only dependent on the uncoupled tool natural frequency, $\bar{\omega}$.

All the simulated results were curve-fitted to give the following equations:

$$C = 7.02 - 2.2(1 - e^{-(\bar{\omega}-29)/18.3})$$
$$k_1 = -0.3712 + 0.3712(1 - e^{-(\bar{\omega}-29)/21})$$
$$k_2 = 0.6302 + 1.3626(1 - e^{-(\bar{\omega}-29)/31.3})$$
$$k_3 = 0.5425 - 0.5425(1 - e^{-(\bar{\omega}-29)/10.3})$$
$$k_4 = -0.3419 - 0.6523(1 - e^{-(\bar{\omega}-29)/51.5}) \qquad (4.4)$$

where

$\bar{\omega}$ = natural frequency of tool assembly (Hz)

These equations are applicable for the following conditions when $K_s$ is greater than $10^8$ N/m:

$$61\text{m/min} < V < 305\,\text{m/min}$$
$$0.127\,\text{mm/rev} < f < 0.88\,\text{mm/rev}$$
$$0.31\,\text{mm} < d < 0.71\,\text{mm}$$
$$0.79\,\text{mm} < r < 2.38\,\text{mm}$$

## 4.3 THE REAL AREA OF CONTACT BETWEEN ROUGH SURFACES

Analytical studies and measurements show that the real area of contact between surfaces occurs at isolated points where the asperities came together. This constitutes a very small fraction of the apparent area for flat surfaces (Fig. 4.3a) or the contour area for curved surfaces (Fig. 4.3b).

For the case of steel on steel flats, the real area of contact is in the order of 0.0001 cm$^2$ per kilogram load. This indicates that pressure on the microcontacts for any combination of materials is constant and is independent of load.

The interactions between the two bodies at the real area are what determines the frictional resistance and wear when they undergo relative

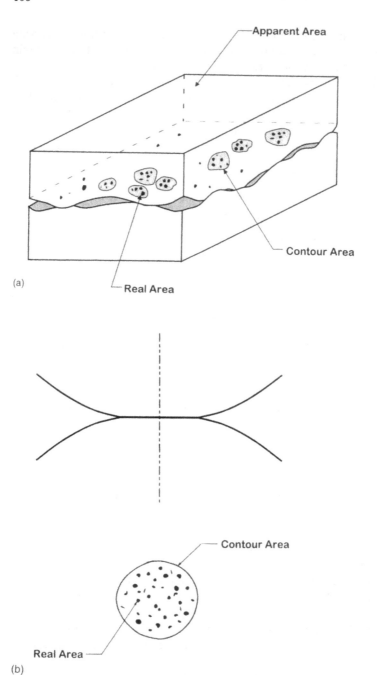

**Figure 4.3**   (a) Contact of flat surfaces. (b) Contact of spherical surfaces.

sliding. Even in the case of a Hertzian contact, the pressure distribution is not continuous. Due to surface roughness it occurs at discrete points, and the force between the bodies is the sum of the individual forces on contacting asperities which constitute the pressure distribution. An interesting investigation of this problem was conducted by Greenwood and Tripp [30]. They analyzed the contact between rough spheres by a physical model of a smooth sphere with the equivalent radius of both spheres pressed against a rough flat surface where the asperity heights follow a normal Gaussian distribution about a mean surface, as illustrated in Fig. 4.4a. They further assumed that the tops of the asperities are spherical with the same radius and that they deform elastically according to Hertz theory. The forces on the asperities constitute the loading on the nominal smooth bodies whose deformation controls the extent of the asperity contacts. The problem is solved iteratively until convergence occurs.

Assuming that the radius of the asperities is $\beta$, the radius of the contact on top of an asperity due to a penetration depth $\omega$ can be calculated as (Fig. 4.4b):

$$\omega = \frac{a^2}{\beta}$$

and the corresponding area of contact

$$A_i = \pi a^2 = \pi \beta \omega$$

From the Hertz theory, the load on the asperity can be calculated as:

$$P_i = \tfrac{4}{3} E_e \beta^{1/2} \omega^{3/2}$$

where

$$\frac{1}{E_e} = \frac{1 - v_1^2}{E_1} + \frac{1 - v_2^2}{E_2}$$

$E_1, E_2$ = elastic moduli for the two materials
$v_1, v_2$ = Poisson's ratios for the materials

If $z$ is the height of the asperity and $u$ is the distance between the nominal surfaces at that location, it can be seen from Fig. 4.2b that $\omega = z - u$. Assuming a height distribution probability function $\phi(z)$, the probability that an asperity is in contact at any location with nominal separation $u$ can be expressed as:

$$\text{prob}(z > u) = \int_u^\infty \phi(z)\, dz$$

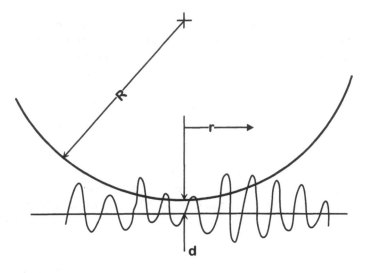

**(a)**

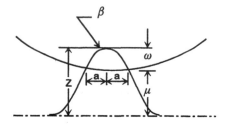

**(b)**

**Figure 4.4** (a) Nominal surfaces and superimposed asperities. (b) Contact between rough surfaces – asperity contact.

The expected force is:

$$\overline{P}_i = \int_u^\infty \tfrac{4}{3} E_e \beta^{1/2} (z - u)^{3/2} \phi(z) \, dz$$

If the asperity density is assumed to be $\eta$, the expected number of asperities to be in contact over an element of the surface $(\partial a)$, where the separation between the nominal surfaces is $u$ can be expressed as:

$$\partial N = \eta(\partial a) \int_u^\infty \phi(z) \, dz$$

The expected area of contact:

$$\partial A = \pi \eta \beta (\partial a) \int_u^\infty (z - u)^{3/2} \phi(z)\, dz$$

and the expected load within $(da)$ is:

$$dP = \tfrac{4}{3} \eta E_e \beta^{1/2} (da) \int_u^\infty (z - u)^{3/2} \phi(z)\, dz$$

Assuming the standard deviation of the asperity heights to be equal to $\sigma$, the following dimensionless relationships were used in developing a general formulation for the problem:

$$\text{Asperity penetration } \omega^* = \frac{\omega}{\sigma}$$
$$\text{Separation of nominal surface at any location } h = \frac{u}{\sigma}$$
$$\text{Minimum separation between the nominal surface } d^* = \frac{d}{\sigma}$$
$$\text{Radial distance in the contact region } \rho = \frac{r}{\sqrt{2R\sigma}}$$
$$\text{Pressure } p^* = \frac{p}{E_e \sqrt{\sigma/(8R)}}$$
$$\text{Height of asperity } S = \frac{z}{\sigma}$$

Accordingly, the equations for the contact conditions within an elementary area $(da)$ can be written as:

$$dN = [\eta(da)]F_0(h)$$
$$\frac{dA}{(da)} = (\pi \eta \beta \sigma)F_1(h) = \frac{\text{real area}}{\text{nominal area}}$$
$$\frac{dP}{(da)} = (\tfrac{4}{3}\eta \beta^{1/2}\sigma^{3/2})F_{3/2}(h) = \text{average pressure}$$

where

$$F_n(h) = \int_{\hat{u}}^\infty (S - h)^n \phi^*(s)\, ds$$
$$\phi^*(s) = \frac{1}{(2\pi)^{1/2}} e^{-s^2/2}$$

The two main independent variables are the total load and the surface roughness. These were used in a dimensionless form as follows to evaluate the contact conditions:

$$T = \text{dimensionless total load} = 2\,P/\sigma\,E_e\sqrt{\sigma R}$$
$$\mu = \text{dimensionless roughness parameter} = \tfrac{8}{3}\eta\sigma\sqrt{2R\beta}$$

Accordingly:

$$T = \int_0^\infty 2\pi\rho\,p^*(\rho)\,dp$$

$a^*$ = dimensionless parameter representing the spread of pressure over the contact region (i.e., affective radius), of the contact area $= \dfrac{1}{2}\left(\dfrac{3T}{2}\right)^{1/3}$

$q_0^*$ = dimensionless maximum value of the average contact pressure

$$= \frac{8a^*}{\pi} = 4\left(\frac{3T}{2}\right)^{1/3}$$

$q_\rho^*$ = dimensionless average contact pressure $= q_0^*\left[1 - \left(\dfrac{\rho}{a^*}\right)^2\right]^{1/2}$

Numerical results are presented in Greenwood and Tripp [30] based on the previous analysis from which the following conclusions can be stated:

1.  Load has remarkably small effect on the mean real pressure on top of the asperities. This is illustrated by the numerical results given in Fig. 4.5.
2.  Consequently the mean real area of contact is approximately linearly dependent on the applied load.
3.  The proportionality constant between the real area and load increases with increased root mean square (r.m.s.) roughness ($\sigma$) decreased asperity density and decreased raidus of the asperities.
4.  The effective radius of the area over which the pressure is spread is considerably larger than the Hertzian contact radius for low loads and approaches the Hertzian contact condition for high loads. Consequently, the average mean pressure is considerably lower than the Hertzian pressure for low loads and approaches it for high loads. This is illustrated in Fig. 4.6.

It is interesting to note that the first two conclusions are the same as those noted by Bowden and Tabor [31] and the electric contact resistance measurements reported by Holm [32]. The constant value of the average pressure on the real area of asperity contact was assumed to be the yield stress at the asperity contacts. However, the analysis presented by Greenwood and Tripp discussed in this chapter provides a rational proce-

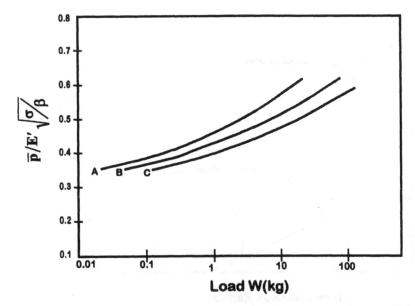

**Figure 4.5** Effect of load on mean real pressure. A: $\eta = 500/mm^2$, $\sigma = 5 \times 10^{-4}$ mm, $\beta = 0.2$ mm; B: $\eta = 940/mm^2$, $\sigma = 5 \times 10^{-4}$ mm, $\beta = 0.2$ mm; C: $\eta = 500/mm^2$, $\sigma = 9.4 \times 10^{-4}$ mm, $\beta = 0.2$ mm. (From Ref. 30.)

dure based on elastic deformation of the asperities for calculating this constant stress value from the surface roughness data and the elastic constants of the surface layer. Later investigations showed that a combination of elastic and plastic asperity contacts can occur for typical surface finishing processes depending on the load and the thickness of the lubricating film. This will be discussed later in the book.

## 4.4 THE INTERACTION BETWEEN ROUGH SURFACES DURING RELATIVE MOTION

It has been shown in the last section that the contact between elastic bodies with rough surfaces occurs at discrete points on the top of the asperities. The interaction takes place at surfaces covered with thin layers of materials, which have different chemical, physical, and thermal characteristics from the bulk material. These surface layers which unite under pressure due to the influence of molecular forces, are damaged when the contact is broken by relative movement. During the making and breaking of the contacts, the

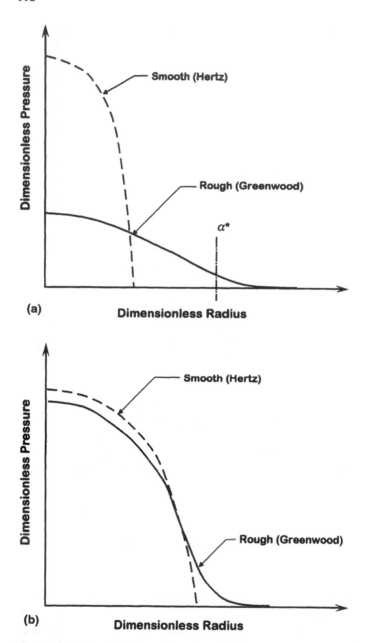

**Figure 4.6**  Comparison of pressure distribution for rough and smooth surfaces: (a) low load; (b) high load.

underlying material deforms. The forces necessary to the making and breaking of the contacts, in deforming the underlying material constitute the frictional resistance to relative motion. It can therefore be concluded that friction has a dual molecular–mechanical nature. The relative contribution of these two components to the resistance to movement depends on the types of materials, surface geometry, roughness, physical and chemical properties of the surface layer, and the environmental conditions in which the frictional pair operates.

## 4.5   A MODEL FOR THE MOLECULAR RESISTANCE

Molecular resistance or adhesion between surfaces is a function of the real area of contact and molecular forces which take place there. A theoretical relationship describing the effect of the molecular forces can be given as:

$$F = \frac{\pi^2 hc}{240 l^4} \left( 1 - 7.2 \sqrt{\frac{m}{n}} \frac{c}{el} \right)$$

where

$h$ = Planck's constant = $6.625 \times 10^{-27}$ erg-sec

$c$ = speed of light

$m, n, e$ = mass, charge, and volume density of electrons in the solid

$l$ = distance between the contacting surfaces

Adhesive forces are generally not significant in metal-to-metal contacts where the surfaces generally have thin chemical or oxide layers. It can be significant, however, in contacts between nonmetals or metals with thin wetted layers on the surface as well as in the contacts between micromachined surfaces.

## 4.6   A MODEL FOR THE MECHANICAL RESISTANCE

The role of roughness in the frictional phenomena has been a central issue since Leonardo da Vinci's first attempt to rationalize the frictional resistance. His postulation that frictional forces are the result of dragging one body up the surface roughness of another was later articulated by Coulomb. This rationale is based on the assumption that both bodies are rigid and that no deformation takes place in the process.

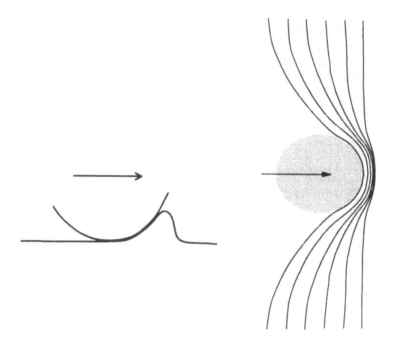

**Figure 4.7** Surface waves generated by asperity penetration.

A modern interpretation of the mechanical role of roughness is based on the elastic deformation of the contacting surfaces due to asperity penetration. The penetrating asperity moving in a tangential direction deforms the underlying material and gives rise to a semi-cylindrical bulge in front of the identor which is lifted up and also spreads sideways as elastic waves. This is diagrammatically illustrated in Fig. 4.7. The size of the bulge depends on the relative depth of penetration $\omega/\beta$, where $\omega$ is the penetration depth and $\beta$ is the radius of the asperity. The process is analogous to that of the movement of a boat creating waves on the water surface. According to this theory, the energy dissipated in the process of deforming the surface is the source of the mechanical frictional resistance and the surface waves generated are the source of frictional noise.

## 4.7  FRICTION AND SHEAR

Both friction and shear represent resistance to tangential displacement. In the first case, the traction resistance is on the surface or "external" to the

**Table 4.2** Friction and Shear

|  | Traction force | Contact | Direction of material displacement | Characteristic of displacement |
|---|---|---|---|---|
| Friction | External | Discrete | Perpendicular to the movement | Sinusoidal waves |
| Shear | Internal | Continuous | Parallel to the movement | Laminar |

body. In the case of shear, the resistance is "internal" in the bulk material. A comparison between the two phenomena can be summarized in Table 4.2.

It should also be noted that friction occurs when the strength of the surface layers is lower than the underlying layers. On the other hand, if the surface layers are harder to deform than the underlying layers, it is expected that shear would occur. In other words, friction can be associated with a "positive gradient" of the mechanical properties with depth while shear can be associated with a "negative gradient" of the material properties with depth below the surface. As illustrated in Fig. 4.8, the former causes gradual destruction of the surface layer with severity depending on the number of passes that one surface makes on the other. A negative gradient of the strength of the surface layer would result in rapid destruction of the bulk material which occurs at the depth where the strength of the material is below what is necessary to sustain the tangential load.

## 4.8 RELATIVE PENETRATION DEPTH AS A CRITERION FOR THE CONTACT CONDITION

An indentor with spherical top is assumed in order to develop a qualitative criterion for the effect of the depth of penetration on the stress condition on

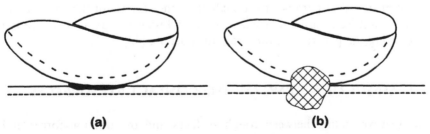

**(a)**          **(b)**

**Figure 4.8** Effect of shear strength gradient on surface damage. (a) $d\tau/dh > 0$, destruction of surface layer. (b) $d\tau/dh < 0$, destruction of bulk material.

the surface. The model can be applied on a microscale where the identor is an asperity, or a macroscale where the indentor is a cutting tool with a spherical radius. Assuming homogeneous materials and applying the Hertz theory, the following relationships can be written:

$$h = \text{penentration depth} = \frac{a^2}{R}$$

$$a = 0.88 \sqrt[3]{\frac{P}{E}} \, R$$

$$q_0 = 0.66 \sqrt[3]{\frac{P}{R^2}} \, E^2 = \text{maximum contact pressure}$$

where

$a$ = radius of contact area
$R$ = radius of the identor
$P$ = applied load
$E$ = effective modulus of elasticity

The relative penetration depth can therefore be expressed as:

$$\frac{h}{R} = \frac{a^2}{R^2} = \frac{(0.7744)P^{2/3}R^{2/3}}{R^2 E^{2/3}}$$

Substituting for $P^{2/3}$ from:

$$P^{2/3} = \frac{q_0^2 R^{4/3}}{0.4356 E^{4/3}}$$

$$\frac{h}{R} = \frac{0.7744 q_0^2 R^{4/3} R^{2/3}}{0.4356 E^{4/3} E^{2/3} R^2} = 1.78 \left(\frac{q_0}{E}\right)^2$$

The above equation shows that the relative depth of penetration can be used as a dimensionless parameter for evaluating the severity of the contact and its transition from elastic to plastic to cutting. Figure 4.9 gives an illustration of utilizing the penetration ratio for this purpose [32].

## 4.9 EFFECT OF SLIDING ON THE CONTACTING SURFACES

The relative sliding between rough surfaces and the traction forces and frictional energy generated in the process result in a change in the temperature and properties of the surface and the layers beneath it. High thermal

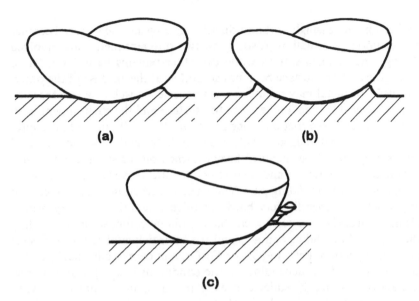

**Figure 4.9**   Effect of relative penetration of severity of contact. (a) Elastic contact: $h/R < 0.01$ ferrous metals; $h/R < 0.0001$ nonferrous metals. (b) Plastic contact: $h/R < 0.1$ dry contact; $h/R < 0.3$ lubricated contact. (c) Microcutting: $h/R > 0.1$ dry contact; $h/R > 0.3$ lubricated contact.

flux can be expected at the asperity contacts for high sliding speeds, and the corresponding thermal gradients can produce high thermal stresses in the asperity and material layers near the surface. Because of its importance, the thermal aspects of frictional contacts will be discussed in greater detail in the next chapter.

The changes in the surface properties that occur include those caused by deformation and strain of the surface layer, by the increase in surface temperature and by the chemical reaction with the environment.

Deformation at the surface may produce microcracks in the surface layer and consequently reduce its hardness. The combination of compressive stress and frictional force and interaction with the environment can cause structural transformation in the surface material known as mechanochemistry. Also, a marked degree of plasticity may occur, even in brittle materials, as a result of the nonuniform stress or strain at the surface. High microhardness may also occur immediately below the surface as a result of sliding. Its depth varies with the parameters contributing to the work-hardening process.

It should be noted that if the contact temperature exceeds the recovery temperature (i.e., the recrystallization temperature of the alloy), the surface layers become increasingly soft and ductile. As a result, the surface becomes smoother upon deformation. Also, when two different metals are involved in sliding, one of them softens while the other remains hard. Transfer of metal occurs and one surface becomes smoother at the expense of the other. The transfer of metal may occur on a microscale, as well as a macroscale.

The chemical interaction between the surface and the environment is an important result of the frictional phenomenon. It is well known that appreciably deformed materials are easily susceptible to oxidation and chemical reactions in general. The chemical layers formed on the surface can significantly influence the friction and wear characteristic, as well as the transfer of frictional heat into the sliding pair. The chemical reaction can produce thin layers, which are generally very hard, on thick layers that are very brittle. Oxide films formed on the surface can have different compositions depending on the nature of the sliding contacts and the environmental conditions. Steel surfaces may produce FeO, $Fe_3O_4$, or $Fe_2O_3$, and copper alloy surfaces can produce $Cu_2O$ or $CuO$ depending on the conditions [32, 33]. For example, hard $Fe_2$) oxides (black oxide) can exist in the sliding contacts between rubber (or soft polymers) and a hard steel shaft in water pump seals. They are known to embed themselves into the soft seal and cause severe abrasive wear to the hard shaft. On the other hand, conditions can cause $Fe_3O_4$ (red oxide) to be formed which is known to act as a solid lubricant at the interface.

## REFERENCES

1. Albrecht, A. B., "How to Secure Surface Finish in Turning Operations," Am. Machin., 1956, Vol. 100, pp. 133–136.
2. Chandiramani, K. L., and Cook, N. H., "Investigations on the Nature of Surface Finish and Its Variations with Cutting Speed," Trans. ASME, 1970, Vol. 86, pp. 134–140.
3. Olsen, K. V., "Surface Roughness in Turned Steel Components and the Relevant mathematical Analysis," Prod. Engr, 1968, pp. 593–606.
4. Solaja, V., "Wear of Carbide Tools and Surface Finish Generated in Finish Turning of Steel," Wear, 1958, Vol. 2, pp. 40–58.
5. Ansell, C. T., and Taylor, J., "The Surface Finish Properties of a Carbide and Ceramic tool," Advances in Machine Tool Design and Research, Proceedings of 3rd International MTDR Conference, Pergamon Press, New York, NY, 1962, pp. 235–243.
6. Taraman, K., "Multi-Machining Output-Multi Independent Variable Turning Research by Response Surface Methodology," Int. J. Prod. Res., 1974, Vol. 12(2), pp. 233–245.
7. Wu, S. M., "Tool Life Testing by Response Surface Methodology Parts I and II," Trans. ASME, 1964, Vol. 86, pp. 105.

8.  Hasegawa, M., Seireg, A., and Lindberg, R. A., "Surface Roughness Model for Turning," Tribol. Int., 1976, pp. 285–289.
9.  Sata, T., "Surface Roughness in Metal Cutting," CIRP, Ann. Alen Band, 1964, Vol. 4, pp. 190–197.
10. Kronenberg, M., *Machining Science and Application*, Pergamon Press, London, England, 1967.
11. Tobias, S. A., *Machine Tool Vibration*, Blackie and Son, London, England, 1965.
12. Kondo, Y., Kawano, O., and Soto, H., "Behavior of Self Excited Chatter due to Multiple Regenerative Effect," J. Eng. Indust., Trans. ASME, 1965, Vol. 103, pp. 447–454.
13. Sisson, T. r., and Kegg, R. L., "An Explanation of Low Speed Chatter Effects," J. Eng. Indust., Trans. ASME, 1969, Vol. 91, pp. 951–955.
14. Armarego, E. J. A., and Brown, R. H., *The Machining of Metals*, Prentice Hall, Englewood Cliffs, NJ, 1969.
15. Rakhit, A. K., Sankar, T. S., and Osman, M. O. M., "The Influence of Metal Cutting Forces on the Formation of Surface Texture in Turning," MTDR, 1970, Vol. 16, pp. 281–292.
16. Sankar, T. S., and Osman, M. O. M., "Profile Characterization of Manufactured Surfaces Using Random Function Excursive Technique," ASME J. Eng. Indust., 1975, Vol. 97, pp. 190–195.
17. Rakhit, A. K., Osman, M. O. M., and Sankar, T. S., "Machine Tool Vibrations: Its Effect on Manufactured Surfaces," Proceedings 4th Canadian Congress Appl. Mech., Montreal, 1973, pp. 463–464.
18. Wardle, F. P., Larcy, S. J., and Poon, S. J., "Dynamic and Static Characteristics of a Wide Speed Range Machine Tool Spindle," Precis. Eng., 1983, Vol. 83, pp. 175–183.
19. Nassipour, F., and Wu, S. M., "Statistical Evaluation of Surface Finish and Its Relationship to Cutting Parameters in Turning," Int. J. Mach. Tool Des. Res., 1977, Vol. 17, pp. 197–208.
20. Zhang, G. M., and Kapoor, S. G., "Dynamic Generation of Machined Surfaces," J. Indust. Eng., ASME, Parts I and II, 1991, Vol. 113, pp. 137–159.
21. Olgac, N., and Zhao, G., "A Relative Stability Study on the Dynamics of the Turning Mechanism," J. Dyn. Meas. Cont., Trans. ASME, 1987, Vol. 109, pp. 164–170.
22. Jemlielniak, K., and Widota, A., "Numerical Simulation of Non-Linear Chatter Vibration in Turning," Int. J. Mach. Tool Manuf., 1989, Vol. 29, pp. 239–247.
23. Tlusty, J., *Machine Tool Structures*, Pergamon Press, New York, NY, 1970.
24. Kim, K. J., and Ha, J. Y., "Suppression of Machine Tool Chatter Using a Viscoelastic Dynamic Damper," J. Indust. Eng., Trans. ASME, 1987, Vol. 109, pp. 58–65.
25. Nakayama, K., and Ari, M., On the Storage of Data on Metal Cutting Forces," Ann. CIRP, 1976, Vol. 25, pp. 13–18.

26. Rao, P. N., Rao, U. R. K., and Rao, J. S., "Towards Improved Design of Boring Bars Part I and II," Int. J. Mech. Tools Manuf., 1988, Vol. 28, pp. 34–58.
27. Tlusty, J., and Ismail, F., "Special Aspects of Chatter in Milling," ASME Paper No. 18-Det-18, 1981.
28. Skelton, R. C., "Surface Produced by a Vibrating Tool," Int. J. Mech. tools Manuf., 1969, Vol. 9, pp. 375–389.
29. Jang, D. Y., and Seireg, A., "Tool Natural Frequency as the Control Parameter for Surface Roughness," Mach. Vibr., 1992, Vol. 1, pp. 147–154.
30. Greenwood, J. A., and Tripp, J., "The Elastic Contact of Rough Spheres," J. Appl. Mech., Trans. ASME, March, 1976, pp. 153–159.
31. Bowden, F. P., and Tabor, D., *Friction and Lubrication of Solids*, Oxford University Press, 1954.
32. Kragelski, I. V., *Friction and Wear*, Butterworths, Washington, 1965.
33. Holm, R., *Electric Contacts Handbook*, Springer, Berlin, 1958.

# 5

# Thermal Considerations in Tribology

## 5.1 INTRODUCTION

This chapter gives a brief review of the fundamentals of transient heat transfer and of some of the extensive literature on the subject. Some representative results and equations are given to illustrate the effect of the different parameters on the transient temperatures generated between rubbing surfaces.

## 5.2 THERMAL ENVIRONMENT IN FRICTIONAL CONTACT

The severe thermal environment which may occur in frictional contacts [1] due to the combination of high pressures and sliding speeds is one of the main factors in the malfunctioning of machine elements such as gears, bearings, cams, brakes, and traction drives. The flash temperature [2], which represents the maximum rise in surface temperature inside the contact zone, has long been used as a design limit against scoring in gears. The thermocracking or warping of the sliding components and the desorption of the protective boundary lubrication film are among several failure mechanisms associated with the high flash temperature. On the other hand, the thermal environment inside the contact zone may precipitate the creation of a beneficial chemical film which protects against asperity interactions for certain material–lubricant combinations and temperature levels [3]. Furthermore, both film temperature and contact pressure control the glass transition for lubricant and consequently the high-slip traction in elastohydrodynamic lubrication [4].

Blok [5] and Jaeger [6] developed the theoretical foundation for the flash temperature prediction in lubricated dry rubbing solids. The thermal solution for the lubricating film temperature was developed much later by Cheng and Sternlicht [7], and Dowson and Whitaker [8]. Jaeger's formula [6] for the fast-moving heat source along a semi-infinite plane was utilized in these studies as a boundary condition along the moving solid surfaces. The possible modification in heat partition among the moving solids due to the fluid film existence was not considered. Both the convected heat by lubricant and the conducted heat in the direction of motion are assumed negligible. Manton et al. [9] investigated the temperature distribution in rolling/sliding contacts lubricated by a Newtonian oil using the finite difference method. They carried out their solution for identical steel disks to compare the temperature distribution resulting from using different oil grades. More recently, the temperature distribution for a rheological fluid was studied by Conry [10]. Wang and Cheng [11] introduced the limiting shear stress concept in their solution for the temperature in spur gear teeth contacts. In both studies the solution is based on the "constant strength" moving heat source theory [6] which may not be valid at the starting and at the end of the engagement cycle in gears where the surface curvature changes rapidly with time.

It has been long recognized that perfectly clean sliding surfaces would not function as a tribological pair. Accordingly, it is generally accepted that friction and wear can be considerably influenced by controlling both physical and chemical properties of surface films and the art of antiwear additives is an integral part of lubrication technology. Another technique used to reduce friction and wear is the coating of one or both of the rubbing surfaces [12, 13] with appropriate layers of different materials. There are several experimental observations which suggest that a reduction in lubricant film thickness in elastohydrodynamic lubrication accompanies the chemical film formtion [14]. These chemical films act as thermal screens on the surfaces, which might add to the inlet viscous heating and the compressibility effect in reducing the film thickness. A drop in the traction coefficient has also been reported to occur with the appearance of these chemical films in lubricated concentrated contacts [14, 15]. The high temperature rise inside the contact zone and the difference in thermal expansion between the material of the layers and the material of the friction pair may cause cracking or complete destruction of the layer. A modification in the heat partition between two rubbing solids may also be related to the existence of oxide films on their surface [16]. In cutting tool technology it is found that coating a tool with a layer of low thermal conductivity gives a significantly longer tool life [17]. Other related studies can be found in Refs 18–22.

Because of the considerable interest in the subject and its practical importance, a comprehensive analysis of the temperature distribution and heat partition for layered rolling/sliding solids was undertaken by Rashid and Seireg [23]. A computer-based simulation is described in Part I of the paper which can be used to study temperature distribution in lubricated layered contacts. The simulation is utilized in the accompanying paper to generate dimensionless relationships, which can be easily used to predict heat partition and maximum temperatures in the contacting surfaces and in the lubricating film for different system parameters. Dimensionless relationships are also developed for lubricated unlayered contacts and dry layered contacts. Because of the recent interest in tribological surface coating, the latter can be utilized to evaluate heat partition and temperature rise in the contact under different coating parameters and operating conditions. Some of the results of the study are presented later in this chapter.

## 5.3 AN INTRODUCTORY TREATMENT OF TRANSIENT HEAT TRANSFER

This section presents some of the fundamental concepts and relationships on which transient heat transfer is based. It deals primarily with one-directional conduction of heat and gives some design equations and illustrative examples of how these relationships are applied. The objective is to give the designer of tribological systems a basic understanding of the phenomenon and how it is influenced by the different system parameters, rather than a rigorous treatment of the subject.

### 5.3.1 Heat Penetration Depth

If the surface of a conductivity material is subjected to a temperature rise $\Delta T_0$, as shown in Fig. 5.1, the depth of heat penetration $d$, at any time $t$ after the initiation of the heat flow process, can be calculated from:

$$d = \sqrt{5Kt} \tag{5.1}$$

where

$K$ = thermal diffusivity = $\dfrac{k}{\rho c}$

$k$ = thermal conductivity

$\rho$ = density

$c$ = specific heat

Equation (5.1) is plotted in Fig. 5.1 for steel.

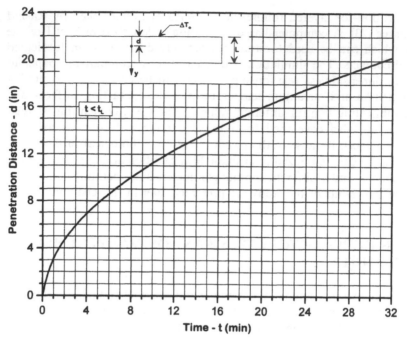

**Figure 5.1**   Penetration distance for steel prior to full penetration ($t < t_L$).

### 5.3.2   Time for Penetrating the Full Thickness of a Heated Slab

For this condition, $d = L$, the thickness of the slab, Eq. (5.1) can be rewritten as:

$$t_L = 0.08L^2 \text{ min} \qquad (L \text{ in inches})$$

where

$t_L$ = full penetration time for steel

### 5.3.3   Temperature Distribution in the Thickness for $t < t_L$

The temperature distribution in the penetrated depth $d$ can be calculated from:

$$\frac{\Delta T}{\Delta T_0} = \left(1 - \frac{y}{d}\right)^2 \qquad\qquad (5.2)$$

The normalized temperature distribution in the penetrated layer is plotted in Fig. 5.2 for different ratios of the thickness of the layer to the thickness of the slab.

After full penetration the temperature at the other surface begins to rise. Assuming no convection of heat from that surface the temperature rise $(\Delta T_L)$ at any time $t > t_L$ can be calculated from:

$$\frac{\Delta T_L}{\Delta T_0} = 1 - \exp\left[-0.214\left(\frac{t}{t_L} - 1\right)\right] \tag{5.3}$$

This relationship is plotted in Fig. 5.3 in a dimensionless form. The time scale is also plotted in the figure for steel slabs with thickness 1, 2, 5, 10, and

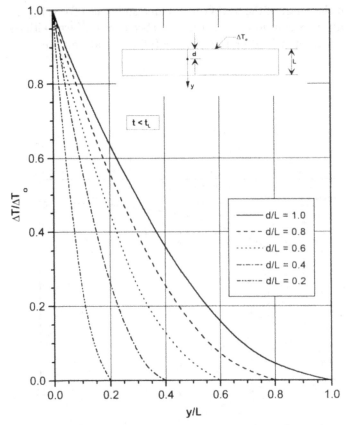

**Figure 5.2** Temperature distribution below the surface prior to full penetration $(t < t_L)$.

20 in., respectively. The graph also shows the penetration time $t_L$ in minutes for the above conditions.

### 5.3.4  Temperature Distribution for $t > t_L$

The temperature distribution across the thickness for $t > t_L$ can be calculated as:

$$\frac{\Delta T}{\Delta T_0} = \left(1 - \frac{\Delta T_L}{\Delta T_0}\right)\left(1 - \frac{y}{L}\right)^2 + \frac{\Delta T_L}{\Delta T_0} \tag{5.4}$$

This relationship is plotted in Fig. 5.4 for different values of $\Delta T_L/\Delta T_0$. It can be seen from Figures 5.2 and 5.4 that the maximum temperature gradients in this case occur at the onset of the heat flow. It can also be seen from Fig. 5.3 that the temperature distribution in the entire slab becomes uniform with less than 5% of deviation of $\Delta T_0$ after 15 times the full penetration time.

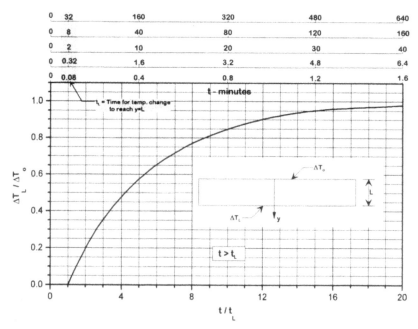

**Figure 5.3**  Temperature rise at the bottom surface of a steel slab due to a step temperature, $\Delta T_0$, applied at the top surface.

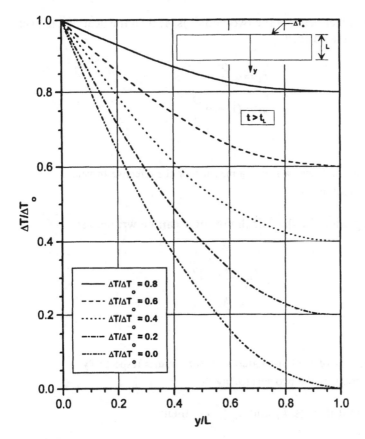

**Figure 5.4** Temperature distribution through the thickness of a steel slab for $t < t_L$.

## 5.4 TEMPERATURE RISE DUE TO HEAT INPUT

The frictional heat input at the interface can be calculated as:

$$\dot{Q} = \frac{\mu PV}{J} \text{ BTU/sec}$$

where

$\mu$ = coefficient of friction
$V$ = sliding velocity
$J$ = Joule equivalent of heat

If the area of contact between the two bodies is $A$, then the heat flux in each body can be calculated from:

$$\dot{q}_1 = \alpha \, \frac{\dot{Q}}{A}$$

$$\dot{q}_2 = (1 - \alpha) \frac{\dot{Q}}{A}$$

where

$\alpha$ = heat partition coefficient which is a measure of the effective thermal resistance of each body

A simplified equation for the calculation of $\alpha$ can be written as:

$$\frac{\alpha}{1 - \alpha} = \frac{\dfrac{1}{k_1} + \dfrac{1}{k_{l_1}}}{\dfrac{q}{k_2} + \dfrac{1}{k_{l_2}}}$$

where

$k_1, k_2$ = conductivity of the bulk material of both bodies respectively
$k_{l_1}, k_{l_2}$ = conductivity of the two surface layers

It can be seen that if $k_{l_1} \gg k_1$ and $k_{l_2} \gg k_2$, then:

$$\frac{\alpha}{1 - \alpha} = \frac{k_2}{k_1}$$

On the other hand, if $k_{l_1} \ll k_1$ and $k_{l_2} \ll k_2$, then:

$$\frac{\alpha}{1 - \alpha} = \frac{k_{l_2}}{k_{l_1}}$$

and in the case of steel bodies separated by an oil film or insulative surface layers of the same composition $k_{l_1} \approx k_{l_2}$. Therefore, $\alpha = (1 - \alpha) = 1/2$, and the heat flow is equally partitioned between the two bodies. This simple relationship is used for qualitative illustration of the concept of heat partition. A procedure for rigorous treatment of the partition is discussed later in this chapter.

If a sustained uniform heat flux, $\dot{q}$, is applied to the surface of a semi-infinite solid, the temperature rise on the surface can be calculated as follows:

$$\Delta T = \frac{1.12 \dot{q} \sqrt{t}}{\sqrt{k \rho c}}$$

For a slab of depth $L$, subjected to a constant heatflux $\dot{q}$, the above equation can be rewritten in dimensionless form as:

$$\frac{\Delta T}{\dot{q} L/k} = 1.12 \frac{k}{L} \frac{\sqrt{t}}{\sqrt{k \rho c}} = 1.12 \sqrt{\frac{kt}{L^2}} \approx \frac{1}{3} \sqrt{\frac{t}{t_L}}$$

$L$ is used as a reference dimension and time is normalized w.r.t. the corresponding full penetration time $t_L$. This relationship is plotted in Fig. 5.5. The temperature rise due to a constant flux $\dot{q}$, applied for a finite period $t_0$, can therefore be calculated as the superposition between a positive $\dot{q}$ input at $t = 0$, and a negative $\dot{q}$ at $t = t_0$. Accordingly:

$$t = 0 \rightarrow t_0 \qquad \frac{\Delta T}{\dot{q} L/k} = \frac{1}{3} \sqrt{\frac{t}{t_L}}$$

$$t = t_0 \rightarrow t_L \qquad \frac{\Delta T}{\dot{q} L/k} = \frac{1}{3} \left( \sqrt{\frac{t}{t_L}} - \sqrt{\frac{(t - t_0)}{t_L}} \right)$$

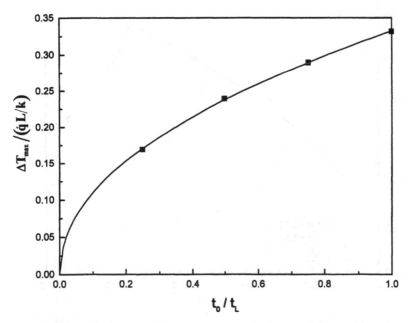

**Figure 5.5** Maximum surface temperature rise in a steel slab subjected to a constant heat flux $\dot{q}$ for a period $t_0$.

Figure 5.6 shows a plot of these relationships for $t_0/t_L = 0.25$, 0.5, and 0.75 respectively.

The principle of superposition can be applied to determine the surface temperature rise for any given function of heat flux. It can be determined either as a summation or an integration of the effect of incremental step inputs that convolute the given function. An illustration of this is the study of the effect of applying the same total quantity of heat input with a triangular flux with the same maximum value but with different slopes, $S_1$ and $S_2$, during the increase and decrease phases respectively. Using the integration approach, therefore:

$$t = 0 \rightarrow t_0, \qquad \Delta T(t) = \int_0^t 1.12 S_1 \frac{\sqrt{\tau}}{\sqrt{k\rho c}} d\tau$$

$$t = t_0 \rightarrow t_L, \qquad \Delta T(t) = \int_0^{t_0} 1.12 S_1 \frac{\sqrt{\tau}}{\sqrt{k\rho c}} d\tau - \int_{t_0}^t 1.12 S_2 \frac{\sqrt{\tau}}{\sqrt{k\rho c}} d\tau$$

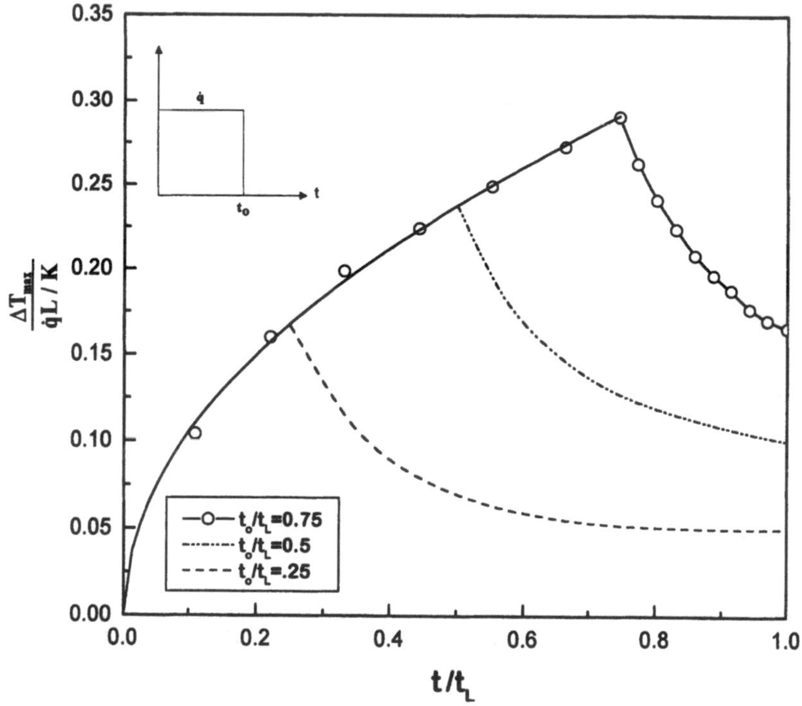

**Figure 5.6**  Temperature rise as a function of time on the surface of a steel slab subjected to constatn flux $\dot{q}$ for different periods $t_0$.

Figures 5.7–5.11 show plots of the surface temperature history for $S_1/S_2 = 9$, 4,1, $\frac{1}{4}$, and $\frac{1}{9}$, respectively.

Figure 5.12 shows the maximum surface temperature as a function of $S_1/S_2$ for the same total heat flow $q_{max}$. It can be seen from Fig. 5.12 that the highest surface temperature rise occurs as the ratio $S_1/S_2$ decreases.

If the heat flux function is determined from experimental data and is difficult to integrate, the temperature rise can be obtained by summing the effects of incremental steps that are constructed to convolute the function as illustrated in Fig. 5.13. Better accuracy can be obtained as the number of steps increases.

The temperature rise at the surface for this case can be calculated as:

$$t = t_1 \rightarrow t_2, \qquad \Delta T = \frac{1.12}{\sqrt{k\rho c}} \left( \dot{q}_1 \sqrt{t - t_1} \right)$$

$$t = t_2 \rightarrow t_3, \qquad \Delta T = \frac{1.12}{\sqrt{k\rho c}} \sum \dot{q}_1 \sqrt{t - t_1} + (\dot{q}_2 - \dot{q}_1)\sqrt{t - t_2}$$

$$t = t_3 \rightarrow t_4, \qquad \Delta T = \frac{1.12}{\sqrt{k\rho c}} \sum \dot{q}_1 \sqrt{t - t_1} + (\dot{q}_2 - \dot{q}_1)\sqrt{t - t_2} + (\dot{q}_3 - \dot{q}_2)\sqrt{t - t_3}$$

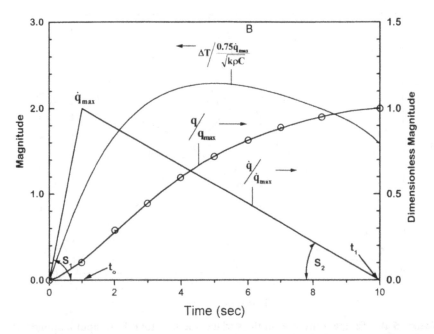

**Figure 5.7** Temperature rise on the surface of a steel slab due to a total heat input $q$ applied at different rates ($t_0/t_1 = 0.1$).

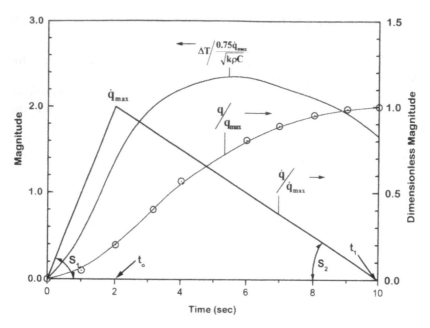

**Figure 5.8** Temperature rise on the surface due to a total heat input $q$ applied at different rates ($t_0/t_1 = 0.2$).

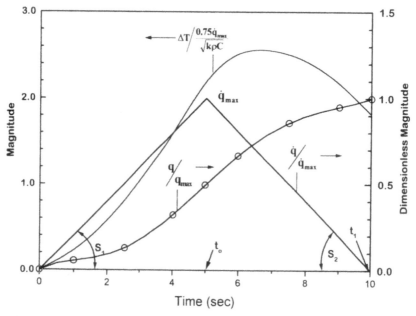

**Figure 5.9** Temperature rise on the surface due to a total heat input $q$ applied at different rates ($t_0/t_1 = 0.5$).

*132*

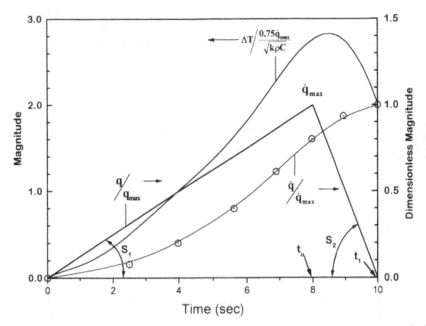

**Figure 5.10** Temperature rise on the surface due to a total heat input $q$ applied at different rates ($t_0/t_1 = 0.8$).

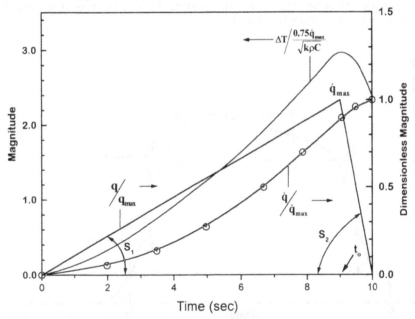

**Figure 5.11** Temperature rise on the surface due to a total heat input $q$ applied at different rates ($t_0/t_1 = 0.9$).

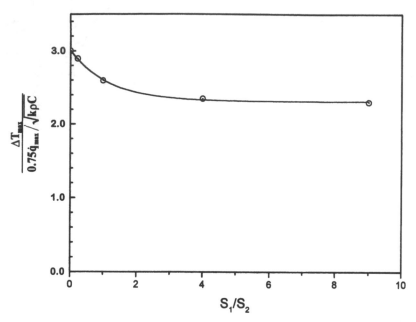

**Figure 5.12**  Dimensionless surface temperature rise as a function of the rate of the slope ratio for the triangular heat input.

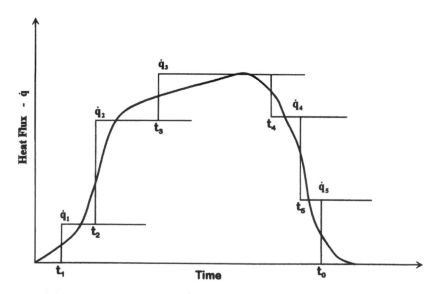

**Figure 5.13**  Convolution integration for a general heat input function.

$$t = t_4 \rightarrow t_5, \quad \Delta T = \frac{1.12}{\sqrt{k\rho c}} \sum \dot{q}_1 \sqrt{t - t_1} + (\dot{q}_2 - \dot{q}_1)\sqrt{t - t_2} + (\dot{q}_3 - \dot{q}_2)\sqrt{t - t_3}$$
$$+ (\dot{q}_4 - \dot{q}_3)\sqrt{t - t_4}$$

$$t = t_5 \rightarrow t_0, \quad \Delta T = \frac{1.12}{\sqrt{k\rho c}} \sum \dot{q}_1 \sqrt{t - t_1} + (\dot{q}_2 - \dot{q}_1)\sqrt{t - t_2} + (\dot{q}_3 - \dot{q}_2)\sqrt{t - t_3}$$
$$+ (\dot{q}_4 - \dot{q}_3)\sqrt{t - t_4} + (\dot{q}_5 - \dot{q}_4)\sqrt{t - t_5}$$

$$t > t_0, \quad \Delta T = \frac{1.12}{\sqrt{k\rho c}} \sum \dot{q}_1 \sqrt{t - t_1} + (\dot{q}_2 - \dot{q}_1)\sqrt{t - t_2} + (\dot{q}_3 - \dot{q}_2)\sqrt{t - t_3}$$
$$+ (\dot{q}_4 - \dot{q}_3)\sqrt{t - t_4} + (\dot{q}_5 - \dot{q}_4)\sqrt{t - t_5} + (0 - \dot{q}_5)\sqrt{t - t_0}$$

## 5.5 HEAT PARTITION AND TRANSIENT TEMPERATURE DISTRIBUTION IN LAYERED LUBRICATED CONTACTS

This section briefly describes a generalized and efficient computer-based model developed by Rashid and Seireg [23], for the evaluation of heat partition and transient temperatures in dry and lubricated layered concentrated contacts. The program utilizes finite differences with the alternating direction implicit method.

The program is capable of treating the transient heat transfer problem in lubricated layered contacts with any arbitrary distribution of layer properties and thicknesses. It takes into consideration the time variation in speeds, load, friction coefficient, fluid film thickness between surfaces, and the effective radius of curvature of contacting solids. It calculates the surface temperature distribution in the layered solids in lubricant film. Also, the role of the chemical layer on surface and film temperatures in lubricated concentrated contacts can be evaluated. Furthermore, the transient operating conditions, which are associated with the performance of such systems, are incorporated in temperature calculations.

A general model for the contact zone in sliding/rolling conditions can be approximated by two moving semi-infinite solids separated by a lubricant film, as shown in Fig. 5.14. The heat generation distribution inside the lubricant film is controlled by the rheological behavior of the lubricant under different pressures, temperatures, and rolling and sliding speed.

In many concentrated contact problems, the moving solids may have different thermal properties, speeds, bulk temperatures and different chemical layers on their surface. All these variables are introduced in the model, as well as any considered heat generation condition in the lubricant film.

The boundary conditions for this problem are based on the fact that the temperature gradient diminishes away from the heat generation zone.

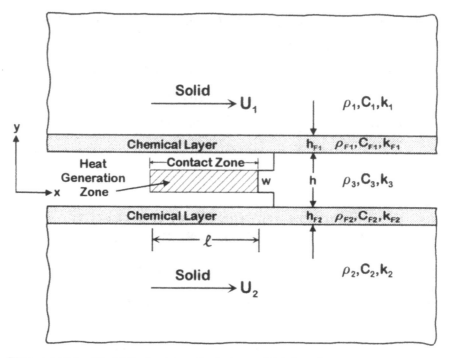

**Figure 5.14** Model for heat transfer in layered lubricated contacts.

The temperature field around the contact zone is represented by a rectangular grid containing appropriately distributed nodal points in the two solids and the lubricant film normal to the flow and in the flow direction.

The size of each division can be changed in such a manner that the boundary conditions can be satisfied for a particular problem without increasing the number of nodes. A larger number of divisions are used across the lubricant film to accommodate the rapid change in both temperature and velocity across the film.

The mesh size is progressively expanded in each moving solid with the distances of the node from the heat generation zone.

The developed program has the following special features:

1. The use of finite difference with the alternating direction implicit method provides considerable modeling flexibility and computing efficiency.
2. It is capable of handling transient variations in geometry, load, speed, and material properties.

3. It can treat dry or lubricated multilayered contacts with relatively small layer thicknesses.
4. Because the program is developed for modeling transient conditions, it can be used for predicting traction characteristics for layered or unlayered solids by incorporation of a proper rheological model for the lubricant. Starting from the ambient temperature conditions, the lubricant properties can be iteratively evaluated from the computed temperatures for any particular operating condition.

The program as developed would be useful in investigating the effect of the different parameters on the temperature distribution in line contacts. It can provide a valuable guide for performing experimental studies to generate empirical design equations for layered surfaces. It can also be utilized to develop empirical equations for lubricated layered contacts applicable to specific regimens of materials and operating conditions. The results for any application can be considerably enhanced by incorporating an appropriate rheological model for the lubricant. This would enable the prediction of traction, velocity, profile in the film, and the heat generation distribution in the contact zone.

### 5.5.1 Numerical Results

Numerical solutions are carried out to illustrate the capabilities of the program utilizing under the following assumptions:

1. The heat source distributed inside the contact zone follows the dry contact pressure distribution (Hertzian pressure). Then the rate of heat generation distribution per unit volume can be calculated as:

$$Q = 2Q_{max} \sqrt{\left(\frac{x}{\ell}\right) - \left(\frac{x}{\ell}\right)^2}$$

where

$$Q_{max} = \frac{4}{\pi} \frac{f W_0 (U_1 - U_2)}{\ell_w}$$

$f$ = coefficient of friction
$W_0$ = load per unit length

2. The solids are homogeneous with no cracks or inclusions.
3. The chemical reaction heat sources are negligible compared to frictional heat sources.

4.  The heat of compression in the lubricant film and the moving solids has a negligible effect on the temperature rise inside the contact zone.

5.  Because the lubricant film thickness and the Hertzian contact with (x-direction) are small in comparison to the cylinder width (z-direction), the temperature gradient in the z-direction is expected to be small in comparison with those across and along the film. Therefore, the conduction in the z-direction is neglected.

The thermal properties of the surface layers in lubricated contacts cover a wide material spectrum. There are some cases where the surface layer has low thermal conductivity in comparison with the lubricant film (for example, paraffinic and the organic surface layers as compared with oil). At the same time, there are some types of coatings, like silicon carbide (SiC), which are much more conductive than any common lubricant. The thermal resistance at the interface between the surface layer and the bulk solid should also be taken into consideration if the thermal boundary layer penetrates the surface layer inside the solid.

The developed program is utilized to study the variation in maximum film temperature versus oil film thickness for several surface layer thicknesses. The attached surface layer to each moving solid is assumed to be identical and the distribution of heat generation is assumed to be uniform across the film ($w = h$). For the considered example:

$$\frac{K_3}{K_F} = 6$$

where

$K_3, K_F$ = thermal conductivities for the lubricant film and the surface layer respectively

The maximum film temperature is expected to be strongly dependent on the layer thickness because of its low thermal conductivity in comparison with the lubricant film. The results as plotted in Fig. 5.15 show a gradual reduction in the influence of the layer thickness on the lubricant film temperature as the film thickness increases for the same friction heat level, as demonstrated by the upper two curves in the figure. All the temperature curves for the layered contacts have the tendency to converge to a common level as the lubricant film thickness increases in magnitude.

This is represented in more detail in Fig. 5.16, which shows the dependency of the maximum film temperature upon a wider range of surface layer

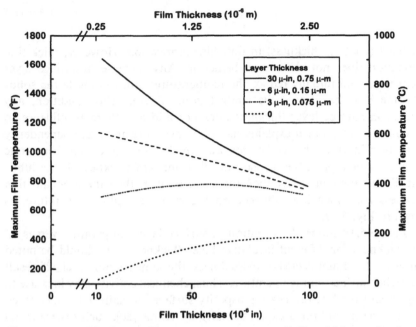

**Figure 5.15** Maximum film temperature versus oil film thickness for layered lubricated contacts (insulative layers, $K_3/K_F = 6$).

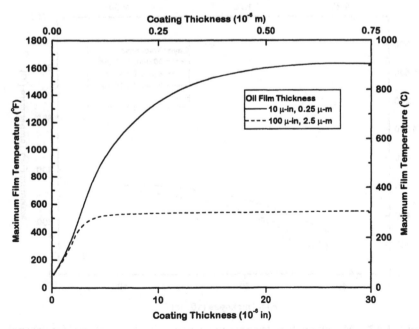

**Figure 5.16** Maximum film temperature versus coating thickness (insulative layers, $K_3/K_F = 6$).

thickness (coating thickness) in thin film lubrication. However, thick film lubrication does not show such behavior. Any increase in surface layer thickness would initially reduce the temperature diffusion inside the solids until the heat flux leaves the contact zone. Beyond this condition, any increase in surface layer thickness does not add any thermal influence to the contact zone, which explains the difference in temperature dependency on surface layer thickness for thin and thick film lubrication. The same argument can explain the increase in lubricant film thickness. If the lubricant film is less conductive than the surface layer, then the lubricant film thickness has much less influence on the maximum film temperature, as shown in Fig. 5.17.

Figure 5.18 shows the variation in surface layer temperature versus oil film thickness for different insulative layer thicknesses. It should be noted that as the fluid film decreases in thickness, the same friction level will result in a higher surface film temperature. Thus, chemical activity may increase to a significant level before bearing asperity surfaces actually achieve contact. This has been confirmed experimentally by Klaus [20]. such experimental

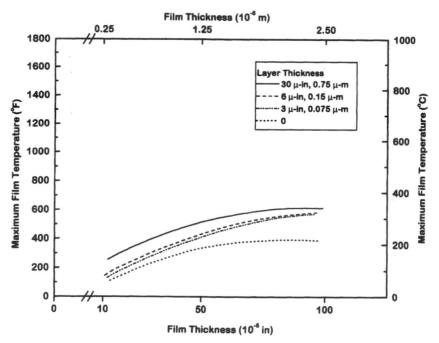

**Figure 5.17** Maximum film temperature versus oil film thickness for layered lubricated contacts (conductive layers, $K_3/K_F = 1/6$).

observation is difficult to perform using the infrared technique [21], because one of the surfaces has to be transparent. If the surface layer is less conductive than the lubricating oil, then the maximum surface layer temperature has a stronger dependency on the lubricant film thickness than in the case of conductive layers, as described in Figs 5.18 and 5.19.

In the case of boundary lubrication, in which the asperity interaction with the solid surfaces plays a major role, the temperature level becomes even more sensitive to surface layer thickness. The small contact width between the asperities generates a shallow temperature penetration across the surface layer, which increases the temperature level even for a very thin layer.

The following can be concluded from the investigated conditions:

1.  In the case of an insulative surface layer, the maximum rise in film temperature is strongly dependent on the surface layer thickness, whereas this is not the case for the conductive surface layer (see Figs 5.15 and 5.17).
2.  In both cases, the surface layer temperature decreases with the increase in lubricant film thickness. This is attributed to the con-

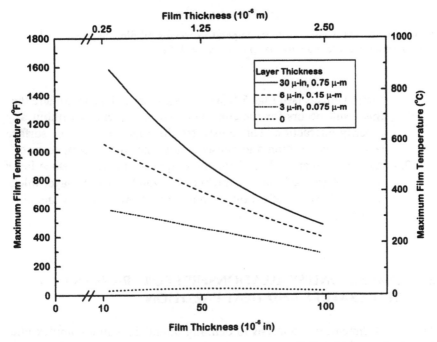

**Figure 5.18** Maximum surface temperature versus oil film thickness for layered lubricated contacts (insulative layers, $K_3/K_F = 6$).

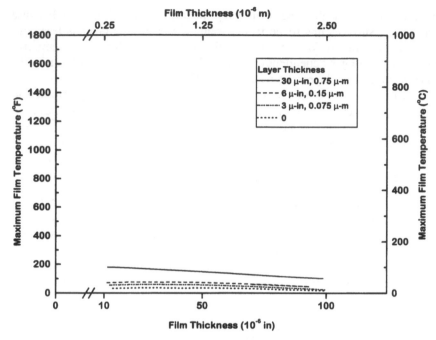

**Figure 5.19** Maximum surface temperature versus oil film thickness for layered lubricated contacts (conductive layers, $K_3/K_F = 1/6$).

vection effects (see Figs 5.18 and 5.19). It should be noted here that this result occurs for the considered smooth surfaces without any asperity interaction. This illustrates the importance of the surface layers on convection and consequently, the surface temperatures.

3. As can be seen in Fig. 5.16, there appears to be a surface layer thickness, for each oil film thickness, beyond which the layer thickness will have no effect on the maximum temperature in the lubricant film.

## 5.6 DIMENSIONLESS RELATIONSHIPS FOR TRANSIENT TEMPERATURE AND HEAT PARTITION

The use of dimensional analysis in defining interactions in a complex phenomenon is a well-recognized art. Any dimensional analysis problem raises two main questions:

1.  The minimum number of the dimensionless groups needed to describe the theoretical analysis;
2.  The physical interpretation of these groups and their most appropriate forms.

Dimensionless relationships for concentrated contact can be of considerable practical importance to the experimentalist and the designer. Most of the theoretical analyses are based on computer solutions and the presentation of the results are generally lacking in presenting generalized trends. The lack of generality is due to the fact that the presentation of the results is usually in the form of discrete examples, there is no provision of insight into the interaction between variables. This section presents dimensionless relationships developed from the computer model described in the previous section which incorporate dimensionless groups representing the system parameters and operating conditions.

## Case 1:   Heat Source Moving over a Semi-Infinite Solid (Fig. 5.20)

This problem is used to check the validity of the modeling approach since an analytical solution by Blok [5] and Jaeger [6] is available for this case. The derived equation for the maximum rise in surface temperature is obtained by using a series approximation as:

$$T_S - T_B = 1.128 \frac{q_t}{\ell} \sqrt{\frac{\ell}{k \rho c U}}$$

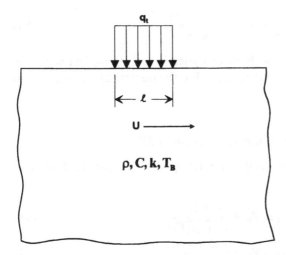

**Figure 5.20**   Moving semi-infinite solid under a stationary heat source.

where

$T_S$ = maximum surface temperature
$T_B$ = bulk temperature

A dimensional analysis is carried out by using the $\pi$ theorem [24, 25] to obtain adequate dimensionless groups for this example. Realizing the fact that the heat input to each material element inside the temperature field is balanced by both conductive and convective modes of heat transfer, it can be concluded that:

$$T_S - T_B = f(U, \rho, c, k, \ell, q_t) \tag{5.5}$$

The final form of the dimensionless equation, as a function of Peclet number, can be derived as:

$$\frac{(T_S - T_B)k}{q_t} = f\left(\frac{\rho c U \ell}{k}\right) \tag{5.6}$$

where

$$\frac{\rho c U \ell}{k} = \text{Peclet number}$$

The log/log plot of the computed data showed a straight line correlation between the two dimensionless number in Eq. (5.6). The equation of this line can be expressed as:

$$\frac{(T_S - T_B)k}{q_t} = 1.03\left(\frac{\rho c U \ell}{k}\right)^{-0.5} \tag{5.7}$$

which is in general agreement with the analytically derived relationship. The differences in the constant can be attributed to the numerical approximation in the computer model.

## Case 2: Sliding/Rolling Dry Contacts (see Fig. 5.21)

The maximum temperatures on the contacting surfaces can be developed from Eq. (5.7) as:

$$T_{S1} = T_{B1} + \frac{1.03\alpha q_t}{\sqrt{k_1 \rho_1 c_1 U_1 \ell}} \tag{5.8}$$

$$T_{S2} = T_{B2} + \frac{1.03(1 - \alpha)q_t}{\sqrt{k_2 \rho_2 c_2 U_2 \ell}} \tag{5.9}$$

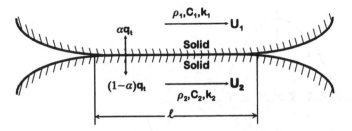

**Figure 5.21** Two cylinders under dry sliding condition.

$T_{S1} = T_{S2}$ in this case, therefore, the heat partition coefficient $\alpha$ can be calculated for equal bulk temperatures as:

$$\alpha = \frac{1}{1 + \sqrt{\dfrac{\rho_2 c_2 U_2 k_2}{\rho_1 c_1 U_1 k_1}}} \tag{5.10}$$

and accordingly:

$$T_{S1} - T_{B1} = \frac{q_t}{\sqrt{\ell}} \frac{1.03}{\sqrt{k_1 \rho_1 c_1 U_1} + \sqrt{k_2 \rho_2 c_2 U_2}} \tag{5.11}$$

Blok [2] derived an identical equation for the flash temperature. The contact in Blok's equation is determined analytically as 1.11 instead of 1.03 determined from the developed program.

### Case 3: Heat Source with a Hertzian Distribution Moving over a Layered Semi-Infinite Solid (Fig. 5.22)

In this case, the relationship for the maximum rise in the solid surface temperature can be obtained using the $\pi$ theorem as:

$$\frac{(T_s - T_{S0})k_0}{q_t} = \left(\frac{U\rho c h_0}{k_0}\right)^{-0.7} \left(\frac{\ell}{h_0}\right)^{-1.17} \left(\frac{k}{k_0}\right)^{-0.7} \left(\frac{U\rho_0 c_0 h_0}{k_0}\right)^{0.78} \tag{5.12}$$

By using the value of the penetration depth $D$ in the solid at the trailing edge [26], Eq. (5.12) can be rewritten as:

$$\frac{(T_s - T_{S0})k_0}{q_t} = 1.137 \left(\frac{D}{\ell}\right)^{1.4} \left(\frac{k_0}{k}\right)^{1.4} \left(\frac{h_0}{\ell}\right)^{0.47} \left(\frac{\ell_e}{h_0}\right)^{0.78}$$

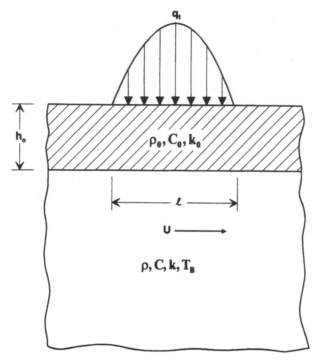

**Figure 5.22**  Layered semi-infinite solid moving under a stationary heat source.

Similarly, the maximum rise in the surface layer temperature is derived as:

$$\frac{(T_0 - T_s)k_0}{q_t} = 1.4\left(\frac{U\rho ch_0}{k_0}\right)^{0.013}\left(\frac{\ell}{h_0}\right)^{-1.003}\left(\frac{k}{k_0}\right)^{0.013} e^{-900\times 10^{-6}\,\frac{U\rho_0 c_0 h_0}{k_0}} \qquad (5.13)$$

or by using the penetration depth concept:

$$\frac{(T_0 - T_s)k_0}{q_t} = 1.164\left(\frac{\ell}{D}\right)^{0.026}\left(\frac{k}{k_0}\right)^{0.026}\left(\frac{h_0}{\ell}\right) e^{-180\times 10^{-6}\,\frac{\ell_e}{h_0}}$$

where

$D = \sqrt{\dfrac{5k\ell}{\rho cU}}$ = temperature penetration depth at the trailing edge

$\ell_e = \dfrac{1}{5}\dfrac{Uh_0^2\rho_0 c_0}{k_0}$ = required entry distance for temperature penetration across the film

$T_{S0}$ = maximum rise in the solid surface temperatue for unlayered semi-infinite
   solids for the same heat input
$T_S$ = maximum rise in the solid surface temperature for unlayered semi-infinite
   solids for the same heat input

### Case 4: Lubricated Rolling/Sliding Contacts

The temperature distribution and heat partition in heavily loaded lubricated contacts is not yet fully understood due to the ill-defined boundary conditions and the modeling complexities in the problem. In this part of the work, a number of dimensionless equations are derived for predicting both the maximum film temperature and the heat partition between the contacting solids.

The model to be analyzed is shown in Fig. 5.23. It represents two rolling/sliding cylinders having different radii, thermal properties, and bulk temperatures, which are separated by lubricant film thickness $h$. Because the lubricant is subjected to extremely high pressures and shear stresses, which only act for a very short time, the assumption that the lubricant behaves as Newtonian liquid is not valid. Experiments demonstrated that typical lubricants exhibit liquid–solid transitions in elastohydrodynamic contacts [4] and that this transition depends on both pressure and temperature. The heat source depth $w$ in the model represents the liquid region where the lubricant undergoes a high shear rate. This region ranges between $0.1$ and $0.4h$. At moderate to high sliding speeds, the magnitude of $w$ is approxiamtely $0.1h$. In order to simplify the derivation of dimensionless equations for this case, $w$ is initially assumed to be equal to zero. Now

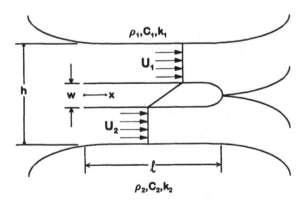

**Figure 5.23** Lubricated, heavily loaded sliding/rolling cylinder.

the partition of heat in lubricated rolling/sliding contacts can be predicted by using Eqs (5.12) and (5.13). In the practical range of different material combinations and bulk temperature difference, the heat generation zone in Fig. 5.23 is assumed to be at the center of the film. Accordingly, by referring to Fig. 5.14, which represents two layered cylinders rubbing against each other, we can assume that:

$$h_{01} = h_{02} = \frac{h}{2}$$

$$k_{01}, \rho_{01}, c_{01} = k_{02}, \rho_{02}, c_{02}$$

By assuming that the amount of heat flowing to the upper semi-infinite layered solid is $\alpha q_t$, then the lower one receives $(1-\alpha)q_t$. Since the maximum temperature rise inside the heat generation zone is the same for the two layered semi-infinite solids, then according to Eq. (5.9), let:

$$\frac{T_{01} - T_{S1}}{\alpha q_t} = B_1 \quad \text{and} \quad \frac{T_{02} - T_{S1}}{(1-\alpha)q_t} = B_2$$

where

$$B_1 = \frac{1.14}{k_0}\left(\frac{U_1\rho_1 c_1 h}{2k_0}\right)^{-0.013}\left(\frac{2\ell}{h}\right)^{-1.003}\left(\frac{k_1}{k_0}\right)^{0.013}\exp{-900 \times 10^{-6}\left(\frac{U_1\rho_0 c_0 h}{2k_0}\right)}$$

Similarly:

$$B_2 = \frac{1.14}{k_0}\left(\frac{U_2\rho_2 c_2 h}{2k_0}\right)^{0.013}\left(\frac{2\ell}{h}\right)^{-1.003}\left(\frac{k_2}{k_0}\right)^{0.013}\exp{-900 \times 10^{-6}\left(\frac{U_2\rho_0 c_0 h}{2k_0}\right)}$$

and from Eq. (5.3), let:

$$\frac{T_{S1} - T_{B1}}{\alpha q_t} = \frac{1.03}{k_1}\left(\frac{\rho_1 c_1 U_1 \ell}{k_1}\right)^{-0.5} = A_1$$

and

$$\frac{T_{S2} - T_{B2}}{(1-\alpha)q_t} = \frac{1.03}{k_2}\left(\frac{\rho_2 c_2 U_2 \ell}{k_2}\right)^{-0.5} = A_2$$

$T_{01} = T_{02}$, therefore:

$$\alpha = \frac{1}{A_1 + B_1 + A_2 + B_2}\left(\frac{T_{B2} - T_{B1}}{q_t} + A_2 + B_2\right) \tag{5.14}$$

Equation (5.14) gives the percentage of heat flowing to the upper layered semi-infinite solid. However, the actual amount of heat flowing to each solid surface is expected to be slightly modified by the lubricant film or surface layer existence. Because the maximum rise in the solid surface temperature is controlled by the amount of heat flow, then from Eq. (5.3) we have:

$$T_{S01} = T_{B1} = \gamma_1 q_t A_1 \tag{5.15}$$
$$T_{S02} = T_{B2} = \gamma_2 q_t A_2 \tag{5.16}$$

But from Eq. (5.12):

$$T_{S01} - T_{B1} = \alpha q_t (A_1 - c_1) \tag{5.17}$$

and

$$T_{S02} - T_{B2} = (1 - \alpha) q_t (A_2 - c_2) \tag{5.18}$$

where

$$c_1 = \frac{1}{k_0} \left( \frac{U_1 \rho_1 c_1 h}{2 k_0} \right)^{-0.7} \left( \frac{2\ell}{h} \right)^{-1.17} \left( \frac{k_1}{k_0} \right)^{-0.7} \left( \frac{U_1 \rho_0 c_0 h}{2 k_0} \right)^{0.78}$$

and

$$c_2 = \frac{1}{k_0} \left( \frac{U_2 \rho_2 c_2 h}{2 k_0} \right)^{-0.7} \left( \frac{2\ell}{h} \right)^{-1.17} \left( \frac{k_2}{k_0} \right)^{-0.7} \left( \frac{U_2 \rho_0 c_0 h}{2 k_0} \right)^{0.78}$$

The percentage of heat flowing to the upper solid can be predicted by substituting Eq. (5.17) into Eq. (5.15) to get:

$$\gamma_1 = \frac{\alpha(A_1 - c_1)}{A_1} \tag{5.19}$$

and from Eqs (5.16) and (5.18), the percentage of heat flowing to the lower solid is

$$\gamma_2 = (1 - \alpha) \frac{(A_2 - c_2)}{A_2} \tag{5.20}$$

The maximum surface layer temperature in this model, Eq. (5.13), is derived without incorporating the influence of the heat source depth $w$ and the percentage of heat flow into the semi-infinite layered solid.

Therefore, the maximum film temperature in elastohydrodynamic lubrication, can be derived by modifying Eq. (5.13) to the following form:

$$T_{01} - T - B1 = \alpha q_t A_1 + \alpha q_t B_1 \exp\left[-0.5\left(\frac{w}{h}\right)\right] \tag{5.21}$$

where $T_{01} = T_3$.

### Film Thickness

The numerical solution for the minimum film thickness in elastochydrodynamic lubrication for compressible, isothermal, smooth, unlayered, and fully flooded cylinders by a Newtonian lubricant was discussed by Hamrock and Jacobson [27]. The equation used for the minimum film thickness in a dimensionless form is written as:

$$H_{min} = 3.07 U_0^{0.71} G_0^{0.57} P_0^{-0.11} \tag{5.22}$$

The dimensionless groups can be defined as follows:

$$\text{Dimensionless film thickness } H_{min} = \frac{h}{R_e}$$

$$\text{Dimensionless speed parameter } U_0 = \frac{\eta_0 UR}{E_e R_e}$$

where

$UR$ = rolling velocity
$R_e$ = effective radius
$\eta_0$ = viscosity at atmospheric pressure

$$\text{Dimensionless materials parameter } G_0 = \alpha_v E_e$$

where

$E_e$ = effective modulus of elasticity
$\alpha_v$ = pressure viscosity coefficient of lubricant
$\eta = \eta_0 e^{\alpha_v P}$
$P$ = pressure
$\eta$ = lubricant viscosity

$$\text{Dimensionless load parameter } P_0 = \frac{W_0}{E_e R_e}$$

Since viscosity is strongly influenced by temperature, thermal effects are expected to have a strong influence on the minimum film thickness. The modification proposed by Wilson and Sheu [28] can be used for a correction factor for the minimum film thickness by considering the thermal build up at the entrance of the contact zone.

The viscosity at atmospheric pressure, $\eta_0$, is based on the average bulk temperatures of the mating solid surfaces. A higher average bulk temperature leads to a lower viscosity and consequently, to a thinner lubricant film. Under extreme conditions, this may result in severe interaction between the rubbing solids. Some work has been devoted to avoid this problem by using different cooling techniques [22, 29].

## Case 5: Parabolic Heat Source Moving on a Metallic Semi-Infinite Solid with Low-Conductivity Surface Layer

The dimensional analysis approach is also used in developing the following dimensionless equations for maximum solid and surface layer temperatures. Let:

$$IC = \frac{1.085}{k_0} \left(\frac{U\rho c h_0}{k_0}\right)^{-0.795} \left(\frac{\ell}{h_0}\right)^{-1.07} \left(\frac{k}{k_0}\right)^{-0.77} \left(\frac{U\rho_0 c_0 h_0}{k_0}\right)^{-0.788} \tag{5.23}$$

then:

$$\frac{T_S - T_{S0}}{q_t} = IC \tag{5.23}$$

and let:

$$IB = \frac{0.135}{k_0} \left(\frac{U\rho c h_0}{k_0}\right)^{-0.21} \left(\frac{\ell}{h_0}\right)^{-0.85} \left(\frac{k}{k_0}\right)^{-0.25} \exp\left[-5750 \times 10^{-6} \left(\frac{U\rho_0 c_0 h_0}{k_0}\right)\right] \tag{5.24a}$$

then:

$$\frac{T_0 - T_S}{q_t} = IB \tag{5.24b}$$

All the variables in Eqs (5.12) and (5.13), and the above equations have identical definitions. However, each set of these equations is valid only for a certain range of thermal properties and surface film thickness.

**Case 6: Parabolic Heat Source Moving on a Low-Conductivity, Semi-Infinite Solid with a Metallic Surface Layer**

By using the same previous procedure, the following equations are derived for this case. Let:

$$CC = 0.676\left(\frac{U\rho ch_0}{k_0}\right)^{-0.831}\left(\frac{\ell}{h_0}\right)^{-0.857}\left(\frac{k}{k_0}\right)^{-0.836}\left(\frac{U\rho_0 c_0 h_0}{k_0}\right)^{0.627} \qquad (5.25a)$$

then:

$$\frac{T_S - T_{S0}}{q_t} = CC \qquad (5.25b)$$

and let:

$$CB = 0.368\left(\frac{U\rho ch_0}{k_0}\right)^{-0.98}\left(\frac{\ell}{h_0}\right)^{-0.84}\left(\frac{k}{k_0}\right)^{-1.0}\left(\frac{U\rho_0 c_0 h_0}{k_0}\right)^{0.8} \qquad (5.26a)$$

then:

$$\frac{T_0 - T_S}{q_t} = CB \qquad (5.26b)$$

Equations (5.23b)–(5.26b) are valid for the following range of operating conditions and thermal properties:

1.  The conductivity ratio which can be applied for each case is:

$$\text{Case 5: } 5 \le \frac{k}{k_0} \le 20$$

$$\text{Case 6: } 0.05 \le \frac{k}{k_0} \le 0.2$$

where

$k_0$ = conductivity of commonly used metallic solids

2.  The speed range is $500 < U < 2000\,\text{in./sec.}$
3.  The limits of the contact zone width are $0.01 \le \ell \le 0.1\,\text{in.}$

4. The film thickness limits are given by $50 \times 10^{-6} \le h \le 200 \times 10^{-6}$ in.
5. The range of $\rho c$ covers all the commonly used materials.

## Case 7: Dry Layered Contacts

The equations derived in Cases 5 and 6 are utilized to develop dimensionless equations for heat partition and maximum temperatures for four different combinations of thermal properties for contacting solids with surface layers, as identified in Table 5.1. The resulting equations are given as follows:

$$\alpha = \frac{1}{A_1 + Z_1 + A_2 + Z_2}\left(\frac{T_{B2} - T_{B1}}{q_t} + A_2 + B_2\right) \tag{5.27}$$

$$T_{S01} - T_{B1} = (1 - \alpha)q_t(A_1 - Z_3) \tag{5.28}$$

$$T_{S02} - T_{B2} = (1 - \alpha)q_t(A_2 - Z_4) \tag{5.29}$$

$$\gamma_1 = \alpha\left(\frac{A_1 - Z_3}{A_1}\right) \tag{5.30a}$$

$$\gamma_2 = (1 - \alpha)\left(\frac{A_2 - Z_4}{A_2}\right) \tag{5.30b}$$

$$T_{01} - T_{B1} = \alpha q_t(A_1 + Z_1)$$
$$T_{01} = T_{02} \tag{5.31}$$

The subscripts for each variable refer to the thermal properties and surface layer thickness of the indicated layered solid in Fig. 5.24.

**Table 5.1**  Different Combinations of Thermal Properties for Layered Dry Contacts

| Combination of thermal properties for the lower layered solid | Combination of thermal properties for the upper layered solid | |
| --- | --- | --- |
| | Case 5 | Case 6 |
| Case 5 | $Z_1 = IB_1$ | $Z_1 = CB_1$ |
| | $Z_2 = IB_2$ | $Z_2 = CB_2$ |
| | $Z_3 = IC_1$ | $Z_3 = CC_1$ |
| | $Z_4 = IC_2$ | $Z_4 = IC_2$ |
| Case 6 | $Z_1 = IB_1$ | $Z_1 = CB_1$ |
| | $Z_2 = CB_2$ | $Z_2 = CB_2$ |
| | $Z_3 = IC_1$ | $Z_3 = CC_1$ |
| | $Z_4 = CC_2$ | $Z_4 = CC_2$ |

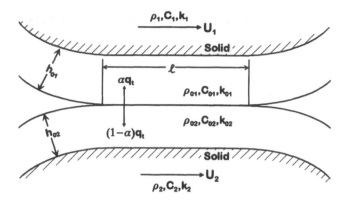

**Figure 5.24** Two rubbing layered cylinders under dry sliding conditions.

Several cases are investigated from Table 5.1 to show various effects for some operating variables on temperature and heat partition. Both denisty and specific heat for all the solid materials in the following examples are assumed to be identical to the corresponding steel properties.

## Numerical Results

Sample conditions are considered to illustrate the results obtained from the dimensionless equations.

Figure 5.25 shows the variation of $T_{max}/q_t$ for different slide/roll ratios. For dry unlayered contacts, the faster and the slower solid surfaces have equal temperatures because there is no reason for a temperature jump across the interface. In the case of dry contacts and constant rolling speed, $T_{max}/q_t$ almost remains constant for different slide/roll ratios, whereas the lubricated contacts show a considerable dependence on this ratio. The slower surface has a higher $T_{max}/q_t$ as compared to the faster solid if there is a film with low thermal conductivity, such as oil, separating the two solids. It can be seen that the film existence would result in a closer heat partition between the two solid surfaces. However, the slower solid has a longer residence time $\ell/U$ under the heat source as compared to the faster solid, therefore, a higher maximum solid surface temprature. On one hand, a thick lubricant film prevents the solids interaction, which eliminates both mechanical and thermal loads between asperities and reduces the friction coefficient. On the other hand, it changes the heat partition in an unfavorable manner.

Figure 5.26 shows a case illustration of the relationships between maximum temperature rise for both lubricant film and solid surfaces and the maximum Hertz pressure. A direct proportionality can be seen with a con-

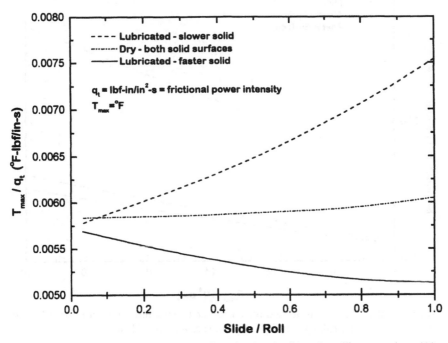

**Figure 5.25** $T_{max}/q_t$ versus slide/roll ratio (steel–oil–steel, rolling speed = 400 in./sec, $R_1 = R_2 = 1$ in., $P_{max} = 100,000$ psi).

siderable difference between the film and solid temperatures. Figure 5.27 represents the case of a metallic solid with an insulative surface layer in contact with another layered solid having an opposite combination of thermal properties. It can be seen from the figure that the heat partition is highly dependent on the ratio $H_d/H_{ave}$ where

$$H_d = h_2 - h_{01} \qquad \text{and} \qquad H_{ave} = \frac{h_{02} + h_{01}}{2}$$

The latter is kept constant to show the main influence of the difference in surface layer thicknesses. The existence of surface layers strongly deviated the heat partition from the dry sliding condition. This phenomenon could be explained by the cooling mechanism in the contact zone by a shallow region near the surface, which mainly incorporates the layer thickness. Figure 5.27 also demonstrates the possibilities for equalizing the heat partition between the moving solids by controlling both thermal properties and thicknesses of surface layers. Negligible sliding is assumed in this case.

Figure 5.28 shows the variation in the maximum temperature rise in the contact zone and the solid surfaces with respect to $H_d/H_{ave}$. The contact

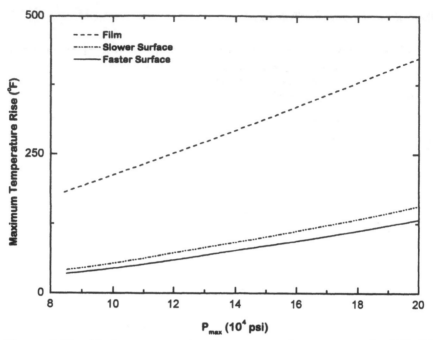

**Figure 5.26** Maximum temperature rise versus maximum pressure for 50% sliding (steel–oil–steel, rolling velocity = 400 in./sec, $R_1 = R_2 = 1$ in.).

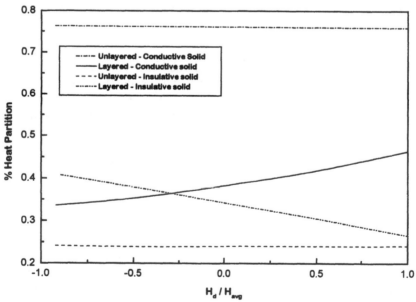

**Figure 5.27** Heat partition versus $H_d/H_{ave}$ (rolling velocity = 2000 in./sec, $l = .02$ in., $K_1 = K_{02}$, $K_{01} = K_2 = .1\ K_1$, $H_{ave} = 10^{-4}$ in.).

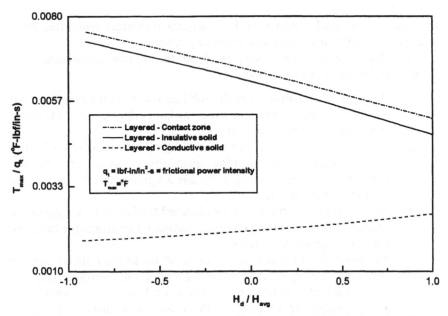

**Figure 5.28** $T_{max}/q_t$ versus $H_d/H_{ave}$(rolling speed = 2000 in./sec, l = .02 in., $K_1 = K_{02}$, $K_{01} = K_2 = .1\ K_1$, $H_{ave} = 10^{-4}$ in.).

zone temperature is almost identical to the solid surface temperature, which carries the conductive surface layer.

The generalized equation for heat partition in lubricated line contact problems, which has been derived for steady-state conditions, is applicable to all metallic solids. It can be deduced from this equation that the deviation in heat partition from that calculated by Jaeger and Blok is highly influenced by the conductivity and thickness of the lubricant film. The existence of the lubricant film tends to equalize heat partition between the rolling/sliding solids independent of their thermal properties and surface speeds. It is interesting to note that the maximum temperature rise for each moving solid is directly proportional to the heat partition coefficient, $\gamma_1$, $\gamma_2$, the ratio of the trailing edge penetration depth to thermal conductivity, $D_1/k_1$, $D_2/k_2$, and the total heat flux, $q_t/\ell$.

The difference between the maximum film and surface temperatures is also controlled by the lubricant film thickness and its conductivity. This can be attributed to the fact that convection is not important in this case.

Although the problem of layered surfaces is appropriately modeled in the developed finite difference program, an evaluation of the effect of the different system parameters on the temperature distribution would be extremely difficult. It was, therefore, imperative to limit the parametric analysis to single layered solids for two specific regimens of thermal properties, layer thicknesses, width of contact and operating speeds.

The following can be concluded from the dimensionless equations developed for the considered cases:

1. The surface temperature of the solid decreases with increasing conductivity of the layers and their thickness due to the convection influence under such conditions.
2. For conductive surface layers, the slide/roll ratio has little influence on the maximum solid surface temperatures, while for insulative surface layers, the slide/roll ratio has a significant influence on the maximum temperature.
3. For conductive surface layers with equal thicknesses, increasing the thickness decreases the maximum surface tempratures for both the solid and the surface layers.
4. For insulative surface layers with equal thicknesses, increasing the thickness slightly decreases the maximum solid surface temperatures and increases the maximum surface layer temperatures.
5. For a conductive solid, $k_1$, with an insulative surface layer, $k_{01}$, contacting an insulative solid, $k_2$, with conductive layer, $k_{02}$, assuming that $k_1 = k_{02}$ and $k_2 = k_{01}$, there is a signficant deviation of the partition from the unlayered case, as shown in Fig. 5.27. It can also be seen that with the above combination of properties it is possible to attain equal heat partition by proper selection of the layer thicknesses (at $H_d/H_{ave} = -0.325$ in this case).

For the above case, the maximum contact temperature is approximately equal to the maximum temperature in the insulative solid with conductive layer for all thickness ratios (Fig. 5.28). The maximum temperature for the conductive solid with insulative layer is significantly lower than the interface temperature.

## REFERENCES

1. Cheng, H. S., "Fundamentals of Elastohydrodynamic Contact Phenomena," International Conference on the Fundamentals of Tribology, Suh, N. and Saka, N., Eds., MIT Press, Cambridge, MA, 1978, p. 1009.
2. Blok, H., "The Postulate About the Constancy of Scoring Temperature," Interdisciplinary Approach to the Lubrication of Concentrated Contacts, P. M. Ku, Ed., NASA SP-237, 1970, p. 153.
3. Sakurai, T., "Role of Chemistry in the Lubrication of Concentrated Contacts," ASME J. Lubr. Technol., 1981, Vol. 103, p. 473.

4.  Alsaad, M., Blair, S., Sanborn, D. M., and Winer, W. O., "Glass Transitions in Lubricants: Its Relation to EHD Lubrication," ASME J. Lubr. Technol., 1978, Vol. 100, p. 404.
5.  Blok, H., "Theoretical Study of Temperature Rise at Surfaces of Actual Contact Under Oiliness Lubricating Conditions," Proc. Gen. Disc. Lubrication, Institute of Mechanical Engineers, Pt. 2, 1937, p. 222.
6.  Jaeger, J. c., "Moving Sources of Heat and the Temperature at Sliding contacts," Proc. Roy. Soc., N.S.W., 1942, Vol. 56, p. 203.
7.  Cheng, H. S., and Sternlicht, B., "A Numerical Solution for the Pressure, Temperature, and Film Thickness Between Two Infinitely Long, Lubricated Rolling and Sliding Cylinders under Heavy Loads," ASME J. Basic Eng., Vol. 87, Series D, 1965, p. 695.
8.  Dowson, D., and Whitaker, A. V., "A Numerical Procedure for the Solution of the Elastohydrodynamic Problem of Rolling and Sliding Contacts Lubricated by a Newtonian Fluid," Proc. Inst. Mech. Engrs, Vol. 180, Pt. 3, Ser. B, 1965, p. 57.
9.  Manton, S. M., O'Donoghue, J. P., and Cameron, A., "Temperatures at Lubricated Rolling/sliding Contacts," Proc. Inst. Mech. Engrs, 1967–1968, Vol. 1982, Pt. 1, No. 41, p. 813.
10. Conry, T. F., "Thermal Effects on Traction in EHD Lubrication," ASME J. Lubr. Technol., 1981, Vol. 103, p. 533.
11. Wang, K. L., and Cheng, H. S., "A Numerical Solution to the Dynamic Load, Film Thickness, and Surface Temperatures in Spur Gears; Part 1 Analysis," ASME J. Mech. Des., 1981, Vol. 103, p. 177.
12. Knotek, O., "Wear Prevention," International Conf. on the Fundamentals of Tribology, Suh, N., and Saka, N., Eds., MIT Press, Cambridge, MA, 1978, p. 927.
13. Torti, M. L., Hannoosh, J. G., Harline, S. D., and Arvidson, D. B., "High Performance Ceramics for Heat Engine Applications," AMSE Preprint No. 84-GT-92, 1984.
14. Georges, J. M., Tonck, A., Meille, G., and Belin, M., "Chemical Films and Mixed Lubrication," Trans. ASLE, 1983, Vol. 26(3), p. 293.
15. Poon, S. Y., "Role of Surface Degradation Film on the Tractive Behavior in Elastohydrodynamic Lubrication Contact," J. Mech. Eng. Sci., 1969, Vol. 11(6), p. 605.
16. Berry, G. A., and Barber, J. R., "The Division of Frictional Heat – A guide to the Nature of Sliding Contact," ASME J. Tribol., 1984, Vol. 106, p. 405.
17. Shaw, M. C., "Wear Mechanisms in Metal Processing," Int. Conf. on the Fundamentals of Tribology, Suh, N., and Saka, N., Eds., MIT Press, Cambridge, MA., 1978, p. 643.
18. Burton, R. A., "Thermomechanical Effects on Sliding Wear," Int. Conf. on the Fundamentals of Tribology, Suh, N., and Saka, N., Eds., MIT Press, Cambridge, MA., 1978, p. 619.
19. Ling, F. F., *Surface Mechanics*, J. Wiley, New York, NY, 1973.

20. Klaus, E. E., "Thermal and Chemical Effects in Boundary Lubrication," Lubrication Challenges in Metalworking and Processing, Proc. 1st Int. Conf., IIT Res. Inst., June, 1978.

21. Winer, W. O., "A Review of Temperature Measurements in EHD Contacts," 5th Leeds–Lyon Symposium, 1978, p. 125.

22. Townsend, D. P., and Akin, L. S., "Analytical and Experimental Spur Gear Tooth Temperature as Affected by Operating Variables," ASME J. Lubr. Technol., 1981, Vol. 103, p. 219.

23. Rashid, M., and Seireg, A., "Heat Partition and Transient Temperature Distribution in Layered Concentrated Contacts. Part 1: Theoretical Model, Part 2: Dimensionless Relationships, ASME J. Tribol., July 1987, Vol. 109, p. 496.

24. Taylor, E. S., *Dimensional Analysis for Engineers*, Oxford University Press, London, England, 1974.

25. David, F. W., and Nolle, H., *Experimental Modeling in Engineering*, Butterworths, London, England, 1974.

26. Arpaci, V. S., *Conduction Heat Transfer*, Addison-Wesley, Reading, MA, 1966, p. 474.

27. Hamrock, B. J., and Jacobson, B. O., "Elastohydrodynamic Lubrication of Line Contacts," ASLE Trans., 1984, Vol. 27, p. 275.

28. Wilson, W. R. D., and Sheu, S., "Effect on Inlet Shear Heating Due to Sliding on Elastohydrodynamic Film Thickness," ASME J. Lubr. Technol., 1983, Vol. 105, p. 187.

29. Suzuki, A., and Seireg, A., "An Experimental Investigation of Cylindrical Roller Bearings Having Annular Rollers," ASME J. Lubr. Technol., 1976, Vol. 98, p. 538.

# 6

# Design of Fluid Film Bearings

## 6.1 HYDRODYNAMIC JOURNAL BEARINGS

Fluid film bearings are a common means of supporting rotating shafts in rotating machinery. In such bearings a pressurized fluid film is formed with adequate thickness to prevent rubbing of the mating surfaces. There are two main types of fluid film bearings: hydrostatic and hydrodynamic. The first type relies on an external source of energy to supply the lubricant with the necessary pressure. In the second type, pressure is developed within the bearing as a result of the relative motion between shaft and bearing. The pressure is influenced by the geometry of the fluid wedge, which is formed to sustain the load, as illustrated in Fig 6.1.

### 6.1.1 Hydrodynamic Equations

The basic equation governing the behavior of the fluid film in the hydro-dynamic case is Reynolds' equation [1]. Assuming isothermal, incompressible flow this equation is derived by consideration of the Newtonian shear – velocity gradient relationship, the equilibrium of the fluid element (Fig. 6.2) in the $x$- and $z$-directions and the continuity equation. Accordingly:

$$\frac{\partial}{\partial x}\left(h^3 \frac{\partial p}{\partial x}\right) + \frac{\partial}{\partial z}\left(h^3 \frac{\partial p}{\partial z}\right) = 6\mu U \frac{\partial h}{\partial x} \qquad (6.1)$$

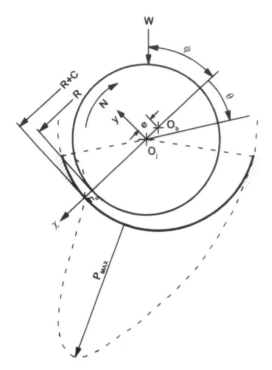

**Figure 6.1** Bearing geometry.

where

$p$ = film pressure
$h$ = film thickness
$\mu$ = oil viscosity (assumed constant throughout the film)
$U$ = tangential velocity in the journal

The solution of this equation for any eccentricity, $e$, would result in expressions for the quantity of flow required, the frictional power loss and the pressure distribution in the oil film. The latter determines the load-carrying capacity of the bearing. A closed-form solution of Reynolds' equation can be obtained with either of the following assumptions:

1. Assume the bearing to be long compared to its diameter. This is generally known as the Sommerfeld bearing. In this case the change in pressure in the axial direction can be neglected compared to the change in the circumferential ($x$) direction. Accordingly:

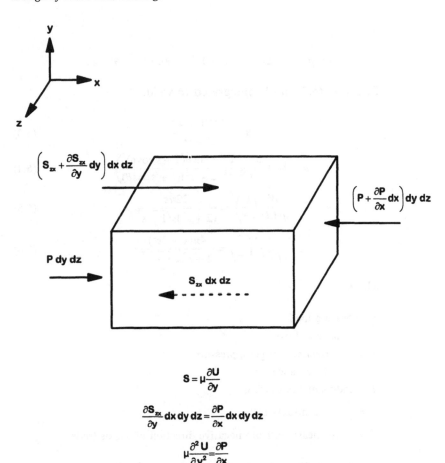

**Figure 6.2**   Equilibrium of fluid element in the *x*-direction.

$$\frac{\partial}{\partial z} \ll \frac{\partial p}{\partial x}$$

and Eq. (6.1) reduces to:

$$\frac{d}{dx}\left(h^3 \frac{dp}{dx}\right) = 6\mu U \frac{dh}{dx} \qquad (6.2)$$

Referring to Fig. 6.1 with the condition:

$$p = p_{\max} \qquad \text{at} \qquad h = h_1$$

and

$$p = p_0 \qquad \text{at} \qquad \theta = 0 \qquad \text{and} \qquad \theta = 2\pi$$

Equation (6.2) can be integrated to yield:

$$h_1 = \frac{2C(1 - \varepsilon^2)}{2 + \varepsilon^2} \tag{6.3}$$

$$p = p_0 + \frac{6\mu UR}{C^2}\left(\frac{\varepsilon \sin\theta(2 + \varepsilon\cos\theta)}{(2 + \varepsilon^2)(1 + \varepsilon\cos\theta)}\right) \tag{6.4}$$

$$\frac{W}{\mu LU}\left(\frac{C}{R}\right)^2 = \frac{12\pi\varepsilon}{(2 + \varepsilon^2)\sqrt{1 - \varepsilon^2}} \tag{6.5}$$

$$\frac{F}{\mu LU}\left(\frac{C}{R}\right) = \frac{4\pi(1 + 2\varepsilon^2)}{2 + \varepsilon^2\sqrt{1 - \varepsilon^2}} \tag{6.6}$$

where

$W$ = bearing load
$F$ = tangential load
$h_1$ = film thickness at peak pressure
$C$ = radial clearance
$R$ = radius of the bearing
$\varepsilon = \dfrac{e}{C}$ = eccentricity ratio
$\theta$ = angle measured from negative direction of $x$-axis (rads)

Equation (6.5) can be rewritten in terms of a dimensionless number, $S$, called the Sommerfeld number:

$$S = \frac{(2 + \varepsilon^2)\sqrt{1 - \varepsilon^2}}{12\pi^2\varepsilon} \tag{6.7}$$

so that

$$S = \frac{\mu UL}{\pi W}\left(\frac{R}{C}\right)^2 = \left(\frac{R}{C}\right)^2 \frac{\mu N}{P} \tag{6.8}$$

where

$N, P$ = journal rotational speed and average pressure, respectively

2. The bearing length is small compared to its diameter. The analysis of this case is called the *short bearing approximation*. The flow in the axial ($z$) direction is assumed to be considerably greater than that in the circumferential direction. Accordingly, Reynolds' equation reduces to

$$\frac{\partial}{\partial z}\left(h^3\frac{\partial p}{\partial z}\right) = 6\mu U\frac{dh}{dz} \tag{6.9}$$

with the boundary conditions:

$$\frac{\partial p}{\partial z} = 0 \qquad \text{and} \qquad z = 0$$

$$p = 0 \qquad \text{at} \qquad z = \frac{L}{2}$$

where

$L$ = bearing length

The solution of Eq. (6.9) yields:

$$p = \left(\frac{3\mu U}{RC^2}\right)\frac{\varepsilon\sin\theta}{(1+\varepsilon\cos\theta)^3}\left(\frac{L^2}{4}-z^2\right) \tag{6.10}$$

$$\frac{W}{\mu LU}\left(\frac{C}{R}\right)^2\left(\frac{D}{L}\right)^2 = \frac{\varepsilon\sqrt{\pi^2(1-\varepsilon^2)+16\varepsilon^2}}{(1-\varepsilon^2)^2} \tag{6.11}$$

$$F = \left(\frac{\mu ULR}{C}\right)\frac{2\pi}{(1-\varepsilon^2)^2} \tag{6.12}$$

$$\tan\phi = \frac{\pi\sqrt{1-\varepsilon^2}}{4\varepsilon} \tag{6.13}$$

where

$\phi$ = attitude angle between the load line and the line of centers
$D$ = journal diameter

Equation (6.11) can be rewritten in terms of the Sommerfeld number, $S$, as:

$$S\left(\frac{L}{D}\right)^2 = \frac{(1-\varepsilon^2)^2}{\pi\varepsilon\sqrt{\pi^2(1-\varepsilon^2)+16\varepsilon^2}} \tag{6.14}$$

### 6.1.2   Numerical Solution

The development of high-speed computers made it possible to obtain numerical solutions for Reynolds' equation (6.1). Accordingly, bearing behavior can be calculated for any given geometry and boundary conditions. The numerical results of Raimondi and Boyd [2–4] are among the most widely known. Their data are presented in the form of design charts with Sommerfeld number as the main parameter. The numerical solutions are based on the assumption of an isoviscous film independent of pressure and temperature variations. The film viscosity is calculated at the mean temperature in the bearing. The bearing performance charts of Raimondi and Boyd are given in Figs 6.3–6.7.

### 6.1.3   Equations for Predicting Bearing Performance

A system of equations, based on these charts, is developed by approximate curve fitting. The equations are utilized in the automated design system described later in this chapter. These equations are:

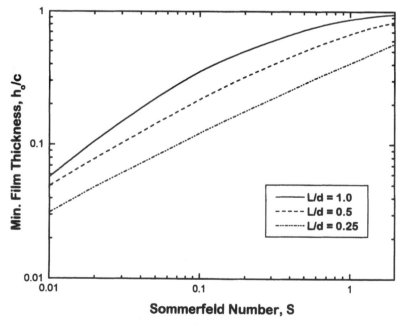

**Figure 6.3**   Minimum film thickness.

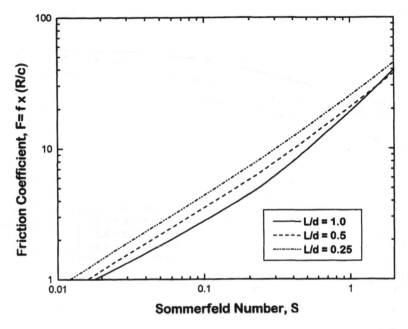

**Figure 6.4** Frictional variable.

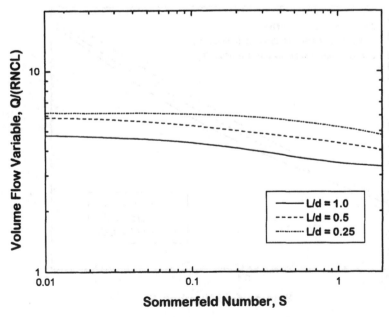

**Figure 6.5** Oil quantity variable.

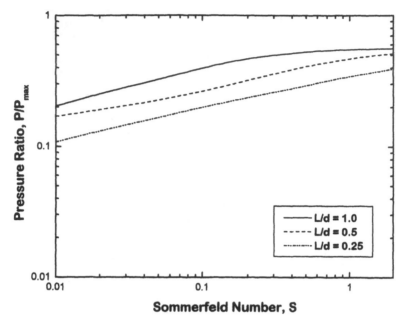

**Figure 6.6**   Peak pressure.

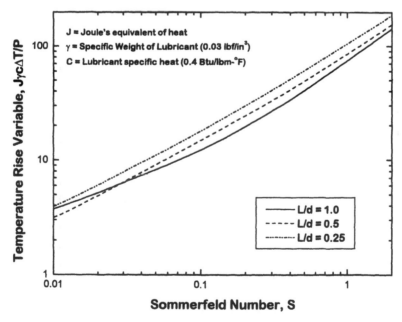

**Figure 6.7**   Temperature variable.

1.0. 
$$S \leq 0.15$$

1.1. 
$$0.25 \leq \frac{L}{D} \leq 0.5$$

$$h_0 = 1.585C\left(\frac{L}{D}\right)^{0.913} (S)^{0.655(L/D)^{0.0922}}$$

$$\Delta t = \frac{0.5}{(L/D)^{0.374}} P(S)^{\frac{0.695}{(L/D)^{0.139}}}$$

$$\frac{P}{P_{max}} = 0.76\left(\frac{L}{D}\right)^{0.62} (S)^{0.24}$$

$$\frac{RNCL}{Q} = 0.128\left(\frac{L}{D}\right)^{0.048} (S)^{0.1(L/D)^{0.47}}$$

$$f\frac{R}{C} = \frac{12.6}{(L/D)^{0.41}} (S)^{\frac{0.62}{(L/D)^{0.1035}}}$$

1.2. 
$$0.5 \leq \frac{L}{D} \leq 1.0$$

$$h_0 = 1.84C\left(\frac{L}{D}\right)^{1.13} (S)^{0.731(L/D)^{0.252}}$$

$$\Delta t = \frac{0.43}{(L/D)^{0.62}} P(S)^{\frac{0.56}{(L/D)^{0.302}}}$$

$$\frac{P}{P_{max}} = 0.76\left(\frac{L}{D}\right)^{0.62} (S)^{0.294(L/D)^{0.292}}$$

$$\frac{RNCL}{Q} = 0.128\left(\frac{L}{D}\right)^{0.048} (S)^{\frac{0.06}{(L/D)^{0.1035}}}$$

$$f\left(\frac{R}{C}\right) = \frac{11.8}{(L/D)^{0.503}} (S)^{\frac{0.62}{(L/D)^{0.1035}}}$$

2.0. 
$$S \geq 0.15$$

2.1. 
$$0.25 \leq \frac{L}{D} \leq 0.50$$

$$h_0 = 1.035C\left(\frac{L}{D}\right)^{0.0673} (S)^{\frac{0.33}{(L/D)^{0.20}}}$$

$$\Delta t = \frac{0.695}{(L/D)^{0.214}} (S)^{0.875(l/D)^{0.042}} (P)$$

$$\frac{P}{P_{max}} = 0.76\left(\frac{L}{D}\right)^{0.62} (S)^{0.24}$$

$$\frac{RNCL}{Q} = 0.128\left(\frac{L}{D}\right)^{0.048} (S)^{0.1(L/D)^{0.47}}$$

$$f\left(\frac{R}{C}\right) = \frac{16.85}{(L/D)^{0.318}} (S)^{0.922(L/D)^{0.087}}$$

The following notation is used in the analysis:

$C$ = radial clearance (in.)

$D$ = journal diameter (in.)

$e$ = journal eccentricity (in.)

$f$ = coefficient of friction

$h_0$ = minimum film thickness (in.)

$L$ = bearing length (in.)

$N$ = journal rotational speed (rps)

$P$ = bearing average pressure (lb/in.$^2$)

$P_{max}$ = maximum oil film pressure (lb/in.$^2$)

$Q$ = quantity of oil fed to bearing (in.$^3$/sec)

$R$ = journal radius (in.)

$t_{max}$ = maximum oil film temperature (°F)

$\Delta t$ = oil temperature rise (°F)

$W$ = bearing load (lb)

$\mu$ = lubricant average viscosity (reyn)

$\phi$ = attitude angle

Dimensionless groups:

$$f \frac{R}{C} = \text{frictional variable}$$

$$\frac{RNCL}{Q} = \text{oil flow variable}$$

$$\frac{L}{D} = \text{length to diameter ratio}$$

$$S = \text{Sommerfeld no.} = \left(\frac{R}{C}\right)^2 \frac{\mu N}{P}$$

### 6.1.4 Whirl and Stability Considerations

The whirl of rotors supported on fluid films is of considerable practical importance and it is no surprise that it has been the subject of numerous investigations. The designer is generally interested in predicting beforehand not only the stability of the rotor operation but the expected peak amplitude of whirl orbit at a given operating speed resulting from a particular amount of unbalance. The safety during the transient start-up phase has to be also insured. Some of the excellent contributions to the analysis of this problem and related bibliographies can be found in Refs 5–12.

A simplified stability criterion, based on an analysis by Lund and Saibel [11], is also given in Fig. 6.8. A modified Sommerfeld number, $\sigma = \pi S(L/D)^2$, and a dimensionless rotor mass, $MC\omega^2/W$, determines the condition for the onset of bearing instability according to the relationship plotted in the figure, where

$$M = \text{rotor mass per bearing}$$
$$C = \text{radial clearance}$$
$$\omega = \text{journal rotational speed (rad/sec)}$$
$$W = \text{bearing load}$$

Mathematical expressions describing the limits of journal stability as a function $\varphi$ of $\sigma$ can be derived from Fig. 6.8 as:

1. $\sigma \le 0.28$: $\varphi(\sigma) = 3/\sigma^{0.55}$
2. $0.28 \le \sigma \le 2.90$: $\varphi(\sigma) = 6.88\sigma^{0.094}$
3. $\sigma > 2.90$: $\varphi(\sigma) = 7.65$

### 6.1.5 The Effect of Rotor Unbalance on Whirl Amplitude and Stability

It is well known that rotor unbalance can have significant effect on whirl and stability. A study by Seireg and Dandage [13] utilized the phase-plane method to simulate the dynamics of an unbalanced rotor supported by an isoviscous film. Response spectra were obtained to illustrate the trend of the influence of the magnitude of unbalance, speed, average film viscosity, load, clearance, and rotor start-up, acceleration on the whirl amplitudes.

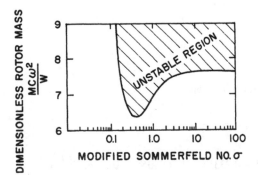

**Figure 6.8** Stability criterion ($\sigma = \pi S(L/D)^2$).

The equations of motion for the rotor under consideration from the steady-state position can be written as:

$$M\ddot{x} + C_{xx}\dot{x} + C_{xy}\dot{y} + K_{xx}x + K_{xy}y = F_x$$
$$M\ddot{y} + C_{yx}\dot{x} + C_{yy}\dot{y} + K_{yx}x + K_{yy}y = F_y$$

which represents a system of coupled nonlinear second-order equations in which:

$F_x$ = force due to the unbalance in the $x$-direction = $mr\omega^2 \cos(\omega t + \phi)$

$\quad - mr \dfrac{d\omega}{dt} \sin(\omega t + \phi)$ in the transient phase

$F_x = mr\omega^2 \cos(\omega t + \phi)$ during the constant speed operation

$F_y = mr\omega^2 \sin(\omega t + \phi) + mr \dfrac{d\omega}{dt} \cos(\omega t + \phi)$ in the transient phase

$F_y = mr\omega^2 \sin(\omega t + \phi)$ during the constant speed operation

where

$\quad m$ = rotor mas per bearing

$\quad mr$ = amount of unbalance

$\quad \omega$ = rotating speed

$\quad \phi$ = phase angle between the unbalance force and the movement of the center of mass of the rotor in the $x$-direction

Expressions for calculating the stiffness and damping coefficients $k_{xx}$, $C_{xx}$, $k_{xy}$, $C_{xy}$, $k_{yy}$, $C_{yy}$, $k_{yx}$, and $C_{yx}$ are given in the following. They are derived by approximate curve fitting from Ref. 14 as a function of $S$, which is evaluated at any instant from the instantaneous eccentricity ratio.

*Stiffness Coefficients*

$0.5 \le L/D \le 1.0, S \le 0.15$:

$$K_{xx} = 0.5979\left(\frac{L}{D}\right)^{-1.0181} S^{-0.8863+0.1927(L/D)}$$

$$K_{xy} = 2.501\left(\frac{L}{D}\right)^{-0.2127} S^{-0.3713+0.1476(L/D)}$$

$$K_{yx} = -0.4816 + 1.7006\left(\frac{L}{D}\right) - 0.9335S + 11.6940S^2 - 16.3368S\left(\frac{L}{D}\right)$$
$$+ 2.2198S^{(L/D)}$$

$$K_{yy} = 2.5181 \left(\frac{L}{D}\right)^{-0.3236} S^{0.4954-0.4007(L/D)}$$

$0.15 \le S \le 1$:

$$K_{xx} = 1.1251 \left(\frac{L}{D}\right)^{-0.6746} S^{-0.8179+0.4691(L/D)}$$

$$K_{xy} = 7.5105 \left(\frac{L}{D}\right)^{0.9778} S^{-0.5584+1.0131(L/D)}$$

$$K_{yx} = 1.463 + 2.044 \left(\frac{L}{D}\right) - 1.290S + 1.053S^2 - 12.272S\left(\frac{L}{D}\right) + 4.378S^{(L/D)}$$

$$K_{yy} = 2.2202 \left(\frac{L}{D}\right)^{-0.1794} S^{0.3145-0.2771(L/D)}$$

$S \ge 1.0$:

$$K_{xx} = 2.3258 - 1.2120 \left(\frac{L}{D}\right) - 0.3413S + 0.3436S\left(\frac{L}{D}\right)$$

$$K_{xy} = 8.1515 \left(\frac{L}{D}\right)^{1.1442} S^{0.4387+0.4717(L/D)}$$

$$K_{yx} = 12.2356 - 12.4891 \left(\frac{L}{D}\right) + 1.2669S + 0.0756S^2 + 2.2224S\left(\frac{L}{D}\right)$$
$$- 10.9395S^{(L/D)}$$

$0.25 \le L/D \le 0.50, S \le 0.15$:

$$K_{xx} = 0.3532 \left(\frac{L}{D}\right)^{-1.7179} S^{-0.2589-1.0922(L/D)}$$

$$K_{xy} = 1.9165 \left(\frac{L}{D}\right)^{-0.2674} S^{-0.6257+0.6357(L/D)}$$

$$K_{yx} = 16.74 - 26.53 \left(\frac{L}{D}\right) + 13.63S - 38.70S^2 + 43.72S\left(\frac{L}{D}\right) - 20.12S^{(L/D)}$$

$$K_{yy} = 7.4551 \left(\frac{L}{D}\right)^{1.1975} S^{-0.6117+1.8354(L/D)}$$

$0.15 < S \le 1.0$:

$$K_{xx} = 1.1897 \left(\frac{L}{D}\right)^{-0.6076} S^{-0.7152+0.3053(L/D)}$$

$$K_{yx} = 4.4333 - 6.0498 \left(\frac{L}{D}\right) + 3.9980S - 4.2133S^2 + 0.8848S\left(\frac{L}{D}\right) - 4.6921S^{(L/D)}$$

$S \le 0.50$:

$$K_{yx} = -7.027 + 10.473\left(\frac{L}{D}\right) - 1.964S + 2.039S^2 - 19.4675\left(\frac{L}{D}\right)$$
$$+ 8.995S^{(L/D)}$$

$S \geq 0.50$:

$$K_{yy} = 3.4171\left(\frac{L}{D}\right)^{0.4507} S^{-0.2406+0.8095(L/D)}$$

$S > 1.0$:

$$K_{xx} = 1.2702\left(\frac{L}{D}\right)^{-0.5053} S^{-0.8199+0.9723(L/D)}$$

$$K_{xy} = 5.7477\left(\frac{L}{D}\right)^{0.6401} S^{0.2659+0.8173(L/D)}$$

$$K_{xx} = 27.92 - 32.71\left(\frac{L}{D}\right) - 1.21S - 0.404S^2 + 19.34\left(\frac{L}{D}\right) - 22.11S\left(\frac{L}{D}\right)$$

$$K_{yy} = 3.1903\left(\frac{L}{D}\right)^{0.3860} S^{0.1813-0.3994(L/D)}$$


## Damping Coefficients

$0.5 \leq L/D \leq 1.0, S \leq 0.05$:
$$C_{xx} = -8.090 + 30.59\left(\frac{L}{D}\right) - 1647.56S + 7078.46S^2 + 739.86S\left(\frac{L}{D}\right) + 219.81S^{(L/D)}$$

$$C_{xy} = -21.11 + 27.88\left(\frac{L}{D}\right) - 705.62S + 2327.28S^2 + 353.02D\left(\frac{L}{D}\right) + 133.45S^{(L/D)}$$

$$C_{yx} == 22.24 + 28.43\left(\frac{L}{D}\right) - 672.00S + 2129.06S^2 + 337.77S\left(\frac{L}{D}\right) + 132.83S^{(L/D)}$$

$$C_{yy} = -2.872 + 5.397\left(\frac{L}{D}\right) - 204.892S + 833.223S^2 + 106.010S\left(\frac{L}{D}\right) + 31.329S^{(L/D)}$$

$0.05 < S \leq 0.25$:
$$C_{xx} = 26.402 - 19.317\left(\frac{L}{D}\right) - 5.201S + 55.945S^2 + 17.102S\left(\frac{L}{D}\right) - 29.361S^{(L/D)}$$

$$C_{xy} = 1.7126 + 0.1870\left(\frac{L}{D}\right) - 9.7244S + 10.0934S^2 + 5.0982S\left(\frac{L}{D}\right) + 2.8168S^{(L/D)}$$

$$C_{yx} = 1.6125 - 0.201\left(\frac{L}{D}\right) - 5.4768S + 0.1344S^2 + 6.7569S\left(\frac{L}{D}\right) + 2.2069S^{(L/D)}$$

$$C_{yy} = 1.587 - 1.083\left(\frac{L}{D}\right) - 18.793S + 13.865S^2 + 30.244S\left(\frac{L}{D}\right) + 2.021S^{(L/D)}$$

$0.25 < S \leq 1.0$:

$$C_{xx} = 6.357 - 2.298\left(\frac{L}{D}\right) - 17.294S + 3.521S^2 + 19.892S\left(\frac{L}{D}\right) + 5.635S^{(L/D)}$$

$$C_{xy} = 0.4641 + 1.8204\left(\frac{L}{D}\right) - 0.6852S + 0.3214S^2 - 2.1412S\left(\frac{L}{D}\right) + 2.2092\left(\frac{L}{D}\right)$$

$$C_{yx} = 0.4619 + 2.0240\left(\frac{L}{D}\right) - 0.5145S + 0.4168S^2 - 2.1685S\left(\frac{L}{D}\right) + 1.9817S^{(L/D)}$$

$$C_{yy} = -7.852 + 10.414\left(\frac{L}{D}\right) - 15.185S + 3.576S^2 + 10.084S\left(\frac{L}{D}\right) + 13.869S^{(L/D)}$$

$S > 1.0$:

$$C_{xx} = 26.424 - 25.646\left(\frac{L}{D}\right) - 11.235S - 0.044S^2 + 43.391S\left(\frac{L}{D}\right) - 16.949S^{(L/D)}$$

$$C_{xy} = 1.353 + 1.109\left(\frac{L}{D}\right) + 0.376S + 0.027S^2 - 1.341S\left(\frac{L}{D}\right) + 0.510S^{(L/D)}$$

$$C_{yx} = 4.059 - 2.013\left(\frac{L}{D}\right) - 0.231S - 0.004S^2 + 1.839S\left(\frac{L}{D}\right) - 1.450S^{(L/D)}$$

$$C_{yy} = 4.735 - 5.019\left(\frac{L}{D}\right) - 7.506S - 0.089S^2 + 26.089S\left(\frac{L}{D}\right) - 2.897S^{(L/D)}$$

$0.25 \leq L/D < 0.50, S \leq 0.10$:

$$C_{xx} = 151.70 - 220.05\left(\frac{L}{D}\right) + 472.18S - 867.94S^2 + 59.59S\left(\frac{L}{D}\right) - 224.26S^{(L/D)}$$

$$C_{xy} = 9.272 - 14.117\left(\frac{L}{D}\right) - 16.803S - 21.363S^2 + 44.802S\left(\frac{L}{D}\right) - 1.517S^{(L/D)}$$

$$C_{yx} = 0.250 - 0.0677\left(\frac{L}{D}\right) - 64.573S + 60.894S^2 + 63.711S\left(\frac{L}{D}\right) + 13.874S^{(L/D)}$$

$$C_{yy} = 10.58 - 14.77\left(\frac{L}{D}\right) + 63.23S - 87.31S^2 - 28.93S\left(\frac{L}{D}\right) - 18.22S^{(L/D)}$$

$0.10 < S \leq 1.0$:

$$C_{xx} = 45.534 - 57.339\left(\frac{L}{D}\right) + 5.652S - 7.791S^2 + 61.060S\left(\frac{L}{D}\right) - 38.020S^{(L/D)}$$

$$C_{xy} = 8.035 - 10.785\left(\frac{L}{D}\right) - 0.134S - 0.8367S^2 + 6.204S\left(\frac{L}{D}\right) - 2.591S^{(L/D)}$$

$$C_{yx} = 5.348 - 6.560\left(\frac{L}{D}\right) - 2.161S - 0.105S^2 + 5.534S\left(\frac{L}{D}\right) - 0.277S^{(L/D)}$$

$$C_{yy} = 1.000 - 0.6411\left(\frac{L}{D}\right) - 0.7971S + 0.4845S^3 + 5.3996S\left(\frac{L}{D}\right) + 1.6987S^{(L/D)}$$

$S > 1.0$:

$$C_{xx} = 33.5977 - 36.630\left(\frac{L}{D}\right) - 6.205S - 0.0315S^2 + 38.344S\left(\frac{L}{D}\right) - 20.881S^{(L/D)}$$

$$C_{xy} = 4.872 + 5.657\left(\frac{L}{D}\right) - 0.391S + 0.007S^2 + 0.576S\left(\frac{L}{D}\right) + 0.201S^{(L/D)}$$

$$C_{yx} = 5.145 - 4.047\left(\frac{L}{D}\right) - 0.577S - 0.016S + 2.744S\left(\frac{L}{D}\right) - 1.615S^{(L/D)}$$

$$C_{yy} = 3.657 - 7.447\left(\frac{L}{D}\right) - 3.246 + 0.174S^2 + 11.932S\left(\frac{L}{D}\right) + 1.951S^{(L/D)}$$

## Illustrative Examples

The equations of motion are numerically integrated for sample conditions in order to illustrate the dynamic behavior of the rotor for different values of the bearing design parameters. The phase-plane is used for the integration [13].

In order to analyze the motion from the start of the rotation until the final uniform speed is reached, the start-up velocity pattern can be incorporated in the integration.

Two sets of coordinates are used in the analysis. The first set is a Newtonian frame for the dynamic analysis. The second set is attached to the shaft and is used to define the film geometry and corresponding dynamic film characteristics at any instant. The necessary transformations between the two frames are continuously performed throughout the simulation.

Unless otherwise specified, the following parameters for the bearing rotor system are used in the calculations:

$$D = \text{bearing diameter} = 2.5\,\text{in.}\,(6.35\,\text{cm})$$
$$c = \text{radial clearance} = 0.0063\,\text{in.}\,(0.016\,\text{cm})$$
$$W = Mg = \text{rotor weight} = 100\,\text{lb}\,(45.4\,\text{kg})$$
$$L = \text{bearing length} = 2.5\,\text{in.}\,(6.35\,\text{cm})$$
$$\text{Unbalance mr} = 0 \text{ to } 0.0007\,\text{lb-sec}^2\,(0 \text{ to } 0.000318\,\text{kg-sec}^2)$$

The oil used is SAE 10 at 150°F (65.5°C) average temperature (corresponding to an average viscosity of $1.76 \times 10^{-6}$ rens). Speeds $= 0$ to 7000 rpm.

Examples of typical computer-plotted whirl orbits, and time history of the eccentricity ratio are shown in figs 6.9–6.14. The results given in Figs 6.9–6.11 are for a perfectly balanced rotor and very high start-up acceleration, $(T_0 = 0)$. It can be seen that at very low speeds, the balanced rotor gradually reaches steady-state equilibrium at a fixed eccentricity (Fig. 6.9). If

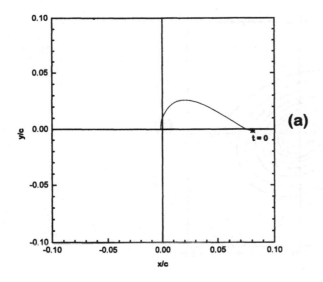

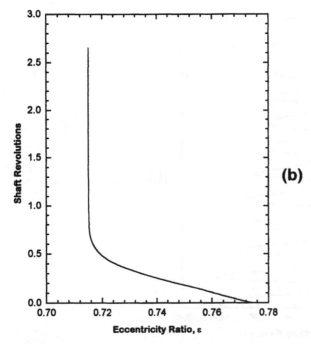

**Figure 6.9** (a) Whirl orbit for balanced rotor at 1000 rpm. (b) Eccentricity–time plot for balanced rotor at 1000 rpm.

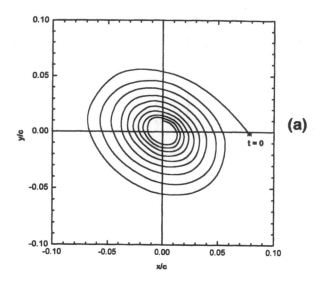

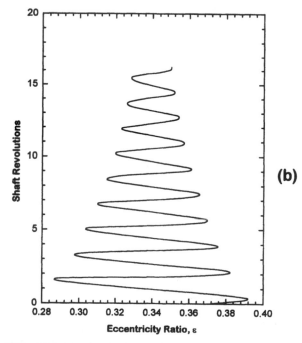

**Figure 6.10** (a) Condition of minimum eccentricity for balanced rotor at 5000 rpm. (b) Eccentricity–time plot at 5000 rpm.

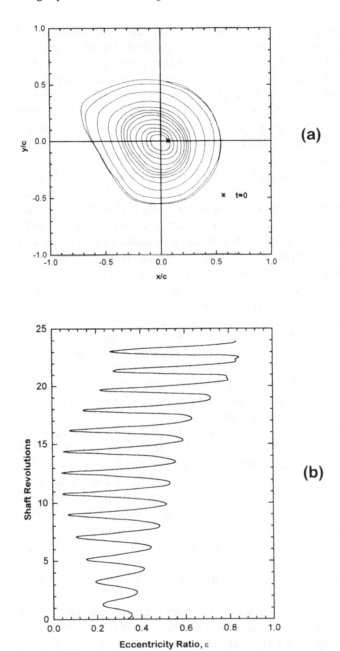

**Figure 6.11** (a) Nonsynchronous whirl for balanced rotor at 6000 rpm. (b) Eccentricity–time plot for balanced rotor at 6000 rpm.

disturbed from that position, the whirl orbit will gradually decay to the equilibrium point. The minimum condition is reached, for this case, at a speed of 5000 rpm (Fig. 6.10). Any increase in the speed beyond this value would produce limit cycle orbits (Fig. 6.11), with increasing amplitudes. Finally, at a speed of 6200rpm, the orbit becomes large enough to consume all the bearing clearance, producing contact between the shaft and the sleeve (neglecting the effect of large amplitudes and near-wall operation on the dynamic characteristics of the film).

Three different whirl conditions were found to occur, also, for the unbalanced rotor as illustrated in Figs 6.12–6.14 for an unbalance $mr = 0.0001$ lb-sec$^2$ (0.0000455kg-sec$^2$) and $T_0 = 0$. At low speeds, the unbalance produces synchronous whirl with relatively high maximum eccentricity, as shown in Fig. 6.12. The maximum eccentricity of the orbit decreases with increasing speed until a minimum condition is reached at a speed of 4900 rpm corresponding to this unbalance. Higher rotor speeds beyond the minimum condition begin to produce nonsynchronous whirl with increasing maximum eccentricities (Fig. 6.13). Finally, at a speed of 6100 rpm, the orbit continues to increase until contact with the sleeve occurs (Fig. 6.14). A summary plot for these different orbit conditions as affected by the magnitude of unbalance is given in Fig. 6.15a. Isoeccentricity ratio lines are plotted from the steady-state orbits to illustrate the effect of speed and unbalance on the type and magnitude of the rotor vibration. A similar plot is given in Fig. 6.15b for the peak eccentricity occurring during the rotor operation. These eccentricities generally occur during the transient phase before steady-state orbits are attained.

Of particular interest is the minimum peak eccentricity locus shown in broken lines in Fig. 6.15a. Also of interest is the sleeve contact curve. Although this curve is obtained with simplifying assumptions, it serves to illustrate the expected trend for the upper speed limit of rotor operation. Both conditions impose a reduction on the corresponding speed as the magnitude of unbalance increases. It is also interesting to note that there appears to be practically a constant speed range of approximately 1200 rpm between the minimum peak eccentricity condition and the sleeve contact conditions.

The following results illustrate the influence of some of the main parameters on the rotor whilr.

Figure 6.16 illustrates the effect of increasing the rotor weight on the whirl. The results show that increasing the rotor weight from 45.5 kg to 142 kg increases the speed for the instability threshold. It also significantly reduces the whirl amplitude.

Changes in the whirl conditions can be seen in Fig. 6.17, when the bearing clearance is changed from 0.0063 in. to 0.01 in. (0.016 to

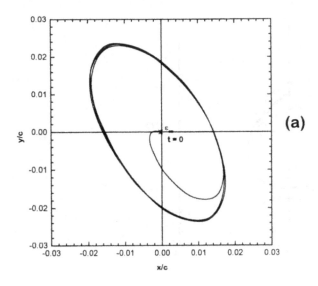

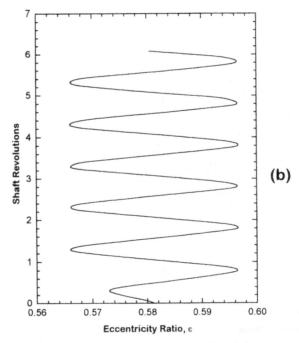

**Figure 6.12** (a) Synchronous whirl for balanced rotor at 1750 rpm. (b) Eccentricity–time plot for unbalanced rotor at 1750 rpm.

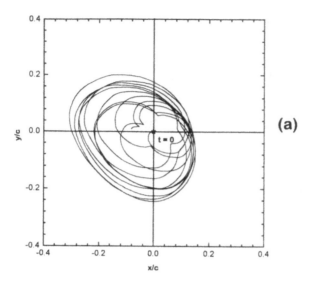

**(a)**

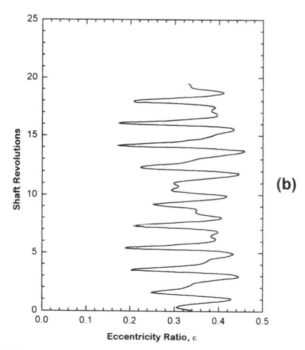

**(b)**

**Figure 6.13** (a) Nonsynchronous whirl for balanced rotor at 5500 rpm. (b) Eccentricity–time plot for unbalanced rotor at 5500 rpm.

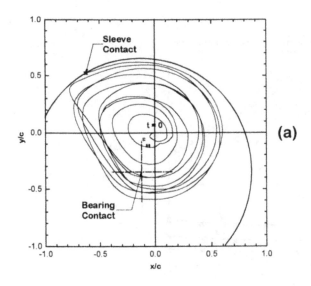

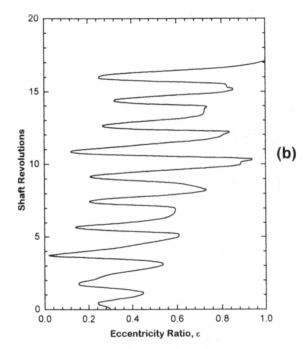

**Figure 6.14** (a) Whirl of unbalanced rotor at sleeve contact condition (6100 rpm). (b) Eccentricity–time plot for unbalanced rotor at 6100 rpm.

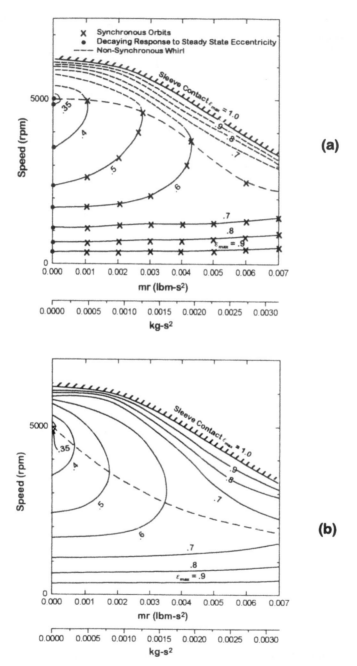

**Figure 6.15** (a) Spectrum of steady-state peak eccentricity for unbalanced rotor. (b) Spectrum of transient peak eccentricity for unbalanced rotor.

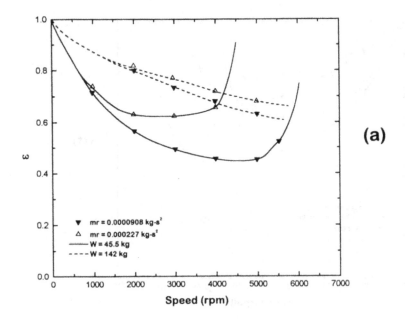

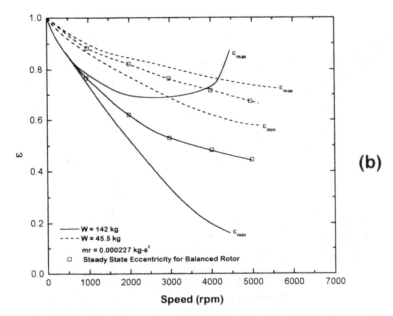

**Figure 6.16** (a) Effect of rotor weight on steady-state peak eccentricity for different unbalanced magnitudes. (b) Effects of rotor weight on amplitude of whirl ($mr = 0.000227$ kg-sec$^2$).

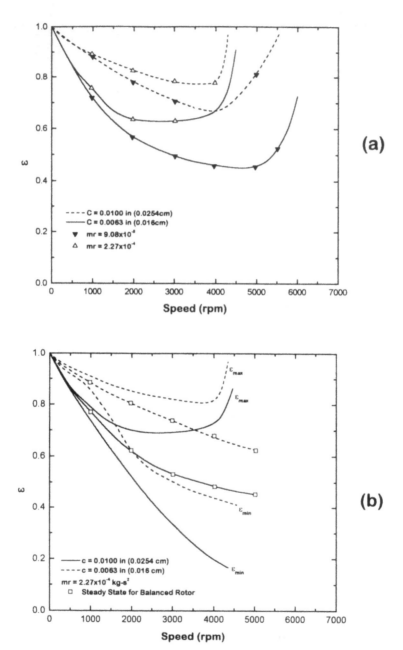

**Figure 6.17** (a) Effect of clearance on steady-state peak eccentricity for different unbalanced magnitudes. (b) Effect of clearance on amplitude of whirl ($mr = 0.000227\,\text{kg-sec}^2$).

0.0254 cm). Figure 6.17a shows that increasing the clearance causes a reduction in the instability threshold. Figure 6.17b on the other hand, shows little effect on the actual whirl orbit amplitude due to the clearance change with $0.000227 \, \text{kg-sec}^2$ unbalance.

Two opposite effects of changing the average film temperature are shown in Fig. 6.18. In the first example, with $W = 100 \, \text{lb}$ (45.5 kg), $C = 0.0063 \, \text{in.}$ (0.016 cm), and $mr = 0.005 \, \text{lb-sec}^2$ (0.000227 kg-sec$^2$), increasing the average film temperature from 37.8°C to 94°C resulted in a considerable reduction in the threshold speed, as well as an increase in the whirl amplitude (Fig. 6.18a). On the other hand, the second example, $W = 1000 \, \text{lbf} = 455 \, \text{kg}$ and $C = 0.016 \, \text{cm}$, shows that considerable reductions in the whirl amplitude resulted from the same increase in the average film temperature (Fig. 6.18b).

The case of a rigid rotor on an isoviscous film considered in this illustration provided a relatively simple model to approximately investigate the effect of rotor unbalance and film properties on the rotor whirl.

The developed response spectrum shown in Fig. 6.15a gives a complete view of the nature of the rotor whirl as affected by the speed and the unbalance magnitude. Of particular interest is the existence of a rotational speed for any particular unbalance where the peak eccentricity is minimal. Nonsynchronous whirl, with increasing amplitudes and eventual instability or rotor sleeve contact, occurs as the speed is increased beyond that condition. It should be noted here that results associated with large whirl amplitudes and those near bearing walls represent qualitative trends rather than accurate evaluation of the whirl in view of the assumptions made.

Investigation of the influence of system parameters on whirl for the considered cases showed, as expected, that improved rotor performance can be attained by increasing the load and reducing the clearance. Increasing the average film temperature showed that an increase or a reduction in the whirl amplitude may occur depending on the particular system parameters.

Although a relatively simple model is used in this study, the technique can be readily adapted to the analysis of more complex rotor systems and film properties.

## 6.2 DESIGN SYSTEMS

### 6.2.1 Procedure Based on Design Graphs

This is an illustration of graph-aided design for journal bearings. The graphs are constructed in such a manner as to enable the designers to

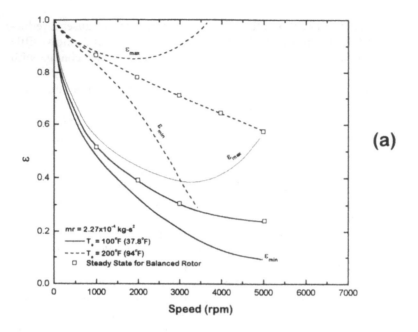

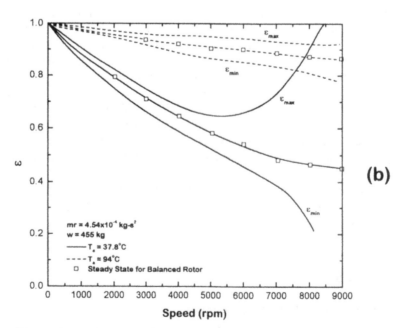

**Figure 6.18** Effect of average film temperature on rotor whirl: (a) $W = 45.5\,\mathrm{kg}$; (b) $W = 455\,\mathrm{kg}$.

select the bearing parameters, which meet their objective with a minimum of calculations.

In constructing the graphs, the main parameters influencing the bearing behavior were divided into two groups.

The first deals with the bearing geometry $(L, D, R, C)$, load $W$, and speed $N$. The second deals with the oil, and its temperature–viscosity characteristics. Because many types of oil can be used in the same bearing, the basic approach in the design graphs given here is to construct separate graphs for the different bearings and oils.

The bearing graphs represent a plot of temperature rise, $\Delta t$, versus average viscosity for a bearing with a known characteristic number, $K = (R/C)^2 N$, length-to-diameter ratio, $L/D$, and average pressure $P$. They are constructed by assuming the average viscosity, calculating the Sommerfeld number and the corresponding $\Delta T$.

Such plots are based on the numerical results of Raymondi and Boyd [2–4] and are shown in Figs 6.19–6.22 for average pressure values of 100,

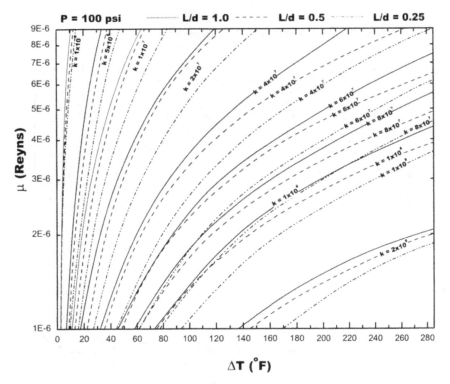

**Figure 6.19** Bearing chart for $P = 100$ psi.

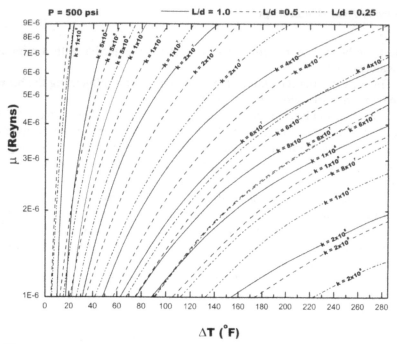

**Figure 6.20** Bearing chart for $P = 500$ psi.

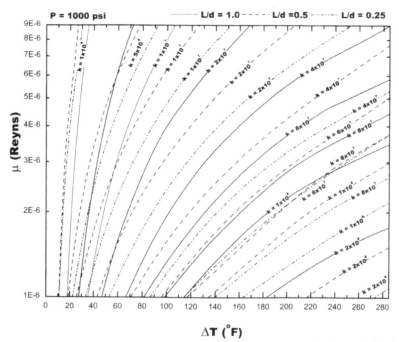

**Figure 6.21** Bearing chart for $P = 1000$ psi.

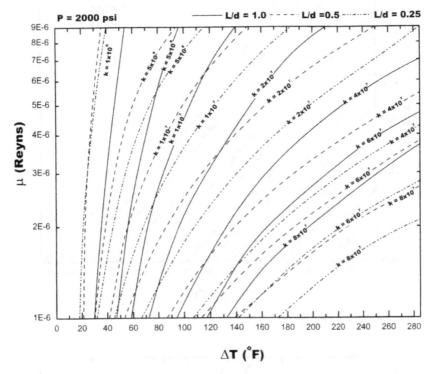

**Figure 6.22** Bearing chart for $P = 2000$ psi.

500, 1000, and 2000 psi, respectively. The length-to-diameter ratios $L/D$ considered are 0.25, 0.50, and 1.0.

The graphs for the lubricants represent the change of average viscosity with temperature rise for any particular initial temperature. Figures 6.23–6.25 represent such plots for SAE 10, 20, and 30 oils, respectively. These graphs give a convenient means of analysis, as well as the design of bearings, as explained in the following section.

### Analysis Procedure

For a bearing with a given geometry, load, and speed, a characteristic number, $K = (R/C)^2 N$, can be readily calculated. As can be seen from Eq. (6.8), this number represents the Sommerfeld number for a particular value of viscosity and average pressure. That is:

$$K = S\left(\frac{P}{\mu}\right)$$

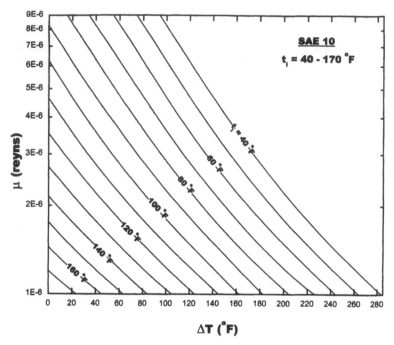

**Figure 6.23** SAE 10 oil chart.

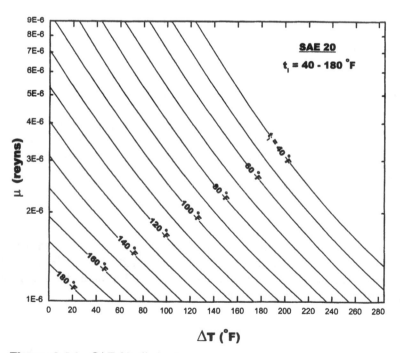

**Figure 6.24** SAE 20 oil chart.

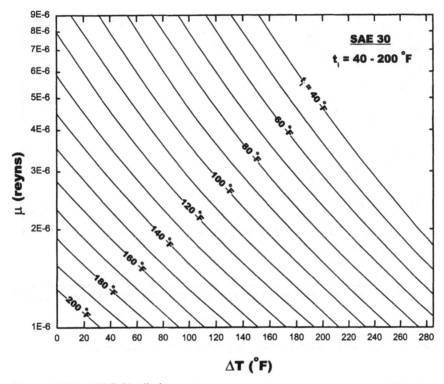

**Figure 6.25** SAE 30 oil chart.

The bearing graph (which represents the relationship between viscosity, $\mu$, versus temperature rise, $\Delta t$, for the particular value of $K$, $L/D$, and $P$), can be readily plotted on a transparent sheet by interpolation from Figs 6.9–6.12. Given the type of oil and its inlet temperature, the oil graph (which represents average viscosity versus temperature rise for the oil), is also plotted on the same sheet from Figs 6.23–6.25. Intersection of the two curves as can be seen in the example illustrated in Fig. 6.26, gives the temperature rise in the bearing and the corresponding average viscosity, $\mu$. The Sommerfeld number for the bearing is then calculated from $S = K\mu/P$.

Consequently, all the behavioral characteristics of the bearing can be read from Figs 6.3–6.8 or calculated from the given bearing performance equations, which are based on the curve fitting of these figures.

## Illustrative Example

The use of the bearing design graphs is illustrated by the following example. It is assumed that a shaft 2 in. in diameter, carrying a radial load of 2000 lb

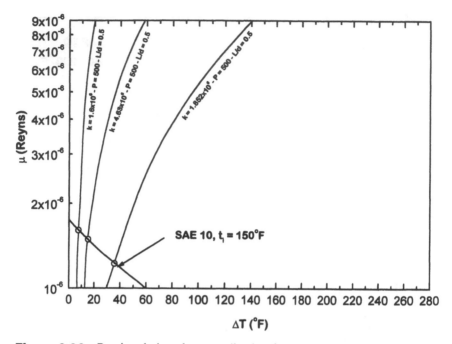

**Figure 6.26** Bearing design chart: application for clearance selection.

at 10,000 rpm is symmetrically supported by two bearings, each of length 1.0 in. The lubricating oil is SAE No. 10 with an inlet temprature of 150°F. The objective is to select a value for the radial clearance, $C$, which minimizes both the oil flow and temperature rise. Because these are conflicting objectives, a weighting factor, $k$, has to be specified to describe their relative importance for a particular bearing application. The design criterion can therefore be formulated as:

$$\text{Find } C, \text{ which minimizes } U = \Delta t + kQ \tag{6.15}$$

where

$\Delta t$ = temperature rise (°F)
$Q$ = oil flow (in.$^3$/sec)

Values of $k = 2$, 5, and 7 are considered to illustrate the influence of the weighting factor on the final design.

The average pressure and the length-to-diameter ratio are first calculated as:

$$P = \frac{W}{LD} = \frac{2000/2}{1 \times 2} = 500 \, \text{psi and} \quad \frac{L}{D} = 0.5$$

Arbitrary values for the design parameter, $C$ are assumed and the corresponding bearing parameter, $k$, is calculated in each case. For example, if $C$ is selected equal to 0.006 in., the corresponding parameter is

$$k = \left(\frac{R}{C}\right)^2 N = \left(\frac{1}{0.006}\right)^2 \left(\frac{10,000}{60}\right) = 4.63 \times 10^6$$

The bearing performance curve, corresponding to this value of $k$ for $P = 500 \, \text{psi}$ and $L/D = 0.5$, can be interpolated from Fig. 6.21 as plotted in Fig. 6.26. The oil characteristic curve for SAE 10 for the 150°F inlet temperature is also traced from Fig. 6.23 as shown in Fig. 6.2. The intersection of the two curves yields the following values for the temperature rise and average viscosity:

$$\Delta t = 15°F \text{ and } \mu = 1.53 \times 10^{-6} \, \text{reyn.}$$

The Sommerfeld number is then calculated:

$$S = \frac{k\mu}{P} = \frac{(4.63 \times 10^6)(1.53 \times 10^{-6})}{500}$$

The quantity of oil flow, $Q$, is readily found from Fig. 6.5 as

$$Q = 5.8 \, RNCL = 5.8 \, \text{in.}^3/\text{sec}$$

The merit value is calculated from Eq. (6.15) for the given weighting factor, $k$.

The process is repeated for different selections of the clearance (0.003 in. and 0.012 in. are tried in this example). The results are listed in Table 6.1 and plotted in Fig. 6.27.

**Table 6.1** Numerical Results for Bearing Design

| $C$ | $\mu$ (reyn) | $C$ | $\Delta t$ (°F) | $Q$ (in.$^3$/sec) | $k = 2$ | $U$ $k = 5$ | $k = 7$ |
|---|---|---|---|---|---|---|---|
| 0.003 | $1.22 \times 10^{-6}$ | 0.0454 | 38 | 2.81 | 43.62 | 52.05 | 57.67 |
| 0.006 | $1.53 \times 10^{-6}$ | 0.0142 | 15 | 5.81 | 26.6 | 44 | 55.6 |
| 0.012 | $1.61 \times 10^{-6}$ | 0.00375 | 8 | 12.00 | 32 | 68 | 92 |

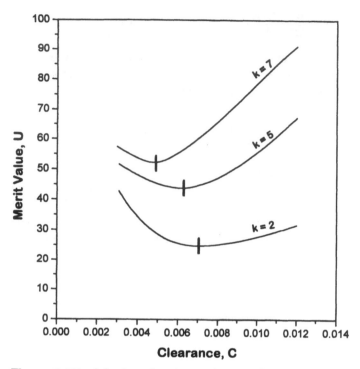

**Figure 6.27** Selection of optimum clearance for the difference objectives.

The optimum clearances can be deduced from the figure for the different values of the weighting factor $k$ as:

$k = 2$: $C^* = 0.005$ in.
$k = 5$: $C^* = 0.006$ in.
$k = 7$: $C^* = 0.007$ in.

### 6.2.2   Automated Design System

This section presents an automated system for the selection of the main design parameters to optimize the performance of the hydrodynamic bearing. In spite of the wealth of literature on the analysis of these bearings, the selection of design parameters in the past has relied heavily on empirical guides. Empiricism was necessary because of complexity of the interaction between the different parameters which govern the behavior of such bearings. The analytical relationships describing the bearing performance are generally based on Reynolds' equation and are, in most cases, numerical solutions of the equation with certain assumptions and approximations.

In this section, the curve-fitted numerical solutions, given in Sections 6.1.3 and 6.1.4, are utilized in a design system that rationally selects the significant parameters of a bearing to optimally satisfy the designer's objective within the constraints imposed on the design. A full journal bearing to operating at a constant speed and supporting a known constant load is considered. The procedure is extended to cover the selection of an optimum bearing for applications where the load and speed may vary from time to time within given bounds.

## System Parameters

The main independent parameters for the problem under consideration are $(D, L, C)$, $\mu$, and $(W, N)$. These parameters, as grouped, describe the bearing geometry, oil characteristics, and load specifications, respectively. In formulating the problem, it will be assumed that $D$, $N$, $W$ are given inputs for the bearing design. The design parameters are therefore $L/D$, $C$, $\mu$. The constraints on the design are:

$$h_0 \geq h_{min}$$
$$t_{max} \leq t_{max\,u}$$
$$P_{max} \leq P_{max\,u}$$
$$\mu \geq \mu_{min}$$
$$L_{min} \leq L \leq L_{max}$$
$$\frac{CM\omega^2}{W} \leq \psi(\sigma)$$

The first five of these inequality constraints represent the limit on the oil film thickness, temperature rise, maximum allowable pressure, minimum oil viscosity, and bearing length. These limits are dictated by the quality of machining, the characteristics of the material–lubricant pair, and the available space. The sixth constraint describes a condition for bearing stability, as described in Fig. 6.8.

## The Governing Equations

The equations governing the behavior of the bearing in this study are developed by curve fitting from Raimondi and Boyd's numerical solution to Reynolds' equation, Eq. (6.1). These equations, which are given in Section 6.1.3, allow the calculation of the temperature rise, minimum oil film thickness, maximum oil film pressure, oil flow, frictional loss, and so forth. The curve-fitted equations for the stability analysis by Lund and Saibel, given in Section 6.1.4, provide a simplified mathematical relationship for the onset of instability constraint.

## Design Criterion

The selection of an optimum solution requires the development of a design criterion, which accurately describes the designer's objective. The topography of this criterion and its interaction with the boundaries of the design domain (constraint surfaces), have a significant effect on the efficiency and success of the search. In the problem of bearing design, many decision criterion can be envisioned. Some of these are: minimizing the maximum temperature rise of the bearing, minimizing the quantity of oil flow required for adequate lubrication, minimizing the frictional loss, and so forth. The objective may also be composed of a multitude of the previously mentioned factors, and weighing their relative importance requires skilled judgement by the individual designer.

## Search Method

In formulating the problem for automated design, the following factors are considered in developing a search strategy: (1) the nature of the objective function, (2) the design domain and behavior of the constraints, (3) the sensitivity of the objective function to the individual changes in the decision parameters, and (4) the inclusion of a preset criterion for search effectiveness and convergence.

A block diagram describing the search is shown in Fig. (6.28). Arbitrary values of the design parameters within their given constraints are the entry point to the system. These values need not satisfy the functional constraints. The first phase of the search deals with guidance of the entry point into the feasible region. In this phase, incremental viscosity changes, of the order of 10%, and clearance changes, on the order of 0.001 in. per inch radius, proved to be adequate. When the stability constraint is violated, a feasible point may be located by dropping the length-to-diameter ratio to its lower limit, and simultaneously halving the viscosity and the clearance. To avoid looping in this phase, a counter can be set to limit the number of iterations. If a feasible point can not be successfully located, the designer can readjust the entry point according to the experience gained from the performed computations. When a feasible point in the design domain is located, the gradient search is initiated according to:

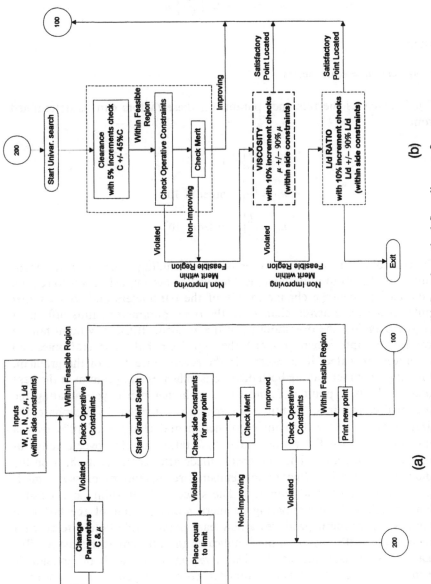

**Figure 6.28** (a) Search method flow diagram 1. (b) Search method flow diagram 2.

$$\mu_{n+1} = \mu_n - \Delta_n \left(\frac{\delta V}{\delta \mu}\right)^n n_\mu^2$$

$$C_{n+1} = C_n - \Delta n \left(\frac{\delta V}{\delta C}\right)^n n_C^2$$

$$L_{n+1} = L_n - \Delta_n \left(\frac{\delta V}{\delta L}\right)^n n_L^2$$

where

$n_\mu, n_C,$ and $n_L$ = scale factors

If $\mu$ is taken as the reference parameter, these factors can be determined from:

$$n_\mu = \left|\frac{\mu}{\mu}\right| = 1$$

$$n_C = \left|\frac{C}{\mu}\right| = \text{order of } 10^3$$

$$n_L = \left|\frac{L}{\mu}\right| = \text{order of } 10^5$$

The control of the step size is exercised by including a provision for changing $\Delta_n$ in the computational logic. One way to accomplish this is to require a specified percentage change in one of the parameters and to set upper limits on the incremental changes in the other parameters, thus offering a safeguard against the overshooting of the gradient. If the new point fails to produce an improvement in merit, the step size is halved several times and the process repeated, if necessary, in the reverse direction of the gradient. Nonimproving merit along both directions indicates an optimum at the base point. If the new point is found to be of higher merit, yet violating the functional constraints, the univariate search is activated. In this phase, the parameters are allowed to undergo incremental changes of 10% over a range of ±90% for the viscosity and $L/D$ ratio, and 5% over a range of 45% for the clearance. The first check, made after each iteration, is on the functional constraints (maximum temperature, maximum pressure, minimum film thickness and stability). The success of returning to a feasible point is followed by comparing its merit value to that of the last base point. A higher merit produces a new base point, while failure to reach a feasible point with improved merit indicates an optimum at the last feasible location. An illustration of the design region, and search progression is given in Fig. 6.29 for a two parameter problem where $L/D$ is assumed constant.

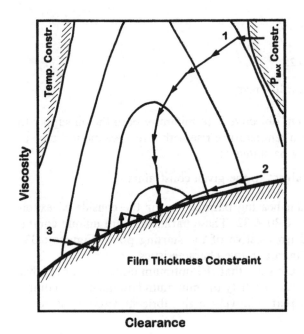

**Figure 6.29** Design region and search progression for the two-parameter problem ($L/D$ = constant).

## Numerical Examples

Two common bearing applications are considered to illustrate the design procedure.

### Design of Bearings for Constant Load and Speed Condition

The inputs are taken as:

$$W = 2000, 1000, 500, 250 \text{ lb, respectively}$$
$$N = 16.66, 33.33, 83.33, 166.66, 250, 333.33 \text{ rps, respectively}$$
$$D = 2 \text{ in.}$$

The constraints are:

$$h_{\min} = 5 \times 10^{-5} \text{ in.}$$
$$t_{\min u} = 300°\text{F} = \text{maximum allowable temperature}$$
$$P_{\max u} = 30,000 \text{ lb/in.}^2 = \text{maximum allowable pressure}$$

$$\frac{L}{D_{max}} = 1.0$$

$$\frac{L}{D_{min}} = 0.25$$

$$\mu_{min} = 1 \times 10^{-7}\,\text{reyn}$$

It is assumed that the design objective is to minimize both the oil supply to the bearing and the oil film temperature rise with a relative merit factor of 5 : 1, respectively. this may be stated as:

Minimize $U = \Delta t + 5Q$ subject to the given constraints

**Results.** The optimum bearing parameters for the considered examples are illustrated in Figs 6.30–6.32. These parameters are unique combinations and are obtained irrespective of the starting point. Figures 6.33–6.37 show the corresponding operative characteristics.

It can be seen from the results that the optimum clearances are higher for high loads and low speeds to satisfy the minimum film thickness requirement. Figure 6.31 shows that relatively high lubricant viscosity is relied

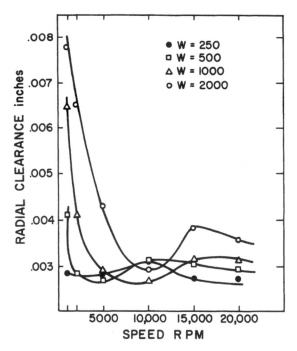

**Figure 6.30**  Optimum clearance.

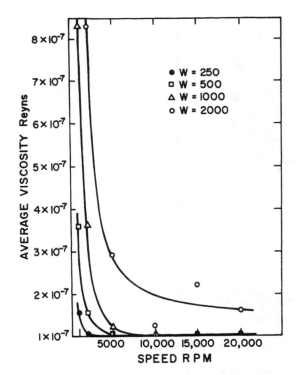

**Figure 6.31** Optimum values of average viscosity.

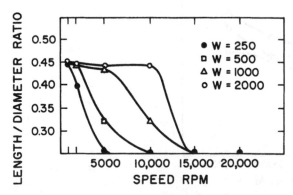

**Figure 6.32** Optimum length/diameter ratio.

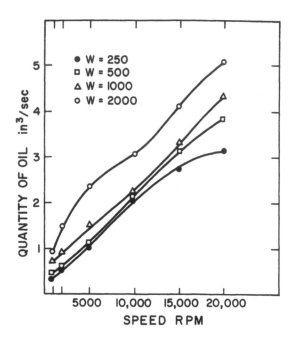

**Figure 6.33**   Temperature rise in optimum bearings.

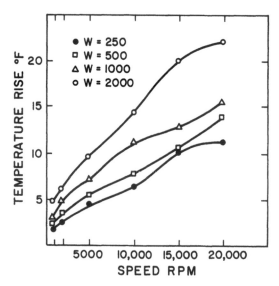

**Figure 6.34**   Oil requirement for optimum bearings.

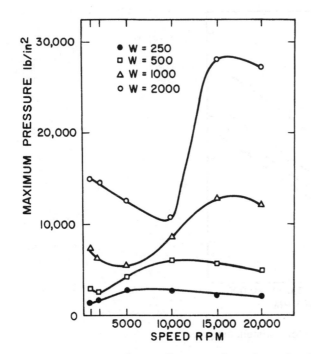

**Figure 6.35** Maximum oil pressure in optimum bearings.

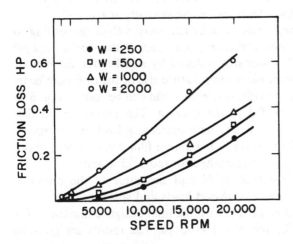

**Figure 6.36** Frictional loss.

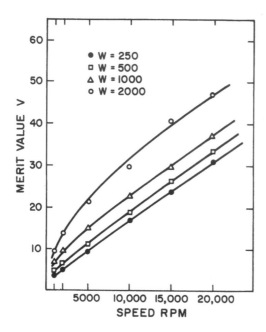

**Figure 6.37** Merit value for optimum designs.

upon at low speeds to maintain the required film thickness. As the speed increases and the journal eccentricity becomes smaller, the optimal viscosities drop in order to satisfy the stability criterion. This trend continues as speed increases until the lower limit set on the viscosity is reached.

Although longer bearings may have higher merit values (according to the design criterion under consideration), the drop in the optimum $L/D$ ratio with increasing speed is primarily induced by stability requirements.

Figure 6.33 shows an increase of temperature rise of the optimum bearing with increased speeds and loads. A similar trend can be seen in Fig. 6.34 for the quantity of oil to be fed to the bearing. The maximum oil film pressure, Fig. 6.35, generally increases with increasing load at any speed. For any particular load, the changes in maximum film pressure with speed are influenced by the corresponding changes in radial clearance.

The frictional power loss (Fig. 6.36) and the value of the objective function (Fig. 6.37) increase with increased loads and speeds.

The effect of the weighting factor, $k$ on the final design is illustrated for the case where $W = 1000$ lb, and $N = 166.66$ rps. The results are given in Table 6.2.

It can be seen that by taking $k = 1$, 5, and 10, respectively, the temperature rise for the optimum bearings are 3.77, 11.0, and 14.44°F, respectively.

**Table 6.2** Effect of Weighting Factor, $k$, on the Optimum Design ($W = 100\,\text{lb}$, $N = 166.6\,\text{rps}$)

| Weighing factor $k$ | $\mu$ | $C$ | $L/D$ | $\Delta t$ (°F) | $Q$ (in.$^3$/ sec) | $P_{max}$ (psi) | HP loss |
|---|---|---|---|---|---|---|---|
| 1 | $1 \times 10^{-7}$ | $3.4 \times 10^{-3}$ | 0.36 | 3.77 | 4.88 | 7600 | 0.147 |
| 5 | $1 \times 10^{-7}$ | $2.55 \times 10^{-3}$ | 0.325 | 11.00 | 2.15 | 8500 | 0.175 |
| 10 | $1 \times 10^{-7}$ | $1.5 \times 10^{-3}$ | 0.313 | 14.44 | 1.67 | 6500 | 0.185 |

The corresponding values of the oil flow are 4.88, 2.15, and 1.67 in.$^3$/sec. It is interesting to note this change in objective produced no change in the required average viscosity since it is already at its lower limit. A small change is necessary however, in the length-to-diameter ratio but the most significant change is in the required clearance.

## Bearings Operating Within a Range of Specified Loads and Speeds

The previous design system is extended so that the optimum parameters, for a bearing operating with equal frequency within a given range of loads and speeds, may be automatically obtained.

In this case, the region under consideration is divided into an array of points, each representing a particular load and speed. A search procedure, similar to that previously mentioned, is adopted. In this case, however, the feasibility of the design at each step is checked for all points in the array. The merit values are also calculated at all points, and the lowest of these values is taken to represent the merit rating of the bearing.

**Results.** Optimum bearing parameters corresponding to several load–speed regions are given in Table 6.3. The input data, constraints, and design criterion are the same as in the previous examples. The regions considered are illustrated in Fig. 6.38. Some of the results shown in the table are obtained for the case where only the corners of the regions (i.e., a point array) are considered. To investigate the effect of grid size on the design, regions 2 and 5 are each divided into a $3 \times 4$ grid. The results, as shown in the table, do not appreciably change with the change of grid size. In all the studied cases, the point of lowest merit is found to be that when both load and speed are highest.

Figure 6.39 shows a comparison of the results from the regional search and those obtained for an optimum bearing designed for the maximum load and speed in region 2. The latter, as expected, shows a higher merit for the load and speed for which it is selected, but its operation is constrained at other parts of the considered region as indicated by the asterisk marker.

**Table 6.3**  Optimum Bearing Parameters and Corresponding Operative Characteristics; $V = 5Q + \Delta t$

| Region no. | Load range (lb) | Speed range (rpm) | Grid points | $\mu$ (reyn) | $C$ (in.) | $L/D$ | Max. $\Delta t$ in region (°F) | Max. $P_{max}$ in region (lb/in.²) | Min. $h_0$ in region (in.) | Max. $Q$ in region (in./sec) | Max. friction loss in region (hp) | Max. merit value in region |
|---|---|---|---|---|---|---|---|---|---|---|---|---|
| 1 | 500–1,000 | 1,000–5,000 | 2 × 2 | $1.3 \times 10^{-7}$ | $1.90 \times 10^{-3}$ | 0.990 | 8.0 | 1,187 | $5.06 \times 10^{-5}$ | 1.519 | 0.115 | 15.73 |
| 2 | 1,000–2,000 | 1,000–5,000 | 2 × 2 | $3.0 \times 10^{-7}$ | $2.55 \times 10^{-3}$ | 0.990 | 12.5 | 2,607 | $5.00 \times 10^{-5}$ | 2.100 | 0.234 | 23.03 |
| 2 | 1,000–2,000 | 1,000–5,000 | 3 × 4 | $3.1 \times 10^{-7}$ | $2.64 \times 10^{-3}$ | 0.998 | 12.3 | 2,592 | $5.07 \times 10^{-5}$ | 2.180 | 0.237 | 23.20 |
| 3 | 500–2,000 | 5,000–10,000 | 2 × 2 | $6.2 \times 10^{-7}$ | $5.90 \times 10^{-3}$ | 0.280 | 17.1 | 28,380 | $5.04 \times 10^{-5}$ | 4.500 | 0.430 | 39.80 |
| 4 | 1,000–2,000 | 10,000–20,000 | 2 × 2 | $3.0 \times 10^{-7}$ | $4.07 \times 10^{-3}$ | 0.275 | 27.6 | 24,750 | $5.14 \times 10^{-5}$ | 6.000 | 0.968 | 57.70 |
| 5 | 500–1,000 | 10,000–20,000 | 2 × 2 | $1.14 \times 10^{-7}$ | $2.95 \times 10^{-3}$ | 0.300 | 17.7 | 9,633 | $5.00 \times 10^{-5}$ | 4.500 | 0.484 | 40.29 |
| 5 | 500–1,000 | 10,000–20,000 | 3 × 4 | $1.4 \times 10^{-7}$ | $3.00 \times 10^{-3}$ | 0.275 | 19.6 | 10,860 | $5.16 \times 10^{-5}$ | 4.450 | 0.523 | 41.80 |

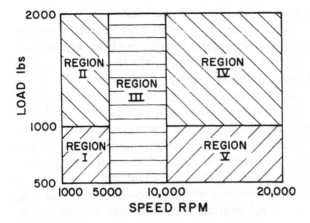

**Figure 6.38** Considered load and speed regions.

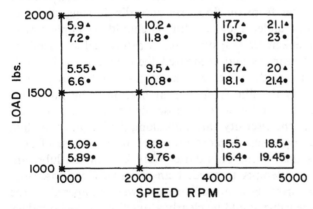

**Figure 6.39** Comparison of designs obtained by regional search and those obtained for the maximum condition of load and speed (Region 2).

## 6.3 THEROMODYNAMIC EFFECTS ON BEARING PERFORMANCE

In the classical hydrodynamic theory presented by Reynolds [1], an isoviscous film is assumed. This assumption is widely used in bearing design, because accounting for the effects of temperature variations along the lubricant film and across its thickness would significantly complicate the analysis.

Many experimental observations, however, show that the isoviscous hydrodynamic theory, alone, does not account for the load-carrying capacity and the temperature rise in the fluid film. McKee and McKee [15], in a series of experiments, observed that under conditions of high speed, the viscosity diminished to a point where the product $\mu N$ remained constant. Fogg [16] found that a parallel-surface thrust bearing can carry higher loads than those predicted by the hydrodynamic theory. His observation, known as the Fogg effect, is explained by the concept of the "thermal wedge," where the expansion of the fluid as it heats up develops additional load-carrying capacity. Shaw [17], Boussages and Casacci [18], Osterle et al. [19], and Ulukan [20] are among the investigators of thermal effects in fluid film lubrication. Cameron [21], in his experiments with rotating disks, suggested that a hydrodynamic pressure is created in the film between the disks arising from the variation of viscosity across the thickness of the film. This variation is generally referred to as the "Cameron effect." Experiments by Cole [22] on temperature effects in journal bearings indicated that at high speeds, severe temperature gradients are set up, both across the film because of heat removal by conduction and in the plane of relative motion because of convective heat transfer from oil flow. He accordingly suggested that constant viscosity theory under such conditions should be applied with caution. Hunter and Zienkeiwicz [23] presented a theoretical study of the heat–energy balance of bearings and compared their findings with Cole's results. They concluded that the effect of temperature, and consequently, viscosity variations across the film in a journal bearing is by no means negligible. Thus pressures were lower than those obtained from a solution which takes into account the viscosity variation along the length of the film only, and the decrease in pressure is more pronounced in the case of non-conducting boundaries than if the boundaries were kept at the lubricant inlet temperature. Their attempts to predict an effective mean viscosity, which would lead to a correct estimate of pressure, were hampered by the fact that such an average value would be clearly a function of the boundary temperature as well as the mean temperature of the oil leaving the bearing. Dowson and March [24] carried out a two-dimensional thermodynamic analysis of journal bearings to include variation of lubricant properties along and across the film. They presented temperature contours in the film, as well as a reasonable estimate of the shaft and bush temperatures.

It was observed during experimental investigations of the pressure distribution in the fluid film developed by rotating an externally supported journal in a sleeve at a predetermined eccentricity that [25–28]:

1. Both the circumferential and axial patterns of pressure distribution normalized to the maximum pressure (Fig. 6.40) are identical to those predicted by the isoviscous hydrodynamic theory (Refs 2–4 for example).

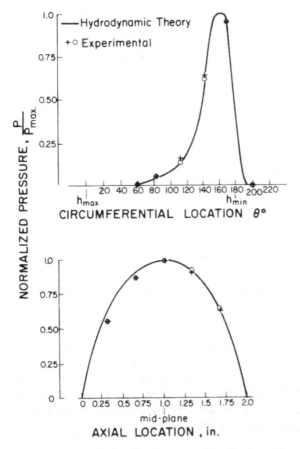

**Figure 6.40** Normalized pressure distribution. (From Ref. 25.)

2. For any particular eccentricity, oil, and inlet temperature, the magnitude of the peak pressure (or the average pressure) in the-film is approximately proportional to the square root of the rotational speed of the journal rather than the approximately linear proportionality predicted by the isoviscous theory (Fig. 6.41).

3. For any particular eccentricity, oil, and inlet temperature, there exists a speed $N^*$ where the isoviscous theory predicts the same magnitude of maximum pressure, $P^*_{max}$ (and consequently, average pressure $P^*_{ave}$) as that measured experimentally (Fig. 6.41).

4. For a given film geometry (fixed eccentricity), oil, and speed, the variation of the maximum pressure (and consequently the average pressure) with inlet temperature is different than that pre-

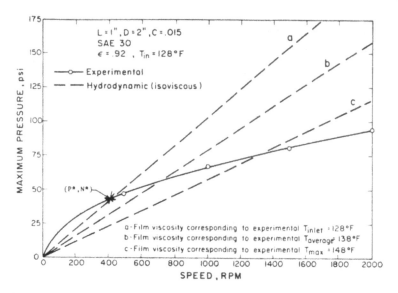

**Figure 6.41** Pressure–speed relationship for fixed geometry bearing. (From Ref. 27.)

dicted by the isoviscous theory. Only at the 0* point can the isoviscous theory predict the film pressures.

5.  For any particular eccentricity, speed, and oil, the 0* condition (where the experimental and predicted isoviscous bearing performances are identical) can be determined according to the following empirical procedure (see Fig. 6.42):

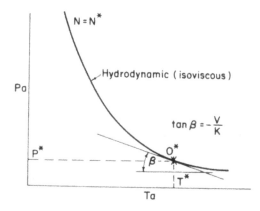

**Figure 6.42** Procedure for determination of the thermohydrodynamic $o^*$ point ($\varepsilon$ = constant).

(a) Construct the curve relating the average pressure, $P_a$, to the average film temperature, $T_a$, based on the isoviscous theory. Since $\varepsilon$ is fixed and $L/D$ is known, the isoviscous Sommerfeld number $S_{iso}$ is constant and can be readily determined by the isoviscous theory. Therefore

$$S_{iso} = \text{constant} = \frac{\mu_a N}{P_a}\left(\frac{R}{C}\right)^2$$

and consequently, for any speed $N$, a curve can be plotted to relate $P_a$ and $T_a$ (which for a given oil defines the average viscosity $\mu_a$).

(b) The 0* condition is found empirically to be the point on that curve where the slope of the tangent is:

$$\tan \beta = \frac{dP_a}{dT_a} = -\frac{V}{K}$$

where

$V$ = volume of the oil drawn into the clearance space in cubic inches per revolution

$K$ = constant which is found empirically (based on the experimental results from Refs 25-32 as detailed in Ref. 33) to be a function of $(R/C)$ as plotted in Fig. 6.43

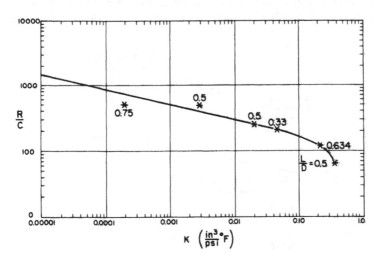

**Figure 6.43**  The empirical factor $k$.

(c) The temperature rise $\Delta T^*$ at the $0^*$ condition can be readily determined based on the isoviscous theory. Consequently, the oil inlet temperature corresponding to this condition can be calculated from:

$$T_i = T_a^* - \frac{\Delta T^*}{2}$$

For a given eccentricity, oil, and inlet temperature, $T_i$, the pressure–speed relationship can be empirically expressed as:

$$\frac{(P_a)_{T_i}}{P_a^*} = \sqrt{\frac{N}{N^*}} \qquad (6.16a)$$

Since the pressure distribution as predicted by the isoviscous theory remains the same (Fig. 6.40), therefore:

$$\frac{(P_{max})_{T_i}}{P_{max}^*} = \sqrt{\frac{N}{N^*}} \qquad (6.16b)$$

This relationship is illustrated in Fig. 6.44 and compared to the corresponding pressure–speed relationship predicted by isoviscous considerations for the same conditions.

### 6.3.1 Basic Empirical Relationships

The objective of the following is to develop, based on experimental findings, a modified Sommerfeld number $S^*$ (and consequently, an effective average

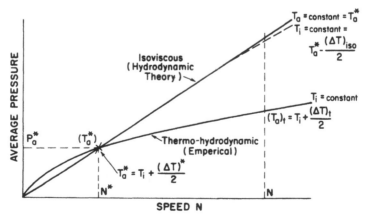

**Figure 6.44** Pressure-speed relationship for a fixed geometry bearing.

viscosity) for any bearing, which accounts for the thermohydrodynamic behavior of the film and can be directly used instead of $S$ to evaluate the performance characteristics for any operating condition, using existing isoviscous analysis and data.

Based on isoviscous hydrodynamic considerations, the Sommerfeld number, $S_{iso}$ for a bearing with a given film geometry is independent of speed, oil, and inlet temperature. Consequently, the relationship between the pressure, $P_a$, speed, $N$, and average viscosity, $\mu_a$, is governed by the condition:

$$S_{iso} = f\left(\varepsilon, \frac{L}{D}\right) = \frac{\mu_a N}{P_a}\left(\frac{R}{C}\right)^2 \tag{6.17}$$

Now defining $S^*$ at the $0^*$ condition as

$$S^* = \frac{\mu_a^* N^*}{P_a^*}\left(\frac{R}{C}\right)^2 \tag{6.18}$$

For the same $\varepsilon$, $S_{iso} = S^*$, from which:

$$(P_a)_{iso} = P_a^* \frac{N}{N^*} \frac{(\mu_a)_{iso}}{\mu_a^*} \tag{6.19}$$

From Eq. (6.18), which is based on the empirical observation for a given film geometry, oil, and inlet temperature, $T_i$:

$$P_a = \text{average pressure in the film with inlet temperature } T_i = P_a^* \frac{N}{N^*}$$

$$\left(\frac{dP_a}{dT_a}\right)^* = -\frac{V}{K} \tag{6.20}$$

where

$$V = \frac{Q}{N} = \left(\frac{Q}{RNCL}\right)RCL$$

$Q/RNCL$ is calculated based on isoviscous considerations from $S_{iso}$ and $L/D$.

From Eq. (6.19):

$$(P_a)_{T_i} = \left((P_a)_{iso} \frac{N^*}{N} \frac{\mu_a^*}{(\mu_a)_{iso}}\right)\sqrt{\frac{N}{N^*}} = (P_a)_{iso}\sqrt{\frac{N^*}{N}} \frac{\mu_a^*}{(\mu_a)_{iso}}$$

Assuming the following viscosity–temperature relationship:

$$\mu = \mu_0 e^{\frac{b}{(T+\theta)}}$$

where $\mu_0$, $\theta$, and $b$ are constants for the given oil (Table 6.4) and $T$ is the temperature of the oil, the isothermal Sommerfeld number $(S)_{iso}$ for the film at a speed $N$ and average temperature $T_a$ can be written as:

$$(S)_{iso} = \frac{\mu_0 N}{(P_a)_{iso}} e^{b/(T_a+\theta)} \left(\frac{R}{C}\right)^2$$

from which:

$$(P_a)_{iso} = \frac{N}{(S)_{iso}} \left(\frac{R}{C}\right)^2 \mu_0 e^{b/(T_a+\theta)}$$

and by differentiation:

$$\left(\frac{dP_a}{dT_a}\right)_{iso} = -\frac{N}{(S)_{iso}} \left(\frac{R}{C}\right)^2 \mu_0 e^{b/(T_a+\theta)} \left(\frac{b}{(T_a^*+\theta)^2}\right) \qquad (6.21)$$

from which $T_a^*$ can be evaluated by iteration. Consequently $(\Delta T)^*$ can be evaluated by isoviscous considerations and the corresponding $T_i$ can be calculated from:

$$T_i = T_a^* - \frac{(\Delta T)^*}{2}$$

**Table 6.4**  Oil Constants

| Oil | $\mu$ (80°F) (reyn-psi) | $\mu$ (140°F) (reyn) | $\mu_0$ (reyn) | $b$ (°F) |
|---|---|---|---|---|
| SAE 10 | $1.18 \times 10^{-5}$ | $2.18 \times 10^{-6}$ | $1.58 \times 10^{-8}$ | 1157.5 |
| SAE 20 | $1.95 \times 10^{-5}$ | $3.15 \times 10^{-6}$ | $1.36 \times 10^{-8}$ | 1271.6 |
| SAE 30 | $3.35 \times 10^{-5}$ | $4.60 \times 10^{-6}$ | $1.41 \times 10^{-8}$ | 1360.9 |
| SAE 40 | $5.50 \times 10^{-5}$ | $6.40 \times 10^{-6}$ | $1.21 \times 10^{-8}$ | 1474.4 |
| SAE 50 | $9.50 \times 10^{-5}$ | $1.05 \times 10^{-5}$ | $1.70 \times 10^{-8}$ | 1509.6 |
| SAE 60 | $1.42 \times 10^{-5}$ | $1.45 \times 10^{-5}$ | $1.87 \times 10^{-8}$ | 1564.0 |

Viscosity at any temperature $T(°F)$ is given by $\mu(T) = \mu_0 e^{b/(T+\theta)}$, where $\theta = 95°F$ and $\mu_0$ = lubricant relative viscosity.

The thermohydrodynamic pressure–speed relationship for the particular film geometry is, according to Eq. (6.16):

$$\frac{(P_a)_{T_i}}{P_a^*} = \sqrt{\frac{N}{N^*}}$$

If it can be assumed that $T_a^*$ is known or can be approximately estimated for a particular oil and film geometry, the corresponding $N^*$ can be computed directly from Eq. (6.21) as:

$$N^* = \frac{V}{K} S^* \left(\frac{C}{R}\right)^2 \frac{1}{\mu_0 e^{b/(T_a^*+\theta)}} \frac{(T_a^*+\theta)^2}{b} = \frac{V}{K}\left[\frac{\mu_a^* N^*}{P_a^*}\left(\frac{R}{C}\right)^2\right]\left(\frac{C}{R}\right)^2 \frac{1}{\mu_0 e^{b/(T_a^*+\theta)}} \frac{(T_a^*+\theta)^2}{b}$$

This equation can be reduced to give:

$$P_a^* = \frac{V}{K} \frac{(T_a^*+\theta)^2}{b} \tag{6.22}$$

which can be readily determined by isoviscous considerations for a given $\varepsilon$, oil, $L/D$, $C/R$, and $T_a^*$.

## 6.3.2 Prediction of Bearing Performance

In a practical situation, the following bearing parameters are usually given: load $W$, diameter $D$, length $L$, radial clearance $C$, journal speed $N$, and oil and inlet temperature $T_i$.

In this section, two empirical procedures will be given for determining a modified Sommerfeld number $S^*$ which can be used instead of the classical Sommerfeld number to determine the bearing performance characteristics using available data and methods based on the isothermal hydrodynamic considerations. In the first procedure, it will be assumed that the temperature rise based on isothermal considerations is approximately the same as the actual thermohydrodynamic temperature rise at the operating condition, as well as at the $0^*$ condition. Under such assumptions:

$$(\Delta T_a)^* = (\Delta T_a)_{T_i} = (\Delta T_a)_{iso} \qquad \text{and} \qquad (T_a)_{T_i} = (T_a)_{iso}$$

This assumption, although approximate, considerably simplifies the determination of the modified Sommerfeld number and consequently the evaluation of the performance characteristics of the bearing. It may be used in situations where the temperature rise is relatively small.

In the second procedure, only the inlet temperature of the oil is considered and the modified Sommerfeld number is obtained by successive iterations. Needless to say, this is the more accurate of the two methods. Design nomograms are also given to facilitate the evaluation of the modified Sommerfeld number $S^*$.

*Empirical Procedure for Obtaining $S^*$ Based on the Assumption that*
$T_a^* \approx (T_a)_{iso} \approx (T_a)_{T_i}$

Sequence of calculations:

1. Compute the isoviscous Sommerfeld number $(S)$ for the given operating conditions (oil, $T_a, N, W, L, D, C$) by using the formula:

$$S = \frac{\mu_a N}{P_a}\left(\frac{R}{C}\right)^2$$

   Note that $P_a = W/(LD)$ and $(P_a)_{iso} = (P_a)_{T_i}$.

2. Corresponding to this $S$, compute $Q/(RNCL)$, the dimensionless quantity of oil flow either from Raimondi and Boyd's charts, or by using the curve-fitted equations given in Section 6.1.3.

3. Estimate the bearing characteristic constant $K$ from Fig. 6.43, and subsequently calculate the value of the parameter $RCL/K$.

4. For the oil and average temperature under consideration, calculate the dimensionless "oil factor" $b\theta/(T_a + \theta)^2$.

5. Calculate the pressure at the $0^*$ condition from Eqs (6.20) and (6.21):

$$P_a^* = \frac{\left(\dfrac{Q}{RNCL}\right)\left(\dfrac{RCL}{K}\right)}{\left(\dfrac{b\theta}{(T_a+\theta)^2}\right)\left(\dfrac{1}{\theta}\right)} \qquad (6.23)$$

6. It is assumed that $\mu_a \approx \mu_a^*$, so the modified Sommerfeld number can be found from the relation:

$$S^* = \frac{\mu_a N^*}{P_a^*}\left(\frac{R}{C}\right)^2$$

$$S = \frac{\mu_a N}{P_a}\left(\frac{R}{C}\right)^2 \qquad \text{and} \qquad \frac{P_a}{P_a^*} = \sqrt{\frac{N}{N^*}}$$

therefore

$$S^* = S\left(\frac{P_a^*}{P_a}\right)$$

7. $S^*$ can then be used in place of $S$ to determine the static and dynamic thermohydrodynamic performance of the bearing using available data and graphs based on isoviscous theory.

**Nomogram for Obtaining $S^*$**

The nomogram given in Fig. 6.45 can be used to determine $S^*$ in this case. To illustrate the procedure for evaluating $S^*$ from $S$, the example shown in the figure by dotted lines and arrows is followed. First, enter the hydrodynamic Sommerfeld number at point A and draw a vertical line AB to the appropriate $L/D$ curve. From B, draw a horizontal line BC to meet the required $RCL/K$ line (note that a given bearing is represented by a particular $RCL/K$ = constant line). The vertical line CD then meets the appropriate $b\theta/(T_a + \theta)^2$ line at D. Point E is the point directly below point A and

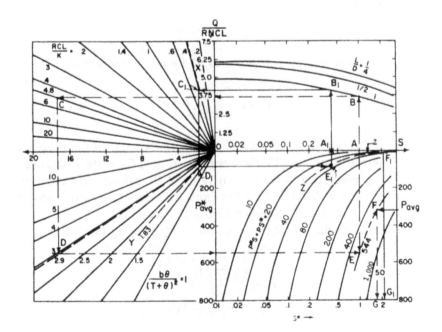

**Figure 6.45** Nomogram for evaluation of $S^*$ from $S$ assuming that the average film temperature is known.

across from D. Point E then defines the curve: $P^*S = PS^* =$ constant. This curve is then followed to point F where the pressure is equal to the actual average pressure $W/(LD)$. The projection of point F on the $S^*$ axis gives the modified Sommerfeld number (point G).

*Empirical Procedure for Obtaining S\* Based on Equal Inlet Temperature*

Sequence of calculations:

1.  Find the bearing characteristic constant $K$ from Fig. 6.43 and evaluate the parameter $RCL/K$ for the bearing.
2.  Calculate the numerical value of the parameter $(N/P_a^2)(R/C)^2$.
3.  Make an initial guess at $T_a^*$ and $P_a^*$, the average temperature and pressure at the $0^*$ condition. (Inlet temperature $T_i$ and average pressure $P_a = W/(LD)$ can be used as initial estimates.)
4.  Compute the dimensionless oil factor $b\theta/(T_a^* + \theta)^2$ corresponding to the current value of $T_a^*$.
5.  compute the average viscosity corresponding to the current value of $T_a^*$:

$$\mu_a^* = Ae^{b/(T_a^* + \theta)}$$

6.  Calculate an approximation to $S^*$ by using the formula:

$$S^* = \mu_a^* \left[ \frac{N}{P_a^2} \left( \frac{R}{C} \right)^2 \right] P_a^*$$

7.  If this approximation to $S^*$ is sufficiently close to the previous approximation to $S^*$, there will be no need for further iteration.
8.  Corresponding to the current value of $S^*$, calculate the quantity of oil flow $Q/(RNCL)$ either from Raimondi and Boyd's charts or by using curve-fitted equations in Section 6.1.3.
9.  Estimate the new approximation to $P_a^*$ by using Eq. (6.23):

$$P_a^* = \frac{\left( \dfrac{Q}{RNCL} \right)\left( \dfrac{RCL}{K} \right)}{\left( \dfrac{b\theta}{(T_a^* + \theta)^2} \right)\left( \dfrac{1}{\theta} \right)}$$

10.  Corresponding to the current values of $S^*$ and $P_a^*$ find the temperature rise $\Delta T^*$ by applying the curve-fitted equations.

11. Revise the estimate for $T_a^*$:

$$T_a^* = T_i + \tfrac{1}{2}\Delta T^*$$

12. Go to step 4.
13. Use the current value of $S^*$ for further analysis of the bearing.

A computer program can be readily developed for performing these calculations.

### Nomogram for Obtaining $S^*$

The nomogram given in Figs 6.46a and b are constructed to facilitate the evaluation of $S^*$ from $S$ by graphical iteration. The classical Sommerfeld number $S$ based on isoviscous hydrodynamic analysis can also be evaluated by graphical iteration from the nomogram given in Fig. 6.47. The procedure is illustrated in detail by numerical examples in the following section.

### 6.3.3   Numerical Examples

The procedure described in this section is a relatively simple method for the determination of a modified sommerfeld number which, when used instead of the classical Sommerfeld number in a standard isoviscous analysis, was found to provide better correlation with the performance of fluid film bearings tested under laboratory conditions.

The modified Sommerfeld number can then be utilized in the standard formulas to calculate eccentricity ratios, oil flow, frictional loss, and temperature rise, as well as stiffness and damping coefficients for full film bearings.

Although no theoretical confirmation is developed for the proposed method, it provides the designer with an alternate method for selecting the main bearing parameters in critical applications. Judgement should be exercised in situations where significant differences exist between the proposed method and existing practices.

Three numerical examples are given in the following to illustrate the different procedures for evaluating a characteristic number for the bearing (Sommerfeld number or modified Sommerfeld number). The first example assumes isoviscous conditions. The second example illustrates the empirical thermohydrodynamic procedure assuming that the average film temperature is known. The third example illustrates the empirical thermohydrodynamic procedure based on the oil inlet temperature.

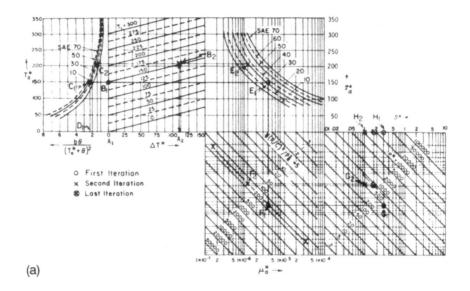

(a)

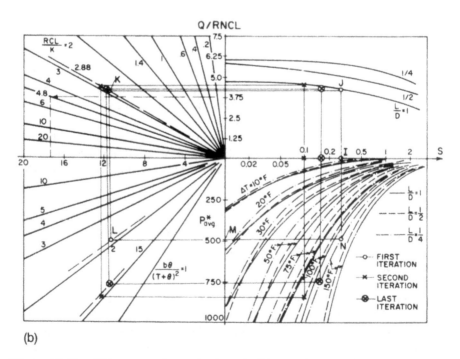

(b)

**Figure 6.46** Nomograms for evaluation of $S^*$ from $S$ based on inlet temperature of oil.

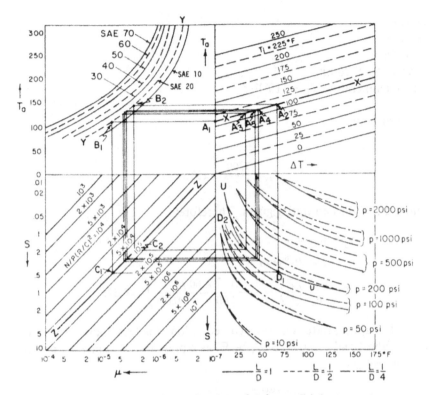

**Figure 6.47** Nomogram for evaluation of $S$ from oil inlet temperature.

## Isoviscous Analysis

Bearing performance characteristics are to be obtained using isoviscous theory when lubricant and the inlet temperature are specified. The main concern in this example is the determination of the average film temperature from the inlet temperature. An iterative procedure is needed. The nomogram (Fig. 6.47) can be used to facilitate the iteration as shown in the following sample problem.

**EXAMPLE 1.** Calculate the Sommerfeld number and the other performance characteristics of a centrally loaded full journal bearing for the following conditions:

$W = 7200\,\text{lb}$

$N = 3600\,\text{rpm}$ (60 rps)

$R = 3$ in.
$L = 6$ in.
$C = 0.006$ in.
Lubricant SAE 20 oil
Average temperature $T_i = 110°F$

Numerical Solution

$$P_a = \frac{7200}{6 \times 6} = 200\,\text{psi}$$

Assume $\Delta T = 0$ as an initial guess. Therefore:

$$T_a = 100°F$$
$$\mu_a = 7 \times 10^{-6}\,\text{reyn}$$
$$S = \frac{\mu_a N}{P_a}\left(\frac{R}{C}\right)^2 = \frac{7 \times 10^{-6} \times 60}{200}\left(\frac{3}{0.006}\right)^2 = 0.525$$

Using the appropriate curve-fitted equation for $\Delta T$ and assuming: $v = 0.03$ lb/in.$^3$ and $c = 0.40$ Btu/(lb-°F) as representative values for a lubricating oil, we get:

$$\Delta T = 842989\left(\frac{L}{D}\right)^{-0.08167} S^{0.85540 + 0.08787(L/D)(P/Jvc)}$$

$$= 842989(1)(0.525)^{0.94327}\,\frac{200}{9336 \times 0.03 \times 0.40} \cong 78.9°F$$

With this new value for $\Delta T$:

$$T_a = 110 + \frac{78.9}{2} \cong 149.5°F$$

The process can now be repeated with this new approximation to $\Delta T$. The results of the first eight iterations are given in Table 6.5. It can be noticed that five or six iterations would give sufficiently close approximation to the final results.

If, instead of assuming $\Delta T$ equal to zero, a better initial guess at $\Delta T$ was made, then only two or three iterations would be needed to reach the final value of $\Delta T$.

**Table 6.5**  Results of the First Eight Iterations

| Iteration | $\Delta T$ | $T_a$ | $\mu$ | $S$ | New $\Delta T$ |
|---|---|---|---|---|---|
| 1 | Initial guess $= 0$ | 110 | $6.7 \times 10^{-6}$ | 0.504 | 78.9 |
| 2 | 79.8 | 149.5 | $2.5 \times 10^{-6}$ | 0.185 | 30.6 |
| 3 | 30.6 | 125.3 | $4.4 \times 10^{-6}$ | 0.328 | 52.5 |
| 4 | 52.5 | 136.3 | $3.3 \times 10^{-6}$ | 0.249 | 40.5 |
| 5 | 40.5 | 130.3 | $3.8 \times 10^{-6}$ | 0.288 | 46.6 |
| 6 | 46.6 | 133.3 | $3.6 \times 10^{-6}$ | 0.268 | 43.4 |
| 7 | 43.4 | 131.7 | $3.7 \times 10^{-6}$ | 0.278 | 45.0 |
| 8 | 45.0 | 132.5 | $3.6 \times 10^{-6}$ | 0.273 | 44.2 |

Nomogram Solution (see Fig. 6.47)

1. A straight line corresponding to $T_i = 110$ is drawn in the first quadrant (line XX).
2. The curve for the SAE 20 oil (second quadrant is identified (curve YY)).
3. The parameter $(N/P)(R/C)^2$ is calculated:

$$\frac{N}{P}\left(\frac{R}{C}\right)^2 = \frac{60}{200}\left(\frac{3}{0.006}\right)^2 = 7.5 \times 10^4$$

   A straight line corresponding to this value is drawn in the third quadrant (line ZZ).
4. In the fourth quadrant, the curve corresponding to $p = 200$ psi and $L/D = 1$ is identified (curve UU).
5. Starting with $\Delta T = 0$ and $T_i = 110°F$ (point $A_1$), a horizontal line $A_1 B_1$ is drawn to the oil curve. The vertical line $B_1 C_1$ meets the $(N/P)(R/C)^2 = 7.5 \times 10^4$ line at $C_1$. A horizontal line from $C_1$ is then drawn to intersect the $(p = 200, L/D + 1)$ curve at $D_1$. A vertical line from $D_1$ meets the line $T_i = 110$ line at $A_2$. Lines $A_2 B_2$, $B_2 C_2$, $C_2 D_2$, and $D_2 A_3$ complete the second iteration. The process is continued until two consecutive "rectangles" are sufficiently close. After five iterations, the range for $\Delta T$ has been narrowed down to 41–46°F. The Sommerfeld number from the last iteration $\approx 0.28$.
6. Once the Sommerfeld number is evaluated, the performance characteristics of the bearing are obtained from Raimondi and Boyd's plots or from the curve-fitted equations.

*Empirical Thermohydrodynamic Analysis*

When Lubricant and Average Film Temperatures are Specified

**EXAMPLE 2.** Calculate the modified Sommerfeld number and the performance characteristics of a centrally loaded journal bearing for the following condition:

$W = 100\,\text{lb}$
$N = 6000\,\text{rpm}\ (100\,\text{rps})$
$R = 1.25\,\text{in.}$
$L = 2.5\,\text{in.}$
$C = 0.0063\,\text{in.}$
Lubricant SAE 10
Average temperature $T_a = 150°F$

Numerical Solution:

First find the numerical value of the bearing characteristic constant $k$ from Fig. 6.43 that for $R/C = 1.25/0063 = 198$, $k = 0.05$, and the parameter $RCL/k$ has the value:

$$\frac{RCL}{k} = \frac{1.25 \times 0.0063 \times 2.5}{0.05} = 0.395$$

The oil parameter $b\theta/(T_a + \theta)^2$ can be evaluated as:

$$\frac{b\theta}{(T_a + \theta)^2} = \frac{1157.5 \times 95}{(150 + 95)^2} = 1.83$$

because for SAE 10, $b = 1157.5$ at $\theta = 95$ (Table 6.4).
   $P_a = W/(LD) = 100/(2.5 \times 2.5) = 16\,\text{psi}$. The average viscosity, $\mu_a = 1.78 \times 10^{-6}$ reyn. Therefore, the Sommerfeld number is:

$$S = \frac{\mu_a N}{P_a}\left(\frac{R}{C}\right)^2 = \frac{1.78 \times 10^{-6} \times 100}{16}(198)^2 = 0.439$$

The quantity of oil flowing in the bearing can be estimated by the curve-fitted equation:

$$\frac{Q}{RNCL} = 3.5251\left(\frac{L}{D}\right)^{-0.2333} S^{-0.1926+0.1149(L/D)} = 3.5251(1)(0.439)^{-0.1926+0.1149}$$

$$= 3.5251 \times 1.066 = 3.76$$

The average pressure at the thermohydrodynamic equilibrium condition can now be calculated:

$$P^* = \frac{\left(\frac{Q}{RNCL}\right)\left(\frac{RCL}{K}\right)\theta}{\frac{b\theta}{(T+\theta)^2}} = \frac{3.76 \times 0.395 \times 95}{1.83} = 77\,\text{psi}$$

The modified Sommerfeld number is obtained by the relation:

$$S^* = S\frac{p^*}{p} = (0.439)\left(\frac{77}{16}\right) = 2.11$$

The eccentricity ratio corresponding to this modified Sommerfeld number is 0.068, whereas the isoviscous theory predicts it to be 0.299. Using $S^*$ in place of $S$, all the performance characteristics for the bearing can be readily calculated based on isoviscous considerations.

Nomogram Solutions (see Fig. 6.45):

The numerical values of the parameters $RCL/k$ and $S$ are calculated to be 0.395 and 0.439, respectively.

The quantity $b\theta/(T+\theta)^2$ can be evaluated by using the upper left portion of Fig. 6.46a to be 1.83. Corresponding to the numerical values of the parameters $RCL/k$ and $b\theta/(T+\theta)^2$, lines OX and OY are drawn in the second and the third quadrant of the nomogram (Fig. 6.45) by interpolation if necessary. The nomogram is then utilized to evaluate $S^*$ from $S$ as follows. Plot point $A_1$ to represent $S$. Draw $A_1B_1$ to the curve $L/D = 1$. Draw the horizontal line $B_1C_1$ to the curve $RCL/k = 0.395$ and the vertical line $C_1D_1$ to the curve $b\theta/(T+\theta)^2 = 1.83$. Read $P^*_{ave}$ on the corresponding scale ($\approx 77\,\text{psi}$). Calculate $S^*$ from $P^*_{ave}S = P_{ave}S^*$ as determined graphically by defining the curve ZZ in the fourth quadrant with $P^*_{ave}S = 77 \times 0.439 = 33.8$, drawing a horizontal line from the $P_{ave}$ scale at 16 psi to intersect it at $F_1$. The vertical line $F_1G_1$ defines the modified Sommerfeld number $S^* \approx 2.1$.

## When Lubricant and Inlet Temperature Are Specified

**EXAMPLE 3.** Calculate the modified Sommerfeld number and the different performance characteristics of a centrally loaded full journal bearing for the following condition:

$W = 947\,\text{lb}$

$N = 4000\,\text{rpm}$ (66.67 rps)
$R = 0.6875\,\text{in.}$
$L = 1.375\,\text{in.}$
$C = 0.0009.15\,\text{in.}$
Lubricant SAE 30 oil
Inlet temperature $T_i = 150°F$

Numerical Solution:

Because $R/C = 751$, the bearing characteristic constant $k = 0.0003$, as can be found from Fig. 6.43. Therefore:

$$\frac{RCL}{k} = \frac{0.6875 \times 0.000915 \times 1.375}{0.0003} = 2.88$$

As an initial guess, assume $\Delta T = 0$, therefore:

$$T_a^* = T_i + \frac{\Delta T^*}{2} = 150°F$$

Consequently:

$$\mu_a^* = 3.6 \times 10^{-6}\,\text{reyn}$$

$$\frac{b\theta}{(T_a^* + \theta)^2} = \frac{1360 \times 95}{(150 + 95)^2} = 2.12$$

Also, because:

$$P_a = \frac{W}{LD} = \frac{947}{1.375 \times 1.375} \cong 500\,\text{psi}$$

the parameter

$$\frac{N}{P_a^2}\left(\frac{R}{C}\right)^2 = \frac{66.67}{(500)^2}(751)^2 \cong 150$$

The average pressure $P_a$ can be used as an initial value for $P_a^*$. Consequently:

$$P_a^* = P_a = 500\,\text{psi}$$

The first approximation for $S^*$ is now computed as:

$$S^* = \frac{\mu_a^* N}{P_a^2}P^*\left(\frac{R}{C}\right)^2 = \frac{3.6 \times 10^{-6} \times 66.7}{(500)^2} \times 500 \times (751)^2 = 0.27$$

The corresponding quantity of oil flow is calculated using the appropriate equation from Section 6.1.3 as $Q/(RNCL) = 3.9$. The new value of $P^*$ can now be calculated.

$$P_a^* = \frac{\dfrac{Q}{RNCL}\dfrac{RCL}{k}\theta}{\dfrac{b\theta}{(T_a^* + \theta)^2}} = \frac{3.9 \times 2.88 \times 95}{2.15} \cong 495\,\text{psi}$$

and the corresponding temperature rise can then be calculated as $\Delta T^* \cong 110°\text{F}$ using $J = 9336\,\text{in.-lb/BTU}$, $v = 0.03\,\text{lb/in.}^3$, and $C = 0.4\,\text{BTU/(lb-°F)}$. With this new value of $\Delta T^*$, a second iteration can be made. The results for the first six such iterations are shown in Table 6.6. Therefore, 0.16 can be taken to be a sufficiently close approximation to the modified Sommerfeld number for the considered example.

Nomogram Solution (see Figs 6.46a and 6.46b):

Only the first, the second, and the sixth (last) iterations are shown. On the first nomogram (Fig. 6.46a), draw the line XX corresponding to:

$$\frac{N}{P^2}\left(\frac{R}{C}\right)^2 = 150$$

and in the second nomogram (Fig. 6.46b), draw the line for $RCL/k = 2.88$.

Now starting with $\Delta T^* = 0$ in Fig. (6.46a) (point $A_1$), draw a vertical line $A_1 B_1$ to intersect the line $T_i = 150$ at point $B_1$. The horizontal line $C_1 B_1$meeting the SAE 30 curve at $C_1$ gives the value of the parameter $b\theta/(T^* + \theta)^2 =$ (point 2.15 $D_1$). Also, the horizontal line $B_1 E_1$ meets with the SAE 30 curve on the right-hand side at $E_1$. A vertical line from $E_1$ meets

**Table 6.6** Results for the First Six Iterations

| Iteration | $\Delta T^*$ (°F) | $T_a^*$ (°F) | $\mu_s^*$ (reyn) | $S^*$ | $Q/RNCL$ | $P^*$ (psi) | $\Delta T^*$ (°F) |
|---|---|---|---|---|---|---|---|
| 1 | Initial guess = 0 | 150 | $3.64 \times 10^{-6}$ | 0.274 | 3.9 | 495 | 110.0 |
| 2 | 110.0 | 205 | $1.32 \times 10^{-6}$ | 0.098 | 4.4 | 847 | 93.4 |
| 3 | 93.4 | 196.7 | $1.50 \times 10^{-6}$ | 0.191 | 4.0 | 722 | 113.8 |
| 4 | 113.8 | 206.9 | $1.28 \times 10^{-6}$ | 0.139 | 4.3 | 848 | 112.4 |
| 5 | 112.4 | 206.2 | $1.29 \times 10^{-6}$ | 0.165 | 4.1 | 779 | 107.0 |
| 6 | 107.0 | 203.5 | $1.35 \times 10^{-6}$ | 0.158 | 4.1 | 767 | 101.0 |

line XX at $F_1$. A horizontal line $F_1G_1$ is then drawn to meet the line $P^* = 500$ at $G_1$. The point $H_1$ vertically above $G_1$ gives the first approximation for $S^*$ to be 0.27.

This value of $S^*$ is then entered as point $I$ in the second nomogram (Fig. 6.46b). Draw a vertical line IJ to meet the curve $L/D = 1$. Then the horizontal line JK meets $RCL/k = 2.88$ line at K. Point L is vertically below point K and on the line $b\theta/(T + \theta)^2 = 2.15$. The horizontal line LMN is such that point N is vertically below point I and point M is on the $P^*_{ave}$ axis. Point M provides the new approximation for $P^* = 495$ psi and point N gives $\Delta T^* \cong 110°F$(for $L/D = 1$).

With $\Delta T = 110°F$, the new entry point $A_2$ in the first nomogram (Fig. 6.46a) is determined and the procedure is continued.

After six iterations, $S^* = 0.16$ can be accepted as the solution for the considered example.

## 6.4 THERMOHYDRODYNAMIC LUBRICATION ANALYSIS INCORPORATING THERMAL EXPANSION ACROSS THE FILM

Viscosity is generally considered to be the single most important property of lubricants, therefore, it represents the central parameter in all lubricant analysis. By far the easiest approach to the question of viscosity variation within a fluid film bearing is to adopt a representative or mean value viscosity. Examples of studies which have provided many suggestions for calculations of the effective viscosity in a bearing analysis are presented By Cameron [34] and Szeri [35]. When the temperature rise of the lubricant across the bearing is small, bearing performance calculations are customarily based on the classical, isoviscous theory. In other cases, where the temperature rise across the bearing is significant, the classical theory loses its usefulness for performance prediction. One of the early applications of the energy equation to hydrodynamic lubrication was made by Cope [36] in 1948. His model was based on the assumptions of negligible temperature variation across the film and negligible heat conduction within the lubrication film as well as into the adjacent solids. The consequence of the second assumption is that both the bearing and the shaft are isothermal components, and thus, all the generated heat is carried out by the lubricant. As indicated in a review paper by Szeri [37]: the belief, that the classical theory on one hand, and Cope's adiabatic model on the other hand, bracket bearing performance in lubrication analysis, was widely accepted for a while.

In 1987, Pinkus [38] in his historical overview of the theory of hydrodynamic lubrication pointed out that one of the least understood and urgent areas of research is that of thermohydrodynamics. In the discussion of parallel surfaces and mixed lubrication, he indicated that the successful operation of centrally pivoted thrust bearings cannot be explained by the hydrodynamic theory. He backs his assertion by reviewing several failed attempts to explain the pressure developed for bearings with constant film thickness.

Braun et al. [39–41] investigated different aspects of the thermal effects in the lubricant film and the thermohydrodynamic phenomena in a variety of film situations and bearing configurations. Dowson et al. [42, 43] adopted the cavitation algorithm proposed by Elord and Adams and studied the lubricant film rupture and reformation effects on grooved bearing performance for a wide range of operating parameters. The comparison of their theoretical results and those presented in a design document revealed that a consideration of more realistic flow conditions will not normally influence the value of predicted load capacity significantly. However, the prediction of the side leakage flow rate will be greatly affected if film reformation is not included in the analysis. Braun et al. [44] also experimentally investigated the cavitation effects on bearing performance in an eccentric journal bearing.

Lebeck [45, 46] summarized several well-documented experiments of parallel sliding or parallel surfaces. The experiments clearly show that as speed is increased, the bearing surfaces are lifted such that asperity contact and friction are reduced. He also suggested that the thermal density wedge, viscosity wedge, microasperity and cavitation lubrication, asperity collisions, and squeeze effects do not provide sufficient fluid pressure to be considered a primary source of beneficial lubrication in parallel sliders. Rohde and Oh [47] reported that the effect of elastic distortion of the bearing surface due to temperature and pressure variations on the bearing performance is small when compared with thermal effects on viscosity.

Most of analytical studies dealing with thermohydrodynamic lubrication utilize an explicit marching technique to solve the energy equation [47, 48]. Such explicit schemes may in some situations cause numerical instability. Also, the effect of variation of viscosity across the fluid film on the bearing performance was acknowledged to be an important factor in bearing analysis [49–53]. The effect of oil film thermal expansion across the film on the bearing load-carrying capacity is not adequately treated in all the published work.

The extensive experimental tests reported by Seireg et al. [25–28] for the steady-state and transient performance of fluid film bearings strongly suggest the need for a reliable methodology for their analysis. This is also a major concern for many workers in the field (Pinkus [38] and Szeri [37]).

Several theoretical studies have been undertaken to address this problem (Dowson and Hudson [54, 55] and Ezzat and Rohde [48, 56]) but were not totally successful in predicting the experimentally observed relationship between speed and pressure for fixed geometry films.

Wang [57] developed a thermohydrodynamic computational procedure for evaluating the pressure, temperature, and velocity distributions in fluid films with fixed geometry between the stationary and moving bearing surfaces. The velocity variations and the heat generation are assumed to occur in a central zone with the same length and width as the bearing but with a significantly smaller thickness than the fluid film thickness. The thickness of the heat generation (shear) zone is developed empirically for the best fit with experimentally determined peak pressures for a journal bearing with a fixed film geometry operating in the laminar regime. A transient thermodynamic computation model with a transformed rectangular computational domain is utilized. The analysis can be readily applied to any given film geometry.

The existence of a thin shear zone, with high velocity gradients, has been reported by several investigators. Batchelor [58] suggested that for two disks rotating at constant but different speeds, boundary layers would develop on each disk at high Reynolds numbers and the core of the fluid would rotate at a constant speed.

Szeri et al. [59] carried out a detailed experimental investigation of the flow between finite rotating disks using a laser doppler velocimeter. Their measurements show the existence of a velocity field as suggeste by Batchelor. More recently [60], the experimental investigation of the flow between rotating parallel disks separated by a fixed distance of 1.27 cm in a 0.029% and 0.053% solutions of polyacrylamide showed the existence of an exceedingly thin shear layer where the velocity gradients are exceedingly high. This effect was found to be most pronounced at higher revolution rates.

Another experimental study by Joseph et al. [61] demonstrates the existence of thin shear layers between two immiscible liquids that have unequal viscosities.

### 6.4.1 Empirical Evaluation of Shear Zone

The computer program described by Wang [67] is used to compute the shear zone ratio, $h_s/h$, which gives the value for the maximum film pressure corresponding to the experimental data. A flow chart of the program is given in Fig. 6.48. The geometric parameters of the bearing (UW-1, UW-2 and UW-3) investigated in this empirical evaluation are listed in Table 6.7. The non-dimensional viscosity–temperature relation used in the computer program is $\bar{\mu}(\bar{T}) = e^{\delta T_{in}(\bar{T}-1)}$, where $\delta$ is the temperature viscosity coefficient. Table 6.8 lists the viscosity coefficients and reference viscosity at 37.8 and 93.3°C for

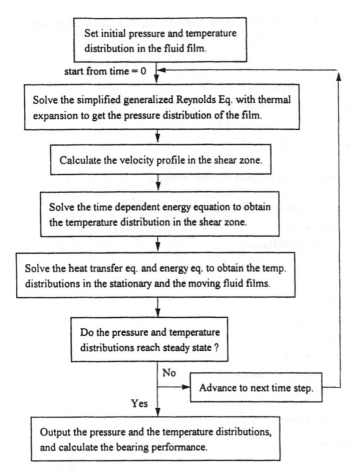

**Figure 6.48** Program flow chart of developed analysis (THD with across film thermal expansion).

**Table 6.7** Geometric Parameters of Investigated Bearings

| Journal bearing | $D$ (mm) | $L$ (mm) | $C$ (mm) | $L/D$ | $R/C$ |
|---|---|---|---|---|---|
| UW-1 | 080.1 | 050.8 | 000.34 | 0.634 | 118 |
| UW-2 | 050.8 | 025.4 | 000.38 | 0.500 | 67 |

| Slider bearing | $B$ (mm) | $L$ (mm) | Slope | $H_{min}$ (mm) | $h_{ave}$ (mm) |
|---|---|---|---|---|---|
| UW-3 | 038.1 | 025.4 | 0.0009 | 0.01524 | 0.02489 |

various grades of oil. In all the cases considered, the properties of the lubricants were taken as follows:

$\rho = 873\,\text{kg/m}^3\,(54.5\,\text{lb/ft}^3)$

$\alpha = 0.000648$ per °C $(0.00036$ per °F$)$

$\beta = 1517 \times 10^6\,\text{N/m}^2\,(31.68 \times 10^6\,\text{lbf/ft}^2)$

$k = 470\,\text{J/(hr-m-°C)}\,(0.0075\,\text{BTU/(hr-ft-°F)})$

$c = 2010\,\text{J/(kg-m-°C)}\,(0.48\,\text{BTU/(lb-°F)}$

The following nomenclature is used in the analysis:

$\alpha = $ lubricant thermal expansion coefficient

$\beta = $ lubricant bulk modulus

$\delta = $ temperature viscosity coefficient

$\varepsilon = $ eccentricity ratio

$\rho = $ lubricant density

$\mu = $ lubricant viscosity

$\bar{\mu} = $ dimensionless viscosity $ = e^{\delta T_{in}(\bar{T}-1)}$

$h = $ thickness of the film

$h_s = $ thickness of the shear layer

$k = $ thermal conductivity of the lubricant

$T = $ fluid film temperature

$\bar{T} = $ dimensionless temperature $ = \left(\dfrac{T}{T_{in}}\right)$

$T_{in} = $ oil inlet temperature

The UW-1 bearing was first selected for evaluating the shear zone due to the fact that extensive test data for various conditions are available for it. The experiment covered eccentricity ratios from 0.6 to 0.9, speeds from 500 to 2400 rpm, lubricating oils from SAE 10 to 50, and inlet temperatures from

**Table 6.8**  Lubricant Viscosity–Temperature Table, $\bar{\mu}(\bar{T}) = e^{\delta T_{in}(\bar{T}-1)}$

| Oil | $\mu$ at 37.8°C [N/(sec-m$^2$)] | $\mu$ at 93.3° [N/(sec-m$^2$)] | $\delta$ |
|---|---|---|---|
| SAE 5  | 0.0155 | 0.00376 | −0.0142 |
| SAE 10 | 0.0318 | 0.0057  | −0.0172 |
| SAE 30 | 0.1059 | 0.00168 | −0.022  |
| SAE 50 | 0.2407 | 0.01975 | −0.025  |

25 to 72°C. The results were used to iteratively evaluate the shear zone ratio, $h_s/h$, which best fits the maximum values of the experimental pressure. The relationship was then applied in the computational program to check the experimental pressure data for the different test conditions of the journal bearing UW-2, as well as the slider bearing UW-3 (see Table 6.7).

## 6.4.2 Empirical Formula for Predicting the Shear Zone

Traditionally, the behavior of an isoviscous fluid film wedge can be characterized based on Newtonian fluid dynamics by a dimensionless number – the Sommerfeld number ($S$). When the transient heat transfer in the fluid wedge for thermohydrodynamic analysis is considered, it stands to reason that another dimensionless number – the Peclet number ($P_e$) should also play an equally significant part. Consequently, the calculated values for $h_s/h$ based on all the considered experimental data (Table 6.9) were plotted versus the product of Sommerfeld and Peclet numbers as shown in Fig. 6.49. It can be seen that a highly correlated curve was obtained. The fitted curve can be defined by the following equations to a high degree of accuracy:

$$SP_e = 0 \to \left(\frac{2}{5\pi} \times 10^6\right) \frac{h_s}{h} = \frac{1}{2\pi}\left(\frac{SP_e}{10^6}\right)^{\frac{7}{2}}$$

$$SP_e = \left(\frac{2}{5\pi} \times 10^6\right) \to \infty \quad \frac{h_s}{h} = \frac{1}{5\pi}\left[1 - \exp{-\frac{\left(\frac{SP_e}{10^6} - \frac{1}{5\pi}\right)}{\left(\frac{2}{\pi}\right)}}\right]$$

**Table 6.9** Summarized Results for Empirical Evaluation of Shear Zone Ratio for Bearings (UW-1)

| Case no. | Speed (rpm) | Eccentricity ratio, $\varepsilon$ | Oil | Oil $T_{in}$ (°C) | $(p_{max})\exp$ ($10^6$ N/m²) | Shear zone ratio, $h_s/h$ |
|---|---|---|---|---|---|---|
| 1 | 525 | 0.90 | SAE 10 | 53.3 | 0.644 | 0.003 |
| 2 | 600 | 0.87 | SAE 50 | 71.1 | 0.700 | 0.008 |
| 3 | 1050 | 0.90 | SAE 10 | 53.3 | 0.931 | 0.010 |
| 4 | 1200 | 0.87 | SAE 50 | 71.1 | 0.994 | 0.020 |
| 5 | 2100 | 0.90 | SAE 10 | 53.3 | 1.274 | 0.025 |
| 6 | 2400 | 0.87 | SAE 50 | 71.1 | 1.414 | 0.038 |
| 7 | 500 | 0.60 | SAE 30 | 41.7 | 0.238 | 0.042 |
| 8 | 2000 | 0.60 | SAE 30 | 41.7 | 0.336 | 0.056 |
| 9 | 1000 | 0.60 | SAE 30 | 41.7 | 0.483 | 0.064 |

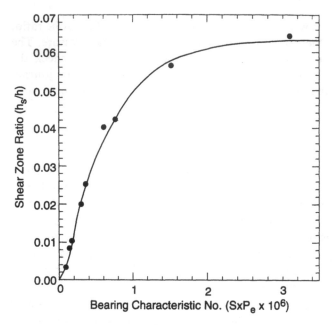

**Figure 6.49**  Fluid film shear zone ratio versus the product of Sommerfeld and Peclet numbers. •, calculated; —, curve-fitted equation.

where the Sommerfeld and Peclet numbers are defined as follows:

$$S = \left(\frac{R}{C}\right)^2 \frac{\mu N}{P_{ave}}$$
$$P_e = \left(\frac{\rho c}{k}\right) \pi R U$$

The following nomenclature is used in the analysis:

$C$ = bearing clearance
$c$ = specific heat of the lubricant
$D$ = bearing diameter
$N$ = rotational speed
$P_{ave}$ = average film pressure
$P_e$ = Peclet number
$R$ = journal radius
$S$ = Sommerfeld number for journal bearing
$L$ = bearing length

The characteristic length of the journal bearing used in the Peclet number is $\pi R$, as this is the length of significance in the transient heat transfer problem.

### 6.4.3 Geometric Analogy Between the Lubricating Film for Journal and Slider Bearings

The film geometry of a journal bearing is considered analogus to that of a generalized slider bearing. The unwrapped journal bearing, assuming the radius of curvature of the bearing is large compared with the film thickness, is basically a slider with a convergent–divergent shape. The divergent portion of the journal bearing cannot be expected to contribute to the load-carrying capacity and consequently the characteristic length of the journal bearing is considered equal to $\pi R$. By using a transformation based on the following geometric analogy:

$$B = \pi R = \text{length of the bearing in the direction of sliding}$$

$$h_{\text{ave}} = \frac{h_{\text{max}} + h_{\text{min}}}{2} = C$$

a characteristic number for a slider bearing, similar to that of a journal bearing, can be obtained. The generally adopted characteristic number (Sommerfeld number, $S$) for journal bearings, which is based on the iso-viscous theory, and the derived characteristic number of slider bearings based on the considered analogy can be written as follows:

$$S = \left(\frac{R}{C}\right)^2 \frac{\mu N}{p} \qquad \text{(journal bearing)}$$

$$S_2 = \frac{1}{2\pi^2}\left(\frac{B}{h_{\text{ave}}}\right)^2 \frac{\mu N}{pB} \qquad \text{(slider bearing)}$$

Based on the same analogy, the Peclet number for slider and journal bearings can be expressed as follows:

$$P_e = \left(\frac{\rho c}{k}\right)\pi R U \qquad \text{(journal bearing)}$$

$$P_e = \left(\frac{\rho c}{k}\right) B U \qquad \text{(slider bearing)}$$

where

$U =$ bearing sliding velocity

The characteristic time constant for thermal expansion, $\Delta t$, for both bearings is defined as follows:

$$\Delta t = 2\pi \left( \frac{C}{U} \right) \qquad \text{(journal bearing)}$$

$$\Delta t = 2\pi \left( \frac{h_{\text{ave}}}{U} \right) \qquad \text{(slider bearings)}$$

and their nondimensional expressions are:

$$\Delta \bar{t} = 2\pi \left( \frac{C}{U} \right) \left( \frac{U}{2\pi R} \right) = \frac{C}{R} \qquad \text{(journal bearing)}$$

$$\Delta \bar{t} = 2\pi \left( \frac{h_{\text{ave}}}{U} \right) \left( \frac{U}{B} \right) = 2\pi \frac{h_{\text{ave}}}{B} \qquad \text{(slider bearing)}$$

The equivalent journal bearing length for a fresh film during one rotation is considered to be equal to $2\pi R$, where

$L = $ length of bearing perpendicular to sliding

To determine the minimum film thickness, based on the isoviscous theory, a transformation is needed between $S_2$ for slider bearings (which is developed based on the journal bearing analogy) and the commonly used characteristic number $S_1$ [62, 63], which is equal to $(1/m^2)([\mu U/(pB)])$. In the formula for $S_1$, $m$ is the slope of the wedge and $p$ is the average pressure of the film. Figure 6.50 shows the relation between $S_1$ and $S_2$ for several $L/B$ ratios. This relationship is obtained by performing an isoviscous calculation to determine the average pressure for slider bearings with different wedge geometries and consequently evaluating the corresponding characteristic number $S_1$.

### 6.4.4 Pressure Distribution Using the Conventional Assumption of Full Film Shear

The pressure distribution in the fluid film is probably the most important behavioral characteristic in the analysis of bearing performance. The other bearing characteristics, e.g., heat distribution, frictional resistance, can be calculated from the pressure field. However, the pressure distribution is strongly influenced by the thermal effects in the fluid film which can not be accurately predicted by previous analytical methods, i.e., isoviscous theory and commonly adopted thermohydrodynamic (THD) analysis [64, 65, 66].

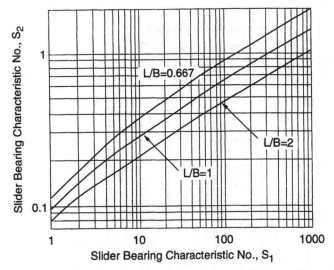

**Figure 6.50** Conversion chart for bearing dimensionless characteristic numbers based on the isoviscous theory.

The isoviscous theory is understandably inadequate in the performance evaluation, due to the lack of consideration of the temperature and viscosity variations in the model. The THD analysis mentioned above are based on the conventional assumption of full film shear and require the simultaneous solution of a coupled system of equations (the energy, momentum, and continuity equations) in the full fluid film. For most cases, however, the solutions predicted from the THD model deviate considerably from experimental results. Figure 6.51 shows a comparison between the solutions obtained from these two theories and the experimental pressure [25]. The latter was obtained by slowly changing the speed of a variable speed motor and a continuous plot of the pressure versus speed is automatically recorded using an *x–y* plotter. All the tests on the UW-1 bearing with different lubricants and eccentricity ratios exhibited a square root relationship between the pressure and speed, as reported by Seireg and Ezzat [25] (Table 6.10). The THD solutions in this figure are obtained by solving the Reynolds equation coupled with the energy equation for the full film without considering the thermal expansion. The boundary of the stationary component is assumed to be thermally insulated, while the moving part surface has the same temperature as the inlet oil. It can be seen that the results based on these theories do not appropriately predict the pressure distribution.

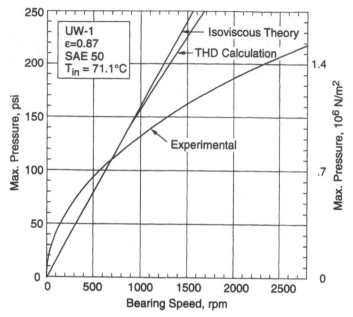

**Figure 6.51** Comparison between the solutions obtained from commonly adopted theories with experimental pressure. The THD solution is obtained by solving the Reynolds equation coupled with energy equation for the full film; no thermal expansion is considered. (From Ref. 25.)

### 6.4.5 Pressure Distribution Using the Proposed Model

Table 6.11 gives a summary of the calculated results for all three bearings (Table 6.10) using the empirical relationship for the shear zone and the computational approach described by Wang and Seireg [67]. The experimental results for the maximum pressure are also given for comparison. It can be seen that the calculated maximum pressures are in excellent correlation with experimental results for all cases.

Figures 6.52a and 6.52b show typical normalized pressure distribution in the film for the UW-1 bearing obtained from the proposed model (which considers thin central shear zone and thermal expansion) and the pressure obtained experimentally [25]. It can be seen from the figures that the calculated pressure distribution correlates well with the experimental pressure. It is interesting to note that the calculated normalized pressure distributions in both the experimental data and to those predicted by the isoviscous theory. The same correlation is found in the case of the UW-2 bearing, as well as the slider bearing UW-3 (Figs 6.53 and 6.54) for the test conditions given in Table 6.10.

**Table 6.10** Test Conditions for the Bearings UW-2 and UW-3

Journal bearing (UW-2)

| Case no. | Speed (rpm) | Eccentricity ratio | Oil | Oil $T_{in}$ (°C) |
|----------|-------------|--------------------|----|-------------------|
| 10 | 1000 | 0.90 | SAE 30 | 37.8 |
| 11 | 2000 | 0.90 | SAE 30 | 37.8 |

Slider bearing (UW-3)

| Case no. | Speed (m/s) | $h_{min}$ (mm) | Slope | Oil | Oil $T_{in}$ (°C) |
|----------|-------------|----------------|-------|-----|-------------------|
| 12 | 0.2286 | 0.01524 | 0.0009 | SAE 5 | 25 |
| 13 | 0.4572 | 0.01524 | 0.0009 | SAE 5 | 25 |

Figures 6.55–6.57 show examples of the pressure–speed characteristics of the bearings investigated under various test conditions. The continuous curves shown in the figures are the best-fit square root relationship between the experimental pressure and speed starting from the origin. The maximum pressure predicted by the isoviscous theory based on the inlet oil temperature is also plotted in each figure for comparison. The calculated maximum

**Table 6.11** Numerical Results Based on the Empirical Formula

| Case no. | $S$ | $P_e \times (10^6)$ | $SP_e \times (10^6)$ | Shear zone ratio | $(p_{max})$exp $\times 10^6$ (N/m²) | $(p_{max})$cal $\times 10^6$ (N/m²) |
|----------|-----|---------------------|----------------------|------------------|-------------------------------------|-------------------------------------|
| 1 | 0.0251 | 3.72 | 0.0933 | 0.0038 | 0.644 | 0.63 |
| 2 | 0.0361 | 4.25 | 0.1534 | 0.0084 | 0.700 | 0.721 |
| 3 | 0.0251 | 7.44 | 0.1867 | 0.0112 | 0.931 | 0.966 |
| 4 | 0.0361 | 8.50 | 0.3069 | 0.0202 | 0.994 | 0.959 |
| 5 | 0.0251 | 14.88 | 0.3734 | 0.0245 | 1.274 | 1.253 |
| 6 | 0.0361 | 17.00 | 0.6137 | 0.0368 | 1.414 | 1.365 |
| 7 | 0.2198 | 3.54 | 0.779 | 0.0430 | 0.238 | 0.252 |
| 8 | 0.2198 | 7.09 | 1.557 | 0.0576 | 0.336 | 0.357 |
| 9 | 0.2198 | 14.17 | 3.115 | 0.0631 | 0.483 | 0.470 |
| 10 | 0.0311 | 2.853 | 0.089 | 0.0036 | 0.455 | 0.42 |
| 11 | 0.0311 | 5.705 | 0.177 | 0.0104 | 0.686 | 0.679 |
| 12 | 0.2649 | 0.117 | 0.031 | 0.0007 | 0.196 | 0.245 |
| 13 | 0.2649 | 0.234 | 0.062 | 0.0020 | 0.322 | 0.343 |

The test conditions for the bearings are given in Tables 6.9 and 6.10 with corresponding case numbers.

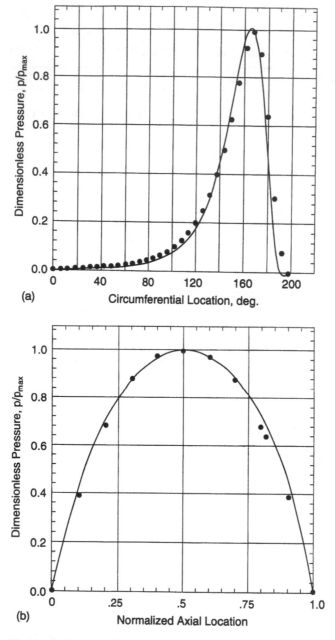

(a)

(b)

**Figure 6.52** (a) Normalized calculated pressure distribution along the centerline of the bearing in the direction of sliding. (b) Normalized calculated pressure distribution at the maximum pressure plane perpendicular to the direction of sliding (UW-1, $\varepsilon = 0.87$, 1000 rpm, SAE 50, $T_{in} = 689°C$). •, THD calculation (with thermal expansion); —, isoviscous theory.

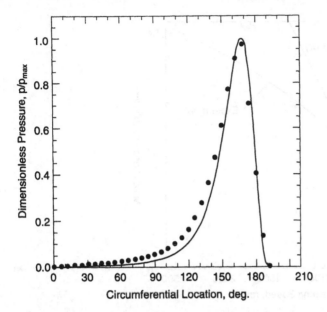

**Figure 6.53** Normalized pressure distribution along the centerline of the bearing in the direction of sliding (UW-2, $\varepsilon = 0.9$, 2000 rpm, SAE 30, $T_{in} = 37.8°C$). •, THD calculation (with thermal expansion); —, isoviscous theory.

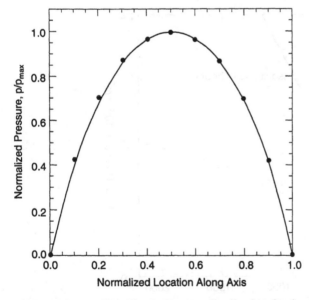

**Figure 6.54** Normalized pressure distribution at the maximum pressure plane perpendicular to the direction of sliding (UW-3, $T_{in} = 25°C$, $h_{min} = 0.01524\,\text{mm}$, slope = 0.0009, 0.4572 m/s, SAE 5). •, THD calculation (with thermal expansion); —, isoviscous theory.

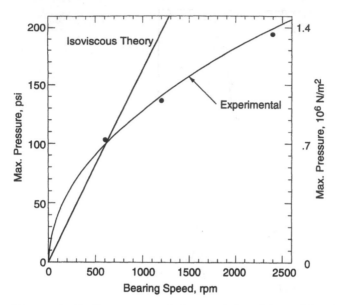

**Figure 6.55** Pressure–speed characteristics of journal bearing (UW-1, $\varepsilon = 0.87$, SAE 50, $T_{in} = 71.1°C$). •, THD calculation (with thermal expansion).

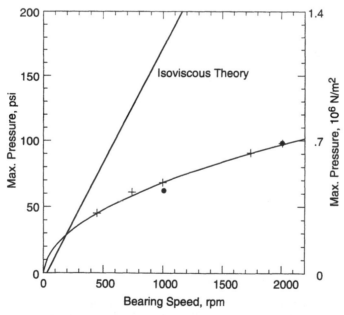

**Figure 6.56** Pressure–speed characteristics of journal bearing (UW-2, $\varepsilon = 0.9$, SAE 30, $T_{in} = 37.8°C$). •, THD calculation (with thermal expansion); +, experimental.

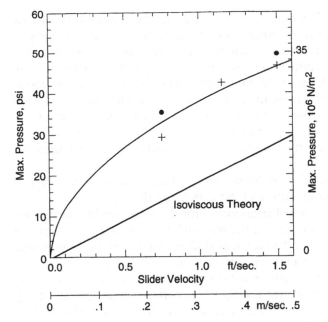

**Figure 6.57** Pressure–speed characteristics of slider bearing (UW-3, $h_{min} = 0.01524 \times 10^{-3}$ m, slope = 0.0009, SAE 5, $T_{in} = 25°C$). •, THD calculation (with thermal expansion); +, experimental.

pressures, based on the developed computational model, show excellent correlation with the experimental data, as well as the square root relation with bearing speed which was found to exist for all the tests on fixed geometry bearings reported by Seireg et al. [25–28, 67].

A numerical solution was also carried out to investigate the pressure–speed characteristics for the case of UW-3 with SAE 5 oil at $T_{in} = 25°C$. A high correlation can be seen between the experimental and the calculated results.

The empirically determined dimensionless effective shear layer thickness, $h_s/h$, which is a function of the film geometry, speed and the thermal properties of the lubricant, is developed based on the experimental data from one bearing, UW-1, and found to be equally applicable to the other journal and slider bearings. The empirical ratio is highly correlated to the product of Sommerfeld and peclet numbers, and reaches an asymptotic value of $1/(5\pi)$, which is the case for bearings operating at high speeds of films with constant thickness.

The predicted ratio, when used in conjunction with the dimensionless characteristic time $\bar{\Delta}t$ in the computational procedure presented by Wang and Seireg [67], was found to accurately evaluate the bearing performance characteristics including the square root relationship between the pressure and speed [68]. It is interesting to note here that experimental tests by Bair et al. [69] with a high-pressure visualization cell on an elastohydrodynamic film revealed a thin hot layer of the lubricant sandwiched between two cooler layers. This hot layer represents a region where the shear deformation is localized. The greatest part of the relative velocity was found to be accommodated within a small fraction (approximately 5%) of the film thickness. This observation adds further credence to the empirical hypothesis discussed in this section.

Theoretical justification of the existence of the central shear zone in thin film lubrication is left for future investigations by rheologists.

The developed, empirical relationship for the dimensionless shear layer thickness is based on test data in the laminar regime with a maximum film temperature rise of 65°C. The calculated thickness has to be viewed with caution if the computed maximum temperature rise is much greater than 65°C. One reason for this caution is the change in the thermal and mechanical properties of the lubricant for high temperature variation may be significant enough to influence the accuracy of the results. This section demonstrates that the pressure and temperature distributions in the film are strongly affected by the thermal properties of the lubricant, especially the thermal conductivity. Bearing designers would benefit from considering the influence of the thermal properties of the lubricants on the bearing performance.

## REFERENCES

1. Reynolds, O., "On the Theory of Lubrication," Phil. Trans. Roy. Soc. (Lond.), 1886, 177, Pt I, p. 157.
2. Raimondi, A. A., and Boyd, J., "A Solution for the Finite Journal Bearing and Its Application to Analysis and Design I," ASLE Trans., 1958, Vol. 1(1), pp. 159–174.
3. Raimondi, A. A., and Boyd, J., "A Solution for the Finite Journal Bearing and its Application to Analysis and Design: II," ASLE Trans., 1958, Vol. 1(1), pp. 175–193.
4. Raimondi, A. A., and Boyd, J., "A Solution for the Finite Journal Bearing and Its Application to Analysis and Design: III," ASLE trans., 1958, Vol. 2(1), pp. 194–209.
5. Lund, J. W., and Orcutt, F. K., "Calculations and Experiments on the Unbalance Response of a Flexible Rotor,"Journal of Engineering for Industry, Trans. ASME, Nov. 1967, Series B, Vol. 89, p. 785.

6. Badgley, R. H., and Booker, J. F., "Rigid-Body Rotor Dynamics: Dynamic Unbalance and Lubrication Temperature Changes," J. Eng. Power, Trans. ASME, Apr. 1971, Vol. 93, p. 279.

7. Rieger, N. F., "Unbalance Response of an Elastic Rotor in Damped Flexible Bearings at Super Critical Speeds," J. Eng. Power, Trans. ASME, Apr. 1971, Vol. 93, p. 265.

8. Someya, T., "Vibrational and Stability Behavior of an Unbalanced Shaft Running in Cylindrical Journal Bearings," VDI-Forsch-Heft 510, p. 5 (in German).

9. Gunter, E. J., Discussion on Reference [3], J. Eng. Power, Trans. ASME, Apr. 1971, Vol. 93, p. 279.

10. Gunter, E. J., "Dynamic Stability of Rotor-Bearing Systems," NASA SP-113, 1966.

11. Lund, J. W., "Stability and Damped Critical Speeds of a Flexible Rotor in Fluid Film Bearings," ASME Trans., J. Eng. Indust., Paper No. 73-DET-103, 1973.

12. Macchia, D., "Acceleration of an Unbalanced Rotor Through the Critical Speed," Paper No. 63-WA-9, Trans. ASME, Winter Annual Meeting, Philadelphia, PA, November 17–22, 1963.

13. Seireg, A., and Dandage, S., "A Phase Plane Simulation for Investigating the Effect of Unbalance Magnitude on the Whirl of Rotors Supported on Hydrodynamic Bearings," Trans. ASME, J. Lubr. Technol., Oct. 1975.

14. Mechanical Technology Incorporated, "Rotor-Bearing Dynamics Design Technology; Part III; Design Handbook for Fluid Film Bearings," Technical Report AFAPL-TR-65-45, Part III, May 1965.

15. McKee, S. A., and McKee, T. R., "Friction of Journal Bearings as Influenced by Clearance and Length," Trans. ASME, Vol. 51, APM-51-15, pp. 161–171.

16. Fogg, A., "Film Lubrication of Parallel Thrust Surfaces," Proc. Inst. Mech. Eng., Vol. 155, pp. 49–67.

17. Shaw, M. C., "An Analysis of the Parallel-Surface Thrust Bearing," Trans. ASME, Vol. 69, pp. 381-387.

18. Boussages, P., and Casacci, S., "Etude sur les pivots a graines paralleles," La Houille Blanche, July–Aug. 1948, pp. 1–9.

19. Osterle, F., Charnes, A., and Saibel, E., "On the Solution of the Reynolds Equation for Slider Bearing Lubrication – IV. Effect of Temperature on the Viscosity," Trans. ASME, Vol. 75, Pt 1, p. 1117.

20. Ulukan, Von Lutfullah, "Thermische Schmierkeilbildung," Bull. Tech. Univ., Istanbul, Vol. 9, pp. 77–101, 1956; Actes Ninth International Congress of Applied Mechanics; Brussels, Vol. 4, p. 303, 1957.

21. Cameron, A., "Hydrodynamic Lubrication of Rotating Discs in Pure Sliding. New Type of Oil Film Formation," J. Inst. Petrol, Vol. 37, p. 471.

22. Cole, J. A., "An Experimental Investigation of Temperature Effects in Journal Bearings," Proc. of the Conference of Lubrication and Wear, 1957 (I. Mech. E.), paper 63, p. 111.

23. Hunter, W. B., and Zienkiewicz, O. C., "Effect on Temperature Variations across the Lubricant Films in the Theory of Hydrodynamic Lubrication," J. Mech. Eng. Sci., 1960, Vol. 2, p. 52.

24. Dowson, D., and March, C. N., "A Thermodynamic Analysis of Journal Bearings," Lubrication and Wear Convention, Plymouth, May, Inst. Mech. Eng., 1967.

25. Seireg, A., and Ezzat, H., "Thermohydrodynamic Phenomena in Fluid Film Lubrication," ASME J. Lubr. Technol., Apr. 1973, Vol. 95.

26. Ezzat, H., and Seireg, A., "Thermohydrodynamic Performance of conical Journal Bearings," Paper H13.1, Proceedings of the World Conference on Industrial Tribology, R. C. Malhotra and J. P. Sharma, editors, Dec. 1972.

27. Seireg, A., and Doshi, R. C., "Temperature Distribution in the Bush of Journal Bearings During Natural Heating and Cooling," Proceedings of the JSLE-ASLE International Lubrication Conference, Tokyo, 1975, p. 105.

28. Seireg, A., Kamdar, B. C., and Dandage, S., "Effect of Misalignment on the Performance of Journal Bearings," Proceedings of the JSLE-ASLE International Lubrication Conference, Tokyo, 1975, p. 145.

29. Dubois, G. B., Ocvirk, F. W., and Wehe, R. L., "Experimental Investigation of Eccentricity Ratio, Friction, and Oil flow of Long and Short Journal Bearings with Load Number Charts," National Advisory Committee for Aeronautics, Technical Note 3491, Sept. 1955.

30. Carl, T. E., "An Experimental Investigation of a Cylindrical Journal Bearing Under Constant and Sinusoidal Loading," Proc. Inst. Mech. Engrs, 1963–1964, Vol. 176, Pt 3N, paper 19, pp. 100–119.

31. Someya, T., "An Investigation into the Stability of Rotors Supported on Journal Bearings. Effects of Unsymmetry and Moments of Inertia of Rotors," Japanese Soc. Lub. Eng., Apr., 1972.

32. Mitchell, J. R., Holmes, R., and Van Ballegooyen, H., "Experimental Determination of a Bearing Oil Film Stiffness," Proc. Inst. Mech. Engrs, 1965–1966, Vol. 180, Pt 3K, pp. 90–96.

33. Seireg, A., and Dandage, S., "Empirical Design Procedure for the Thermohydrodynamic Behavior of Journal Bearings," Trans. ASME, J. Lubr. Technol., 1982, pp. 135–148.

34. Cameron, A., The Principles of Lubrication, Longmans Green and Co., Ltd., 1966.

35. Szeri, A. X., Tribology – Friction, Lubrication and Wear, Hemisphere Publishing Co., 1980.

36. Cope, W., "The Hydrodynamic Theory of Film Lubrication," Proc. Roy. Soc., 1948, Vol. A197, pp. 201–216.

37. Szeri, A. Z., "Some Extensions of the Lubrication Theory of Osborne Reynolds," ASME J. Tribol., 1987, pp. 21–36.

38. Pinkus, O., "The Reynolds Centennial: A Brief History of the Theory of Hydrodynamic Lubrication," ASME J. Tribol., 1987, pp. 2–20.

39. Braun, M. J., Mullen, R. L., and Hendricks, R. C., "An Analysis of Temperature Effect in a Finite Journal Bearing with Spatial Tilt and Viscous Dissipation," Trans. ASLE, 1984, Vol. 47, pp. 405–411.
40. Braun, M. J., Wheeler, R. L., and Hendricks, R. C., "A Fully Coupled Variable Properties Thermohydraulic Model for Hydrostatic Journal Bearing," ASME J. Tribol., 1987, Vol. 109, pp. 405–417.
41. Braun, M. J., Wheeler, R. L., and Hendricks, R. C., "Thermal Shaft Effects on the Load Carrying Capacity of a Fully Coupled Variable Properties Journal Bearing," Trans. ASLE, 1987, Vol. 30, p. 292.
42. Dowson, D., Tayler, C. M., and Miranda, A. A., "The Prediction of Liquid Film Journal Bearing Performance with a Consideration of Lubricant Film Reformation, Part 1: Theoretical Results," Proc. Inst. Mech. Engrs, 1985, Vol. 199(C2), pp. 95–102.
43. Dowson, D., Tayler, C. M., and Miranda, A. A., "The Prediction of Liquid Film Journal Bearing Performance with a Consideration of Lubricant Film Reformation, part 2: Experimental Results," Proc. Inst. Mech. Engrs, 1985, Vol. 199(C2), pp. 103–111.
44. Braun, M. J., and Hendricks, R. C., "An Experimental Investigation of the Vaporous/Gaseous Cavity Characteristics in an Eccentric Journal Bearing," Trans. ASLE, 1984, Vol. 27, pp. 1–4.
45. Lebeck, A. O., "Parallel Sliding Load Support in the Mixed Friction Regime, Part 1: The Experimental Data," ASME J. Tribol., 1987, Vol. 109, pp. 189–195.
46. Lebeck, A. O., "Parallel Sliding Load Support in the Mixed Friction Regime, Part 2: Evaluation of the Mechanisms," ASME J. Tribol., 1987, Vol. 109, pp. 196–205.
47. Rohde, S. M., and Oh, K. P., "A Thermoelastohydrodynamic Analysis of a Finite Slider Bearing," ASME J. Lubr. Technol., 1975, pp. 450–460.
48. Ezzat, H. A., and Rohde, S. M., "A Study of the Thermohydrodynamic Performance of Finite Slider Bearings," ASME J. Lubr. Technol., 1973, pp. 298–307.
49. Hunter, W., and Zienkiewicz, O., "Effects of Temperature Variation Across the Film in the Theory of Hydrodynamic Lubrication," J. Mech. Eng. Sci., 1960, Vol. 2, pp. 52–58.
50. Raimondi, A. A., "An Adiabatic Solution for the Finite Slider Bearing (L/B = 1)," Trans. ASLE, 1966, Vol. 9, pp. 283–298.
51. Hahn, E. J., and Kettleborough, G. E., "The Effects of Thermal Expansion in an Infinite Wide Slider Bearing-Free Expansion," ASME J. Lubr. Technol., 1968, pp. 233–239.
52. Hahn, E. J., and Kettleborough, G. E., "Thermal Effects in Slider Bearings," Proc. Inst. Mechn. Engrs, 1968–1969, vol. 183, pp. 631–645.
53. Boncompain, R., Fillon, M., and Frene, J., "Analysis of Thermal Effects in Hydrodynamic Bearings," ASME J. Tribol., 1986, Vol. 108, pp. 219–224.
54. Dowson, D., and Hudson, J. D., "Thermohydrodynamic Analysis of the Infinite Slider Bearing: Part 1, the Plane Inclined Slider Bearing," Proc. Inst. Mech. Engrs, 1963, pp. 34–44.

55. Dowson, D., and Hudson, J. D., "Thermohydrodynamic Analysis of the Infinite Slider Bearing Part 2, the Parallel Surface Bearing," Proc. Inst. Mech. Engrs, 1963, pp. 45–51.

56. Ezzat, H. A., and Rohde, S. M., "Thermal Transients in Finite Slider Bearings," ASME J. Lubr. Technol., 1974, pp. 315–321.

57. Wang, N. Z., "Thermohydrodynamic Lubrication Analysis Incorporating Thermal Expansion Across the Film," Ph.D, Thesis, University of Wisconsin–Madison, 1993.

58. Batchelor, G. K., "Note on a Class of Solutions of Navier–Stokes Equations Representing Steady Rotationally-Symmetric Flow," Q. J. Mech. Appl. Maths., 1951, Vol. 4, pp. 29–41.

59. Szeri, A. Z., Schneider, S. J., Labbe, F., and Kaufmann, H. N., "Flow Between Rotating Disks, Part 1: Basic Flow," J. Fluid Mech., 1983, Vol. 134, pp. 103–131.

60. Sirivat, A., Rajagopal, K. R., and Szeri, A. Z., "An Experimental Investigation of the Flow of Non-Newtonian Fluids between Rotating Disks," J. Fluid Mech., 1988, Vol. 186, pp. 243–256.

61. Joseph, D. D., Nguyen, K., and Beavors, G. S., "Non-Uniqueness and Stability of the Configuration of Flow of Immiscible Fluids with Different Viscosities," J. Fluid Mech., 1984, Vol. 141, pp. 319–345.

62. O'Connor, J. J., and Boyd, J., *Standard Handbook of Lubrication Engineering*, 1968, McGraw-Hill, New York, NY.

63. Winer, W. O., and Cheng, H. S., "Film Thickness Contact Stress and Surface Temperatures," Wear Control Handbook, ASME, 1980, pp. 81–141.

64. Dowson, D., and Hudson, J. D., "Thermohydrodynamic Analysis of the Infinite Slider Bearing: Part 1, The Plane Inclined Slider Bearing," Proc. Inst. Mech. Engrs, 1963a, pp. 34–44.

65. Dowson, D., and Hudson, J. D., "Thermohydrodynamic Analysis of the Infinite Slider Bearing: Part 2: The Parallel Surface Bearing," Proc. Inst. Mech. Engrs, 1963a, pp. 45–51.

66. Ferron, J., Frene, J., and Boncompain, R., "A Study of the Thermal Hydrodynamic Performance of a Plain Journal Bearing, Comparison Between Theory and Experiments," ASME J. Lubr. Technol., Vol. 105, pp. 422–428.

67. Wang, N. Z., and Seireg, A., "Thermohydrodynamic Lubrication Analysis Incorporating Thermal Expansion Across the Film," (presented at the STLE/ASME Tribology Conference, Oct. 24–27, 1993), Trans. ASME, J. Tribol., Oct. 1994.

68. Wang, N. Z., and Seireg, A., "Empirical Prediction of the Shear Layer Thickness in Lubricating Films," J. Tribol., 1995, Vol. 117, pp. 444–449.

69. Bair, S., Qureshi, F., and Khonsari, M., "Adiabatic Shear Localization in a Liquid Lubricant under Pressure," Trans. ASME, J. Tribol., Oct. 1994, Vol. 116, pp. 705–709.

# 7

# Friction and Lubrication in Rolling/Sliding Contacts

## 7.1 ROLLING FRICTION

The frictional resistance to rolling in dry conditions was extensively investigated by Palmgren [1] and Tabor [2], who concluded that slip is negligible and cannot be considered as the mechanism causing rolling friction. Tabor suggested that rolling friction is a manifestation of the energy loss due to hysteresis in the stressed material at the contact zone during the rolling motion under normal load.

It is difficult to set down quantitative laws of dry rolling friction analogous to those of sliding friction because each of the mechanisms enumerated above has its own, quite different character, and the overall coefficient of friction will depend on which components of the rolling friction force are the most important for the particular system under consideration. Rabinowicz [3] generalized the laws of rolling friction as follows:

1.  The friction force varies as some power of the load, ranging from 1.2 to 2.4. For lightly loaded systems, where the deformation at the contact is primarily elastic, the friction force generally varies as a low power of the load. For heavily loaded systems, where plastic deformation occurs in the contact area, the friction force varies as a higher power of the load.
2.  The friction force varies inversely with the radius of curvature of the rolling elements.
3.  The friction force is lower for smoother surfaces than for rougher surfaces.

4.  The static friction force is generally much greater than the kinetic, but the kinetic is slightly dependent on the rolling velocity and generally drops off somewhat as the rolling velocity is increased.

## 7.2 HYDRODYNAMIC LUBRICATION AND FRICTION

Assuming rigid cylinders (Fig. 7.1), and isoviscous film subjected to a normal load $P_y$ per unit length, the minimum film thickness can be obtained from the solution of Reynolds' equation (Eq. (6.1)) as:

$$\frac{h_0}{R_e} = 4.9 \frac{\eta U}{P_y} \tag{7.1}$$

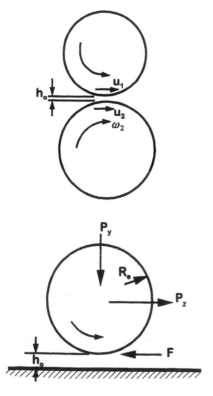

**Figure 7.1** Rolling/sliding contacts.

where

$P_y$ = load intensity (lb/in.)

$\eta$ = viscosity (reyn)

$U = \dfrac{U_1 + U_2}{2}$ = rolling velocity (in./sec)

$R_e$ = effective radius = $\dfrac{1}{\left(\dfrac{1}{R_1} + \dfrac{1}{R_2}\right)}$

If the viscosity change with pressure is taken into consideration, i.e.:

$$\eta = \eta_0 e^{\alpha P}$$

where

$\eta_0$ = viscosity at atmospheric pressure

$\alpha$ = viscosity pressure coefficient

$P$ = pressure

The limit value of the minimum film thickness for $P \to \infty$ can be given by:

$$\left(\frac{h_0}{R_e}\right) \approx 2.3\left(4.0\,\frac{\eta U}{P_y}\right) \tag{7.2}$$

A good approximation for the coefficient of friction can be given as [4]:

$$f = 0.5\frac{1 + 1.546Z}{\left(1 + 0.206\,\dfrac{R_e}{h_0}\right)^{1/2}} \tag{7.3}$$

where

$$Z = \frac{U_1 - U_2}{U_1 + U_2}$$

# 7.3 ELASTOHYDRODYNAMICS IN ROLLING/SLIDING CONTACTS

Hydrodynamic considerations alone as expressed in Eqs (7.1) and (7.2) do not adequately explain the significantly higher film thickness experienced in practice in many mechanical elements such as rolling element bearings, gears and cams. Grubin, in 1949 [5], postulated that this large discrepancy can be attributed to the assumption of rigid cylinders in situations where the applied loads and, consequently, the pressures are of such high magnitude to cause significant deformation in the material. He assumed a Hertzian-type flat area to occur in the contact region, which would significantly change the geometry of the film used in the Reynolds equation. This led to very significant increase in the predicted film thickness and consequently initiated the important area of tribology called elastohydrodynamic (EHD) lubrication.

The formula given for calculating the minimum film thickness in the EHD regime is:

$$\frac{h_0}{R_e} = \frac{1.95(\eta_0 U \alpha)^{8/11} E_e^{1/11}}{R_e^{7/11} P_y^{1/11}} \tag{7.4}$$

where

$\eta_0 =$ viscosity of the lubricant

$\alpha =$ pressure–viscosity coefficient

$R_e =$ effective radius $= \dfrac{1}{\left(\dfrac{1}{R_1} + \dfrac{1}{R_2}\right)}$

$U =$ rolling velocity $= \dfrac{U_1 + U_2}{2}$

$E_e =$ effective modulus of elasticity

$\quad = \dfrac{1}{\dfrac{1}{2}\left(\dfrac{1-\sigma_1^2}{E_1} + \dfrac{1-\sigma_2^2}{E_2}\right)}$

$P_y =$ load per unit length

$\sigma_1, \sigma_2 =$ Poisson's ratio

To illustrate the considerable difference in the results obtained by Eqs (7.1) and (7.4), the following example is assumed for heavily loaded steel rollers, lubricated with mineral oil:

$E_e = 33 \times 10^6\,\text{psi}$

$R_e = 1\,\text{in.}$

$\alpha = 10.9 \times 10^{-6}\,\text{reyn}$

$U = 200\,\text{in./sec}$

$P_y = 15,000\,\text{lb/in.}$

$\dfrac{h_0 \text{ (from equation (7.4))}}{h_0 \text{ (from equation (7.1))}} \approx 104$

which suggests that the elasticity of the rollers causes the minimum film thickness to increase by approximately 100 times.

Dowson and coworkers [4, 6] approached the problem from first principles and simultaneously solved the elasticity and the Reynolds equations. Their formula for the minimum film thickness is given in a dimensionless form as:

$$H = 1.6 \frac{G^{0.6}\overline{U}^{0.7}}{W^{0.13}} \qquad (7.5)$$

where

$$H = \frac{h}{R_e}$$

$$\overline{U} = \text{speed parameters} = \frac{\eta_0 U}{E_e R_e}$$

$$G = \text{material parameter} = \alpha E_e$$

$$W = \text{load parameter} = \frac{P_y}{E_e R}$$

Using the same dimensionless groups suggested by Dowson and Higginson [4], the Grubin solution can be given as:

$$H = 1.95 \frac{(G\overline{U})^{0.73}}{W^{0.091}} \qquad (7.6)$$

What is particularly significant in the EHD theory is the very low dependency of the minimum film thickness on load. The important parameters influencing the generation of the film are the rolling speed, the effective radius of curvature and the oil viscosity. Consequently, Dowson and Higginson suggested the following simplified formula for practical use:

$$h_0 = 5(\eta_0 U R_e)^{1/2} \qquad (7.7)$$

where

$h_0$ = minimum film thickness (in.)
$\eta_0$ = inlet oil viscosity (poise)
$R_e$ = effective radius (in.)
$U$ = rolling speed (in./sec)

## 7.4  FRICTION IN THE ELASTOHYDRODYNAMIC REGIME

The EHD lubrication theory developed over the last 50 years has been remarkably successful in explaining the many features of the behavior of heavily loaded lubricated contacts. However, the prediction of the coefficient of friction is still one of the most difficult problems in this field. Much experimental work has been done [7–21], and many empirical formulas have also been proposed based on the conducted experimental results.

Plint investigated the traction in EHD contacts by using three two-roller machines and a hydrocarbon-based lubricant [14]. He found that roller surface temperature has a considerable effect on the coefficient of friction in the high-slip region (thermal regime). As the roller temperature increases the coefficient of friction falls linearly until a knee is reached. With further increase in temperature the coefficient of friction rises abruptly and erratically and scuffing of the roller surface occurs. He also gave the following equation to correlate all the experimental results, which was obtained from 28 distinct series of tests:

$$f = 0.0335 \log\left(\frac{21,300}{\theta_c + 40}\right) - 44.5b^3 \tag{7.8}$$

where $\theta_c$ is the temperature on the central plane of the contact zone (°C) and $b$ is the radius of the contact zone (inches).

Dyson [15] considered a Newtonian liquid and derived the expression for maximum coefficient of friction as:

$$f_{\max} = \frac{0.66\alpha}{h_0}\left(\frac{8K\eta_0}{\gamma}\right)^{0.5}\frac{e^{0.5\alpha P}}{\alpha} \tag{7.9}$$

where

$\alpha$ = pressure–viscosity coefficient
$K$ = heat conductivity
$P$ = pressure
$\eta_0$ = dynamic viscosity
$h_0$ = minimum oil film thickness
$\gamma$ = temperature–viscosity coefficient

If $\alpha P \gg 1$, the coefficient of friction increases rapidly with pressure.

Sasaki et al. [16] conducted an experimental study with a roller test apparatus. The empirical formula of the friction coefficient $f$ in the region of semifluid lubrication as derived from the tests is given as:

$$f = k \left( \frac{\eta U^4}{w^{1.7}} \right)^{-0.12} \tag{7.10}$$

where

$\eta$ = lubricant dynamic viscosity
$U$ = rolling velocity
$w$ = load per unit width
$k$ = function of the slide/roll ratio

When slide/roll ratio $= 0.31$, $k = 0.037$; when slide/roll ratio $= 1.22$, $k = 0.026$.

Drozdov and Gavrikov [17] investigated friction and scoring under conditions of simultaneous rolling and sliding with a roller test machine. The formula for determination of $f$ at heavy contact loads from more than 10,000 experiments is found to be:

$$f = \frac{1}{0.8 v_0^{0.5} + V_t \varphi(P_{max}, v_0) + 13.4} \tag{7.11}$$

where

$\varphi(P_{max}, v_0) = 0.47 - 0.12 \times 10^{-4} P_{max} - 0.4 \times 10^{-3} v_0$
$\quad v_0$ = kinematic viscosity of the lubricant (cSt) at the mean surface temperature $(T_0)$ and atmospheric pressure
$\quad V_t$ = sum rolling velocity (sum of the two contact surface velocities, m/sec)
$P_{max}$ = maximum contact pressure (kg/cm$^2$)

O'Donoghue and Cameron [18] studied the friction in rolling sliding contacts with an Amsler machine and found that the empirical relation relating friction coefficient with speed, load, viscosity, and surface roughness could be expressed as:

$$f = \frac{S + 22}{35} \frac{0.6}{\eta^{1/8} V_s^{1/3} V_t^{1/6} R^{1/2}} \tag{7.12}$$

where

$S$ = total initial disk surface roughness (µin. CLA)
$V_s$ = sliding velocity (difference of the two contact surface velocities) (in./sec)
$V_t$ = sum rolling velocity (in./sec)
$\eta$ = dynamic viscosity (centipoises)
$R$ = effective radius (in.)

Benedict and Kelley [19] conducted experiments to investigate the friction in rolling/sliding contacts. The coefficient of friction has been found to increase with increasing load and to decrease with increasing sum velocity, sliding velocity, and oil viscosity when these quantities are varied individually. The viscosity was determined at the temperature of the oil entering the contact zone. The results are combined in a formula, which closely represents the data as below:

$$f = 0.0099 \, \frac{1}{1 - \dfrac{S}{45}} \, \log\left(\frac{3.5 \times 10^8 \, W}{\eta_0 \, V_s \, V_t^2 (R_1 + R_2)^2}\right) \tag{7.13}$$

where

$R$ = effective radius (in.)
$S$ = surface roughness (µin. rms)
$V_s$ = sliding velocity (in./sec)
$V_t$ = sum rolling velocity (in./sec)
$W$ = load per unit width (lb/in.)
$\eta_0$ = dynamic viscosity (cP)

The limiting value of $S$ is 30 µin.
    Misharin [20] also studied the friction coefficient and derived the formula:

$$f = \frac{0.325}{(V_s V_t v_0)^{1/4}} \tag{7.14}$$

where

$V_s$ = sliding velocity (m/sec)
$V_t$ = sum rolling velocity (m/sec)
$v_0$ = kinematic viscosity (cSt)

The limiting values are:

$$R: \text{nonsignificant deviation from 1.8 cm}$$
$$\text{slide/roll ratio: } 0.4\text{–}1.3$$
$$\text{contact stress} \geq 2500\,\text{kg/cm}^2$$
$$0.08 \geq f \geq 0.02$$

The accuracy of this empirical formula is reported to be within 15%.

Ku et al. [21] conducted sliding–rolling disk scuffing tests over a wide range of sliding and sum velocities, using a straight mineral oil and three aviation gas turbine synthetic oils in combination with two carburized steels and a nitrided steel. It is shown that the disk friction coefficient is dependent not only on the oil–metal combination, but also on the disk surface treatment and topography as well as the operating conditions. The quasisteady disk surface temperature and the mean conjection–inlet oil temperature are shown to be strongly influenced by the friction power loss at the contact, but not by the specific make-up of the frictional power loss. They are also influenced by the heat transfer from the disk, mainly by convection to the oil and conduction through the shafts, which are dependent on system design and oil flow rate.

For AISI 9310 steel:

$$f = \frac{0.0666}{V_t^{0.5}} + \frac{130}{W\,V_s^{0.6} + 1965} + 0.0009 + 0.0003S \tag{7.15}$$

For AMS 6475 steel:

$$f = \frac{0.0666}{V_t^{0.5}} + \frac{130}{W\,V_s^{0.6} + 1965} - 0.0041 + 0.0003S \tag{7.16}$$

where

$V_t$ = sum rolling velocity (m/sec)
$V_s$ = sliding velocity (m/sec)
$W$ = load (kN)
$S$ = surface roughness (μm CLA)

## 7.5 DOMAINS OF FRICTION IN EHD ROLLING/SLIDING CONTACTS

The coefficient of friction for different slide-to-roll ratio $z$ has three regions of interest as interpreted by Dyson [15]. As illustrated in Fig. 7.2, the first region is the isothermal region in which the shear rate is small and the amount of heat generated is so small as to be negligible. In this region, the lubricant behavior is similar to a Newtonian fluid. The second region is called the nonlinear region where the lubricant is subjected to larger strain rates. The coefficient of friction curve starts to deviate significantly from the Newtonian curve and a maximum coefficient of friction is obtained, after which the coefficient of friction decreases with sliding speed. Thermal effects do not provide an adequate explanation in this region because the observed frictional traction may be several orders of magnitude lower than the calculated values even when temperature effects are considered. The third region is the thermal region. The coefficient of friction decreases with increasing sliding speed and significant increase occurs in the temperature of the lubricant and the surfaces at the exit of the contact.

Almost all the empirical formulas discussed in the previous section are for the thermal regime. Each formula shows good correlation with the test data from which it was derived, as illustrated in Fig. 7.3, but generally none of these formulas correlates well with the others, as shown in Fig. 7.4. This suggests that these formulas are limited in their range of application and that a unified empirical formula remains to be developed.

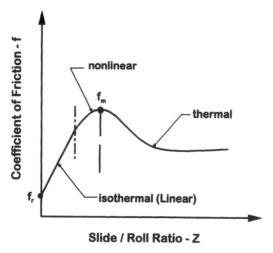

**Figure 7.2**  Friction in rolling/sliding contacts.

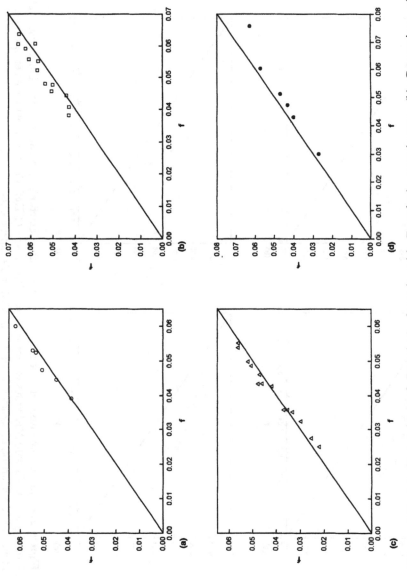

**Figure 7.3** (a) Comparison of Drozdov's formula with Drozdov's experiments. (b) Comparison of Cameron's formula with Cameron's experiments. (c) Comparison of Kelley's formula with Kelley's experiments. (d) Comparison of Misharin's formula with Misharin's experiments.

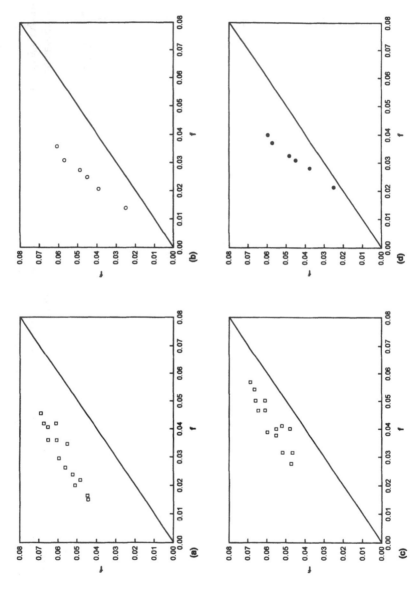

**Figure 7.4** (a) Comparison of Drozdov's formula with Cameron's experiments. (b) Comparison of Drozdov's formula with Misharin's experiments. (c) Comparison of Kelley's formula with Cameron's experiments. (d) Comparison of Kelley's formula with Misharin's experiments.

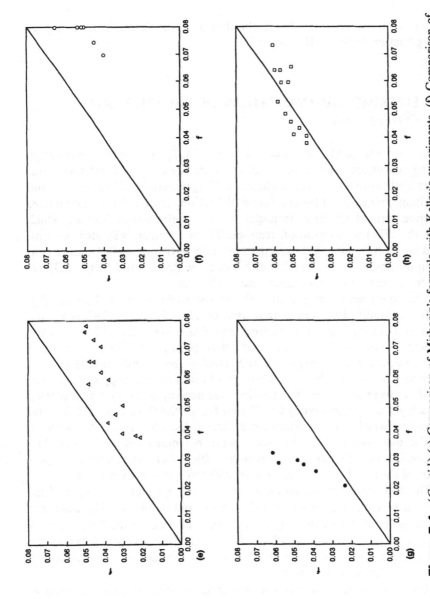

**Figure 7.4** (*Cont'd.*) (e) Comparison of Misharin's formula with Kelley's experiments. (f) Comparison of Misharin's formula with Drozdov's experiments. (g) Comparison of Cameron's formula with Misharin's experiments. (h) Comparison of Misharin's formula with Cameron's experiments.

No formulas are available in the literature for determination of pure rolling friction in the EHD regime.


## 7.6 EXPERIMENTAL EVALUATION OF THE FRICTIONAL COEFFICIENT

An experimental study was undertaken by Li [22] to simulate typical engineering conditions, and explore and evaluate the effects of different parameters such as loads, speeds, slide/roll ratios, materials, oil viscosities, and machining processes on the coefficient of friction. The results were then used to derive general empirical formulas for the coefficient of friction, which cover the different lubrication regimes. These formulas will also be compared with other published experimental data to further evaluate their general applicability. The formulas developed by Rashid and Seireg [23] are used to calculate the temperature rise in the film.

The experimental setup used in this study is schematically shown in Fig. 7.5. It is a modified version of that used by Hsue [24]. The shaft remained unchanged during the tests, whereas the disks were changed to provide different coated surfaces. The shaft was ground 4350 steel, diameter 61 mm, and the disks were ground 1020 steel, diameter 203.2 mm. The coating materials used for the disks were tin, chromium, and copper. Uncoated steel disks were also used. The coating was accomplished by electroplating with a layer of approximately 0.0127 mm for all the three coated disks, and the contact width was 3.175 mm for all the disks. The disk coated with tin and the one coated with chromium were machined before plating. The measured surface roughness is shown in Table 7.1 and the material properties are shown in Table 7.2. A total of 240 series of tests were run.

The disk assembly was mounted on two 1 in. ground steel shafts which could easily slide in four linear ball bearing pillow blocks. The load was applied to the disk assembly by an air bag. This limited the fluctuation of load caused by the vibration which may result from any unbalance in the disk. The frictional signal obtained from the torquemeter was relatively constant in the performed tests.

A variable speed transmission was used to adjust the rolling speed to any desired value. A toothed belt system guaranteed the accuracy of sliding–rolling ratios. This was particularly important for the rolling friction tests.

The lubricant used was 10W30 engine oil with a dynamic viscosity of 0.09 Pa-s at 26°C; the loads were 94,703, 189,406, 284,109, and 378,812 N/m; the slide/roll ratios were 0, 0.08, 0.154, 0.222, 0.345; the rolling speeds varied from 0.3 to 2.76 m/s, and the sliding speeds were in the range 0 to 0.95 m/s.

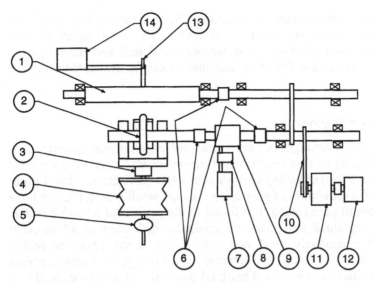

1. Shaft
2. Disk Assembly
3. Load Cell
4. Gas Bag
5. Air Meter
6. Couplings
7. Digital Oscilloscope
8. Amplifier
9. Torque Meter
10. Chains
11. Variable Speed Transmission
12. Motor
13. Oil Valve
14. Oil Container

**Figure 7.5** Experimental setup.

**Table 7.1** Surface Roughness Measurement

| Disk coating material | Surface roughness (μm AA) |
| --- | --- |
| Tin | 0.42 |
| Chromium | 0.38 |
| Copper | 0.17 |
| Steel | 0.20 |

**Table 7.2** Material Properties

| Material | $K$ (W/(m-K)) | $\rho$ (kg/m$^3$) | $C$ (J/(kg-K)) | $E$ (GPa) |
| --- | --- | --- | --- | --- |
| Steel | 43 | 7800 | 473 | 203.4 |
| Copper | 401 | 8930 | 386 | 103 |
| Chromium | 94 | 7135 | 450 | 250 |
| Tin | 67 | 7280 | 222 | 46 |
| 10W30 | 0.145 | 888 | 1880 | – |

The experimental results cover rolling friction, the isothermal regime, the nonlinear regime, and the thermal regime. The variables in the tests include load, speed, slide/roll ratio, surface roughness, and the properties of the coated layer. The following conclusions can be drawn from the test results.

### 7.6.1 Friction Regimes

Although many investigators have conducted experimental investigations on the coefficient of friction, no experimental results have been reported in the literature for the rolling friction with EHD lubrication. This is probably due to the difficulties of measuring the very small rolling friction force to be expected in pure rolling. It is found in the performed tests that rolling friction is very small and increases gradually with load in all cases. It decreases at a relatively rapid rate with rolling speed when the rolling speed is small ($< 1.5\,\text{m/s}$), then decreases at a lower rate at higher rolling speeds. The effects of the coated material properties and surface roughness on rolling friction appear to be insignificant for all the performed tests. Figure 7.6 shows the experimentally determined variations of rolling friction with load and rolling speed.

In the isothermal regime, it is expected that the surface roughness, the modulus of elasticity of the coated and base materials, and the thickness of the coated layers play an important role. On the other hand, the material thermal properties do not appear to have significant influence. Coating layers of soft materials are found to give a higher coefficient of friction. The surface roughness also increases friction. The coefficient of friction is also found to increase with load and decrease with rolling speed. Figures 7.7–7.10 show the variation of coefficient of friction with slide/roll ratio. It can be seen from these figures that the coefficient of friction for steel and copper coating reaches its maximum in the nonlinear regime. For chromium and tin coatings, the coefficient of friction continues to increase, but at much slower rate than in the isothermal and the nonlinear regimes. The magnitude and position of the maximum value of the coefficient of friction are influenced by the surface roughness, material physical properties, load, speed, and viscosity.

In the thermal regime the coefficient of friction is found to decrease slightly with the slide/roll ratio. The thermal properties of the coated and the base materials are found to have significant effect on the coefficient of friction as would be expected. The surface with a high diffusivity $K/(\rho C)$ usually produces a lower coefficient of friction because the surface contact temperature rise is lower, and consequently, the actual oil viscosity is higher, which produces a better lubrication condition. Rough surfaces give higher coefficient friction as in the isothermal and nonlinear regimes. However, the

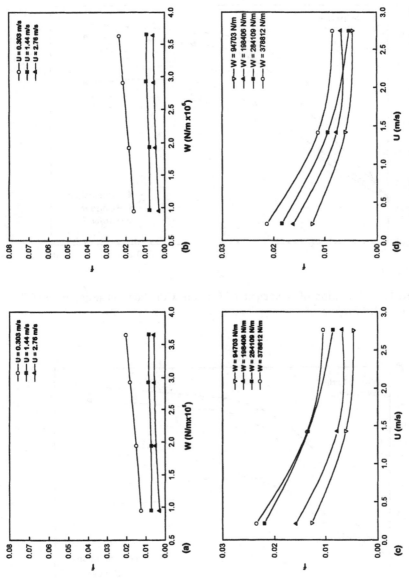

**Figure 7.6** Variation of coefficients of friction: (a) with load, chromium; (b) with load, copper; (c) with rolling speed, tin; (d) with rolling speed, steel.

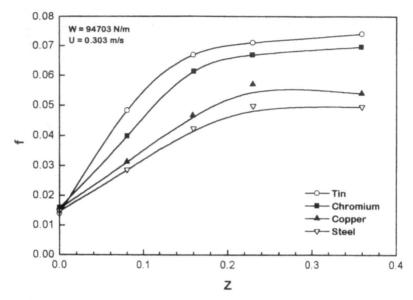

**Figure 7.7** Variation of coefficient of friction with slide/roll ratio; $W = 94,703$ N/m, $U = 0.303$ m/sec.

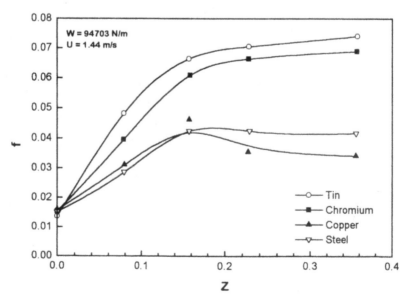

**Figure 7.8** Variation of coefficient of friction with slide/roll ratio; $W = 94,703$ N/m, $U = 1.44$ m/sec.

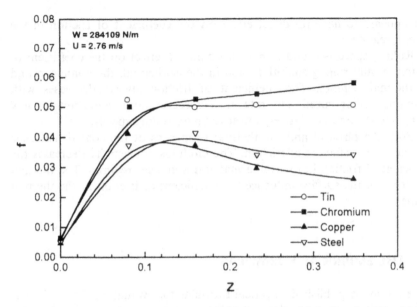

**Figure 7.9** Variation of coefficient of friction with slide/roll ratio; $W = 284,109$ N/m, $U = 2.76$ m/sec.

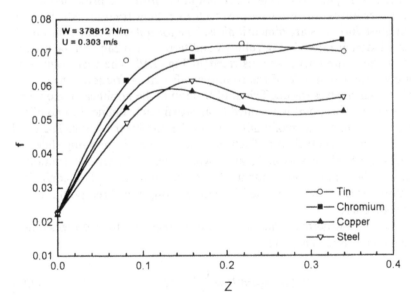

**Figure 7.10** Variation of coefficient of friction with slide/roll ratio; $W = 378,812$ N/m, $U = 0.303$ m/sec.

load appears to have no direct effect on the coefficient of friction in the thermal regime.

Rolling speed is found to have a significant effect on the coefficient of friction in pure rolling conditions and in the isothermal, the nonlinear, and the thermal regimes. The coefficient of friction always decreases with increasing rolling speed. The rate of decrease is more significant for low rolling speeds, and is relatively lower for high rolling speeds.

Both the physical and the thermal properties of the coated materials influence the coefficient of friction. The modulus of elasticity decreases the coefficient of friction in the isothermal and nonlinear regimes. The thermal properties of the surface influence the coefficient of friction in the thermal regime.

## 7.7  THE EMPIRICAL FORMULAS

There are many published empirical formulas for evaluating the coefficient of friction. They were developed by different investigators under different experimental conditions, and therefore, it it no surprise that they do not correlate with each other. All of these formulas are developed from test data in the thermal regime. The generalized empirical formulas presented in this section cover all the three regimes, as well as rolling friction. All the variables in these formulas are dimensionless. The formulas calculate the coefficient of friction at three sliding/rolling conditions which can then be used to construct the entire curve, as illustrated in Fig. 7.10. The first point is $f_r$, which gives the magnitude of the rolling coefficient of friction. The second point is $f_n$, which gives the coefficient of friction in the nonlinear region, and $z^*$, its location. This point is assumed to approximately define the end of the isothermal region or the maximum value in the nonlinear regime. The third one is the thermal coefficient of friction, $f_t$, and the corresponding slide/roll ratio location is chosen as 0.27, after which the coefficient of friction is assumed to be almost independent of the slide/roll ratio. The coefficient of friction curve is then presented by curve fitting the three points by an appropriate curve.

1.   In the isothermal and the nonlinear regimes, four dimensionless parameters are used. They are:

$$\text{Rolling speed } \underline{U} = \frac{U^2 \rho}{E'} \times 10^{10} \tag{7.17}$$

$$\text{Viscosity } \underline{\eta} = \frac{\eta^2}{E' R^2 \rho} \times 10^{11} \tag{7.18}$$

$$\text{Load } \underline{W} = \frac{W}{E'R} \times 10^5 \qquad (7.19)$$

$$\text{Surface roughness } \underline{S} = \frac{S_{ec}}{R} \times 10^6 \qquad (7.20)$$

$S_{ec}$ is calculated according to Eq. (7.25), and $\rho = 0.865$ (which is an approximate value for most lubricating oils used in test conditions). All the other variables are defined in the following notation:

$U = \text{rolling speed} = \dfrac{U_1 + U_2}{2}$

$U_1, U_2 = \text{rolling speeds of rollers 1, 2}$

$R = \text{effective radius} = \dfrac{R_1 R_2}{R_1 + R_2}$

$R_1, R_2 = \text{radii of rollers 1, 2}$

$E' = \text{effective modulus of elasticity} = \dfrac{1}{\dfrac{1}{2}\left(\dfrac{1 - v_1^2}{E_1} + \dfrac{1 - v_2^2}{E_2}\right)}$

$E_1, E_2 = \text{elastic modulii of solids in contact}$

$v_1, v_2 = \text{Poisson's ratio for solids in contact}$

$\eta = \text{dynamic viscosity of oil}$

$W = \text{load per unit length}$

2.  The coefficient of rolling friction is the value at which the sliding speed is equal to 0. It is found to be best fitted for the experimental data by the following equation:

$$f_r = \frac{0.00138}{0.05 + \underline{U}^{0.433}} \underline{W}^{0.367} \qquad (7.21)$$

3.  The transition coefficient of friction $f_n$ can be calculated from:

$$f_n = \alpha \underline{S}^{1/9} \underline{W}^\beta - \gamma \qquad (7.22)$$

where

$\alpha = 0.0191 - 1.15 \times 10^{-4} \sqrt{\underline{\eta}}$

$\beta = 0.265 + 6.573 \times 10^{-3} \underline{\eta}$

$\gamma = (7.778 \times 10^{-3} + 1.778 \times 10^{-3} e^{-\underline{U}/0.0141}) \dfrac{\ln(1.336 \underline{U})}{\sqrt{\underline{W} + 1}}$

and its location $z^*$ is calculated from:

$$z^* = \frac{\alpha'}{1 + \underline{W}^{\beta'}} \, e^{-\sqrt{\underline{U}}/5.5} (2 + e^{-\underline{U}/0.00474}) \, \frac{1}{1 + 4.4 \times 10^{-5} \underline{S}^{2.7}} \qquad (7.23)$$

where

$$\alpha' = 0.219(1 - e^{-\underline{\eta}\sqrt{\underline{S}}/6.368}) + 0.0122$$

$$\beta' = 0.344(1 - e^{-\underline{\eta}\sqrt{\underline{S}}/4.472}) + \frac{0.922}{\sqrt{\underline{S}}}$$

    4.   In the thermal regime, where slide/roll > 0.27:

$$f = f_0 - [a(1 - e^b)] \qquad (7.24)$$

where

$$f_0 = \text{coefficient of friction at } h_0 = 0, \text{ from Fig. 7.11}$$

$$a = 0.0864 - 1.372 \times 10^3 \left( \frac{S_{ec}}{R} \right)_e$$

$$b = -\frac{\dfrac{h_0}{R}}{2.873 \left( \dfrac{S_{ec}}{R} \right)_e - 2.143 \times 10^{-5}} \qquad (7.25)$$

$$S_{ec} = \sqrt{S_e^2 + S_e^2}$$

where

$S_e$ = effective surface roughness, from Fig. 7.12; for $S < 0.05\,\mu\text{m}$
    take $S_e = 0.05\,\mu\text{m}$

$\left( \dfrac{S_{ec}}{R} \right)_e$ = effective surface roughness ratio, from Fig. 7.13

$R$ = effective radius

$h_0$ = oil film thickness calculated by the well-known Dowson–Higginson
    formula:

$$h_0 = 2.65 \, \frac{G^{0.54} \eta_0^{0.7} U^{0.7}}{E'^{0.57} W^{0.13}} \, R^{0.43} \qquad (7.26)$$

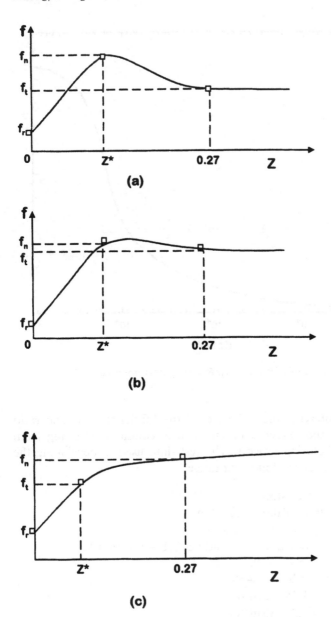

**Figure 7.11** Possibilities for construction of the empirical curves from the calculated three points.

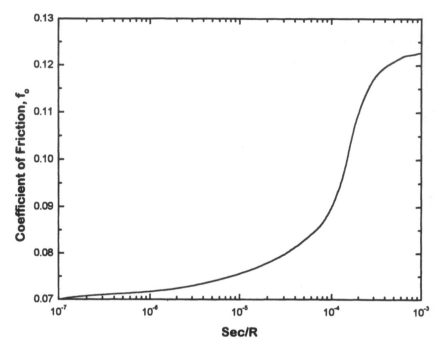

**Figure 7.12**   Coefficient of friction at $h_0/R = 0$ against nominal $S_{ec}/R$.

The ratio $h_0/R$ represents the influence of the lubricant film. The ratio $(S_{ec}/R)_e$ represents the influence of the surface condition resulting from a particular manufacturing process. The test data used in developing the proposed formula cover the following range:

contact surfaces: steel–steel
effective radius $R = 0.0109 - 0.0274$ m
lubricant viscosity $\eta = 2.65 - 2000$ cP
surface processing operation = grinding $(0.1 - 1.6\,\mu m\,AA)$
film/surface roughness $\lambda = 0.21 - 14.31$
slide/roll ratio $z = 0.268 - 0.455$
sliding speed $V_s = 1.35 - 5$ m/sec
rolling speed $U_t = 3.2 - 15$ m/sec
material = EN32 steel cast hardened to 750 VPN to a depth of 0.025 in.
      [17]
SAE 8622 carburized and hardened to Rockwell hardness 60 [18]
Steel 38XMI–OA [19]

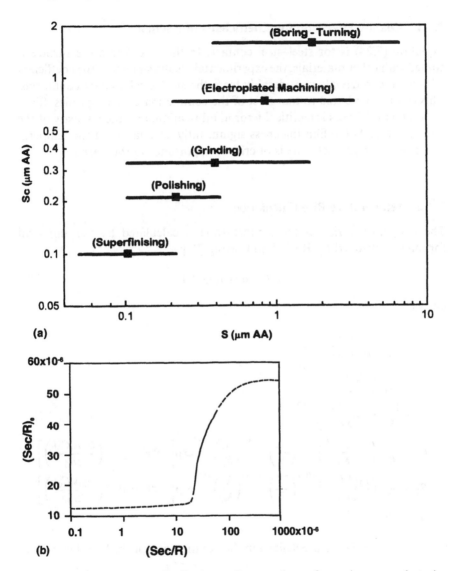

**Figure 7.13** (a) Proposed effective surface roughness for various manufacturing processes. (b) Effective $(S_{ec}/R)_e$ against nominal $S_{ec}/R$.

Grade 12X2H4A steel carburized to a depth of 1–1.5 mm and heat-treated to a Rockwell hardness of 58–60 [20]

load $W = 1.54 \times 10^5 - 20.3 \times 10^5 \, \text{N/m}$

maximum contact stress $\sigma_{\text{max}} = 6724 - 16{,}825 \, \text{kg/cm}^2$

### 7.7.1   Coating Effects on the Coefficients of Friction

Equation (7.24) is for steel–steel contact. In the case where the surface is coated with other materials, the experimental results show that the coefficient of friction can deviate considerably from the steel–steel contact conditions. This can be attributed to the effect of the coating material properties. Since the oil film thickness is a critical fctor in lubrication, and the viscosity of the lubricant affects the film thickness significantly, evaluation of the temperature rise in the contact zone is of critical importance in this case.

### 7.7.2   Temperature Rise Calculation

The temperature rise in the contact zone is calculated by the empirical formulas developed by Rashid and Seireg [23]:

$$\Delta T = \alpha q_t (A_1 + A_2) \tag{7.27}$$

where

$$\alpha = \frac{A_2 + B_2}{A_1 + B_1 + A_2 + B_2}$$

$$A_1 = \frac{1.03}{K_1} \left( \frac{\rho_1 C_1 U_1 L}{K_1} \right)^{-0.5}$$

$$A_2 = \frac{1.03}{K_2} \left( \frac{\rho_2 C_2 U_2 L}{K_2} \right)^{-0.5}$$

$$B_1 = \frac{1.14}{K_0} \left( \frac{\rho_1 C_1 U_1 h}{2K_0} \right)^{0.013} \left( \frac{2L}{h} \right)^{-1.003} \left( \frac{K_1}{K_0} \right)^{0.013} \exp\left[ -900 \times 10^{-6} \left( \frac{U_1 \rho_0 C_0 h}{2K_0} \right) \right]$$

$$B_2 = \frac{1.14}{K_0} \left( \frac{\rho_2 C_2 U_2 h}{2K_0} \right)^{0.013} \left( \frac{2L}{h} \right)^{-1.003} \left( \frac{K_2}{K_0} \right)^{0.013} \exp\left[ -900 \times 10^{-6} \left( \frac{U_2 \rho_0 C_0 h}{2K_0} \right) \right]$$

$$q_t = |U_1 - U_2| Wf$$

where all the variables are defined in the notation except the film thickness $h$:

$$h = \varepsilon h_0$$

$\varepsilon$ is a factor proposed by Wilson and Sheu [25]:

$$\varepsilon = \frac{1}{1 + 0.241[(1 + 14.8\, z^{0.83}) \delta^{0.64}]} \tag{7.28}$$

where

$z$ = sliding/rolling ratio

$$\delta = \frac{\eta_0 \gamma U^2}{K}$$

$\eta_0$ = lubricant viscosity at the entry condition

$\gamma$ = temperature–viscosity coefficient of the lubricant

$U$ = mean rolling velocity

$K$ = heat conductivity of the lubricant

### 7.7.3 Coating Thickness Effects on Temperature Rise

It should be noted here that Eq. (7.27) is derived for the case when the two entire disks have homogeneous properties, i.e., $K_1, C_1, \rho_1, E_1$ for disk 1 and $k_2, C_2, \rho_2, E_2$ for disk 2. In order to use the formula to calculate the temperature rise for coating surfaces, the temperature penetration depth, $D$, is calculated and the result is plotted in Figs 7.14–7.16.

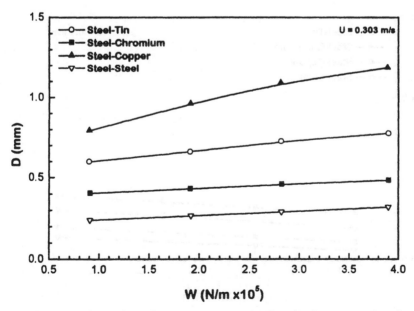

**Figure 7.14** Variation of temperature penetration depth on coated surface for load for $U = 0.303\,\text{m/s}$.

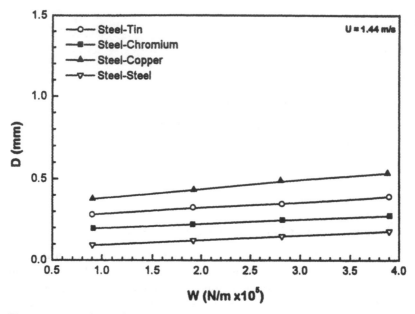

**Figure 7.15** Variation of temperature penetration depth on coated surface with load for $U = 1.44\,\text{m/s}$.

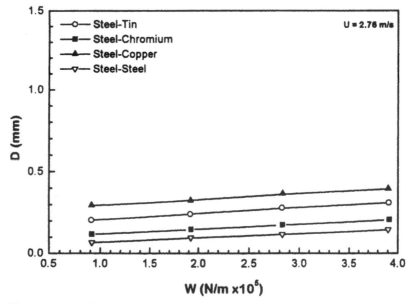

**Figure 7.16** Variation of temperature penetration depth on coated surface with load for $U = 2.76\,\text{m/s}$.

$$D = \sqrt{\frac{5KL}{\rho CU}} \qquad (7.29)$$

From Figs 7.14–7.16 it can be seen that the temperature penetration depth in all cases is much higher than the coating thickness (0.0127 mm). This means that both the coating material properties and the base material properties must be considered during applying Eq. (7.27). Therefore, a coating thickness factor $\beta$ is used to modify the temperature rise calculated with the coating material properties:

$$\beta = 1 - \exp\left(-\frac{h_C}{\left(\frac{D}{\tau}\right)}\right) \qquad (7.30)$$

where $h_C$ is the coating thickness, $D$ is the temperature penetration depth for steel under the corresponding conditions, $\tau$ is a constant with a value of 0.033. Figure 7.17 shows the variation of $\beta$ with the ratio of coating thickness to temperature penetration depth.

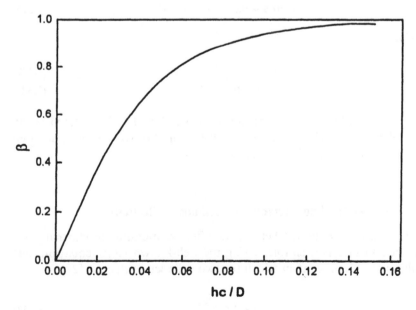

**Figure 7.17** Variation of $\beta$ with $h_c/D$.

### 7.7.4 Effective Viscosity

Using the notation:

$T_b$ = absolute bulk disk temperature (e.g., $T_b = 273.16 + °C$)
$\Delta T_S$ = temperature rise for steel–steel contact
$\Delta T_C$ = temperature rise from Eq. (7.27) using the material properties of the contacting surfaces for steel–coating contact
$\Delta T = \Delta T_C - \Delta T_S$ = temperature rise difference between the steel–coating contact and the steel–steel contact
$\Delta T_e$ = effective temperature rise difference between the steel–coating contact and the steel–steel contact

Then:

$$\Delta T_e = \Delta T \beta \tag{7.31}$$

where $\beta$ is the coating thickness factor from the previous section.

Then $T_e = T_b + \Delta T_e$ is used to calculate the viscosity for that coating conditions, and the viscosity is then substituted into Eq. (7.24) to calculate the corresponding coefficient of friction. The viscosity of 10W30 oil is calculated by the ASTM equation [27]:

$$\log(cS + 0.6) = a - b \log T_e \tag{7.32a}$$

therefore

$$\text{viscosity} = 10^\phi - 0.6 \tag{7.32b}$$

$$\phi = 10^{(a - b \log T_e)} \tag{7.32c}$$

where $T_e$ is the absolute temperature ($K$ or $R$), $cS$ is the kinematic viscosity (centistokes). $a = 7.827$. $b = 3.045$ for 10W30 oil. For some commonly used oil, $a$ and $b$ values are given in Table 7.3.

### 7.7.5 Coating Thickness Effects on Modulus of Elasticity

For the reasons mentioned before, the effective modulus of elasticity, $E_e$, for coated surface is desirable. Using the well-known Hertz equation, one calculates the Hertz contact width for two cylinder contact as [27]:

$$b = 1.6\sqrt{\frac{WR}{E'}} \tag{7.33}$$

**Table 7.3** Values of $a$ and $b$ for Some Commonly Used Lubricant Oils

| Oil | $a$ | $b$ |
|-----|-----|-----|
| SAE 10 | 11.768 | 4.6418 |
| SAE 20 | 11.583 | 4.5495 |
| SAE 30 | 11.355 | 4.4367 |
| SAE 40 | 11.398 | 4.4385 |
| SAE 50 | 10.431 | 4.0319 |
| SAE 60 | 10.303 | 3.9705 |
| SAE 70 | 10.293 | 3.9567 |

where

$$\frac{1}{E'} = \frac{1}{2}\left(\frac{1 - v_1^2}{E_1} + \frac{1 - v_2^2}{E_2}\right)$$

$E'$ and $v$ are the modulus of elasticity and Poisson's ratio.

Coating material properties are used for $E_2$ and $v_2$ because coating thickness is an order greater than the deformation depth (this can be seen later). Therefore, the deformation depth is calculated by (Fig. 7.18):

$$h_d = R\sin\theta\tan\theta$$

$\theta$ is very small, therefore:

$$h_d = R\theta\theta = R\,\frac{b}{R}\,\frac{b}{R} = \frac{b^2}{R} \tag{7.34}$$

The variation of the deformation depth with load is shown in Fig. 7.19. Then the effective modulus of elasticity of the coated surface is proposed as:

$$E_e = E_b + (E_c - E_b)\left(1 - \exp\left(-\frac{h_c}{\left(\frac{h_d}{\tau}\right)}\right)\right) \tag{7.35a}$$

where

$E_b$ = modulus of elasticity of base material
$E_c$ = modulus of elasticity of coating material
$E_e$ = modulus of elasticity of coated surface
$h_c$ = coating film thickness
$h_d$ = elastic deformation depth
$\tau$ = constant (it is found that $\tau = 13$ best fits the test data)

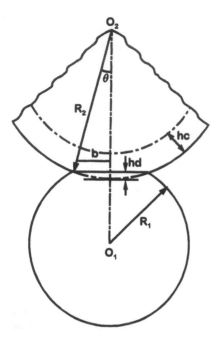

**Figure 7.18**   The contact of the shaft and the coated disk.

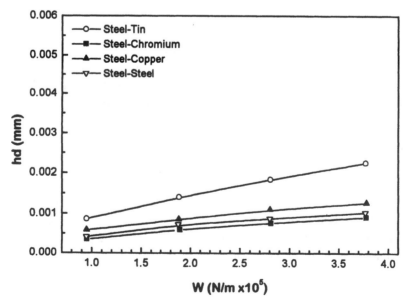

**Figure 7.19**   Variation of deformation depth on coated surface with load.

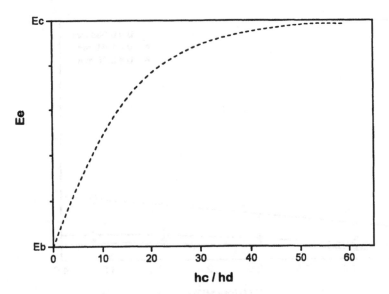

**Figure 7.20**   Variation of $E_e$ with $h_c/h_d$.

Figure 7.20 shows the variation of effective modulus of elasticity of coated surface for different values of $h_c/h_d$. The effective combined modulus of elasticity is therefore calculated by

$$\frac{1}{E'_{ec}} = \frac{1}{2}\left(\frac{1-v_1^2}{E_1} + \frac{1-v_e^2}{E_e}\right) \tag{7.35b}$$

For most metals used in engineering, the variation of $v$ is small, and consequently, the variation in $1 - v^2$ is smaller. Therefore, no significant error is expected from using the Poisson's ratio of the base material or the coating material.

Figures 7.21 and 7.22 show the comparison of the calculated coefficient of friction in pure rolling conditions with the test results for chromium and steel. Figure 7.23 shows the calculated coefficient of friction in the thermal regime compared with test results for tin, steel, chromium, and copper. Figure 7.24 shows the calculated coefficient of friction in the thermal regime compared with the results from Drozdov's [17], Cameron's [18], Kelley's [19], and Misharin's [20] experiments. Figures 7.25–7.28 show sample comparisons of the experimental results with the curves, which are constructed by using the calculated $f_r, f_n,$ and $f_t$, and appropriate curves against slide/roll ratios. Figure 7.29 shows the comparison of Plint's test data with prediction. It can be seen that the correlations are excellent.

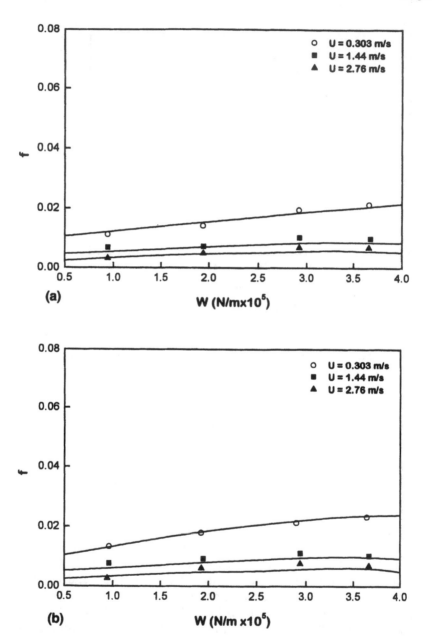

**Figure 7.21**  Comparison of experimental coefficient of rolling friction vs. load with prediction for (a) tin; (b) chromium; (c) copper; (d) steel.

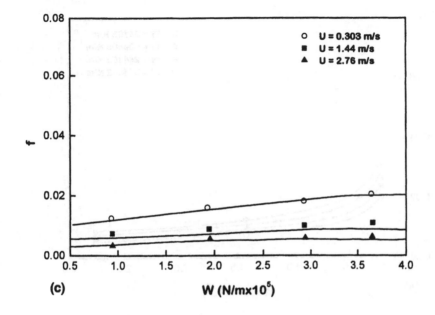

**(c)**

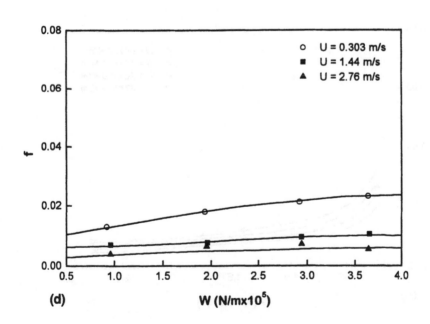

**(d)**

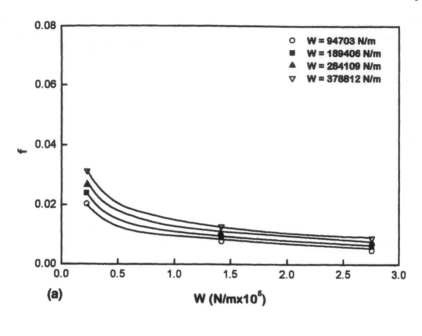

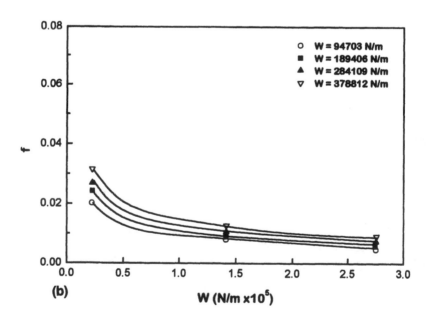

**Figure 7.22**  Comparison of experimental coefficient of rolling friction vs. rolling speed with prediction for (a) tin; (b) chromium; (c) copper; (d) steel.

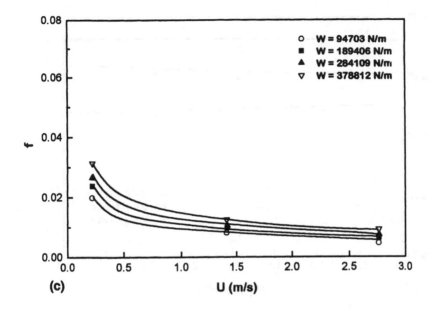

(c)

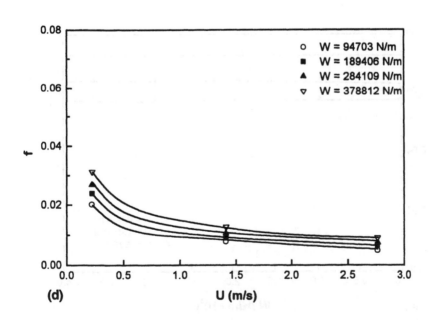

(d)

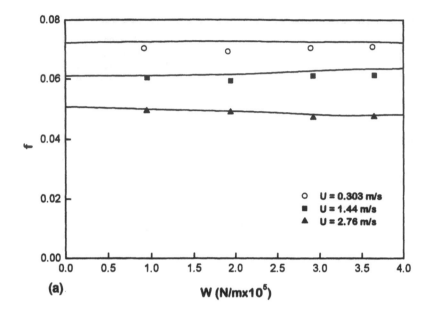

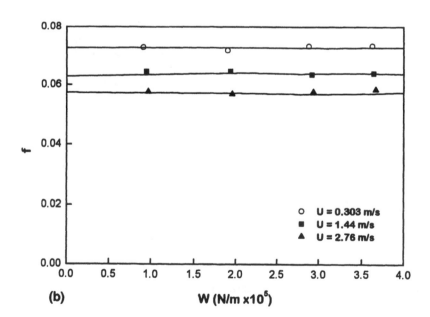

**Figure 7.23** Comparison of experimental coefficient of thermal friction vs. load with prediction for (a) tin; (b) chromium; (c) copper; (d) steel.

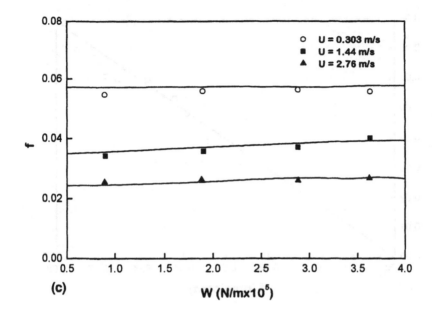

**(c)**

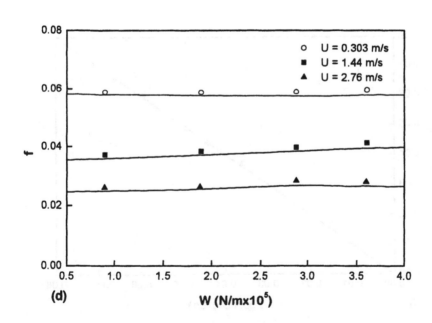

**(d)**

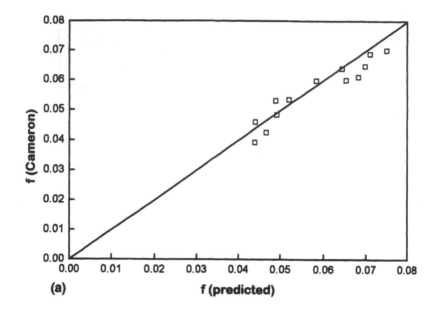

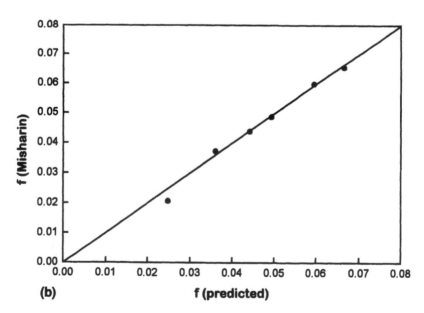

**Figure 7.24** Comparison of test data with prediction: (a) Cameron; (b) Misharin; (c) Kelley; (d) Drozdov.

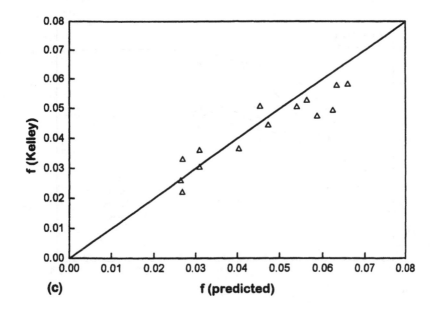

(c)

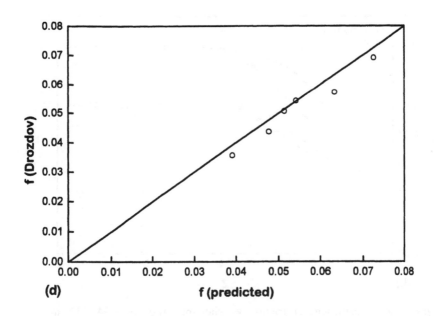

(d)

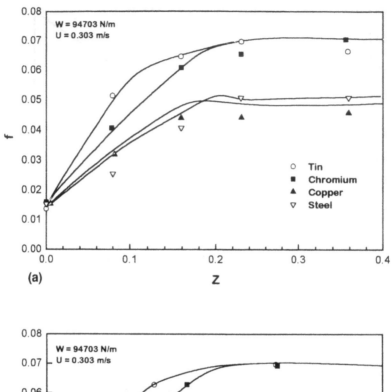

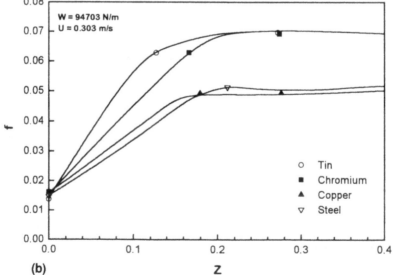

**Figure 7.25** Coefficient of friction vs. slide/roll ratio ($W = 94,703\,\text{N/m}$, $U = 0.303\,\text{m/sec}$): (a) experimental; (b) calculated.

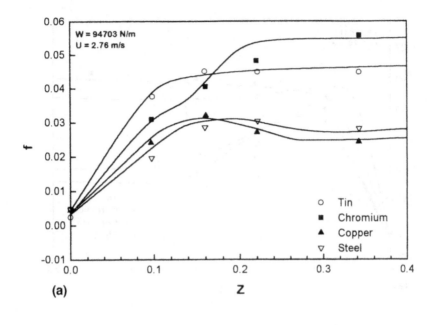

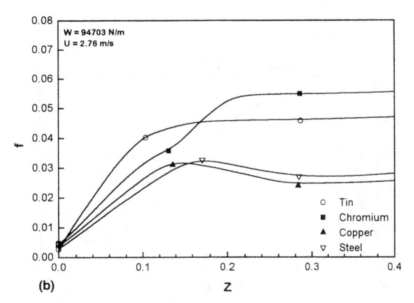

**Figure 7.26** Coefficient of friction vs. slide/roll ratio ($W = 94,703\,\text{N/m}$, $U = 0.303\,\text{m/sec}$): (a) experimental; (b) calculated.

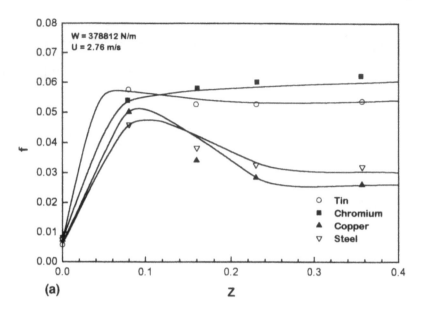

(a)

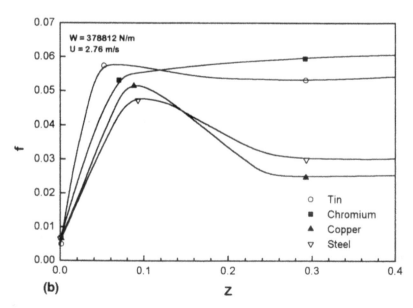

(b)

**Figure 7.27**  Coefficient of friction vs. slide/roll ratio ($W = 378,812\,\text{N/m}$, $U = 2.76\,\text{m/sec}$): (a) experimental; (b) calculated.

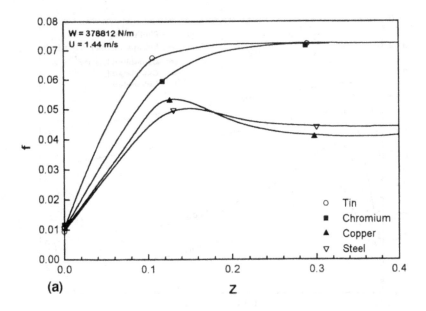

(a)

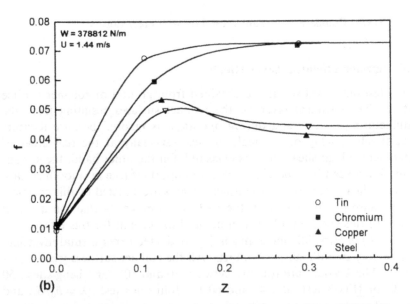

(b)

**Figure 7.28** Coefficient of friction vs. slide/roll ratio ($W = 378,812\,\text{N/m}$, $U = 1.44\,\text{m/sec}$): (a) experimental; (b) calculated.

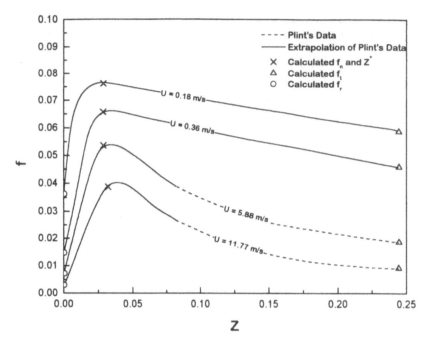

**Figure 7.29** Comparison of Plint's test data with prediction.

### 7.7.6 Surface Chemical Layer Effects

All the test data used so far are obtained from ground or rougher surface contacts. The chemical layer on the contact surfaces resulting from the manufacturing process and during operation is ignored because it wears off relatively quickly on a rough surface, especially at the real areas of contact where high shear stress is expected. On the other hand, the surface chemical layer can be expected to play an important role in smooth surface contacts. The properties of the contact surfaces are affected by this chemical layer because it can remain on the surfaces more easily than on a rough surface. In this case, Eq. (7.27) can be used to account for this effect.

Cheng [7], Hirst [8], and Johnson [9] conducted experimental investigations on very smooth surface contacts. All the contact surfaces were superfinished. The $\lambda$ values are roughly between 80 and 100 for Cheng's test, 50 and 60 for Hirst's test, and 40 and 60 for Johnson's test. ($\lambda = h_0/S_c$, and $S_c = \sqrt{S_1^2 + S_2^2}$, where $S_1$ and $S_2$ are surface roughness of contact surfaces, CLA.)

Because the surface chemical layers usually are very thin, they are assumed to have little effect on the elastic properties of the surfaces.

However, they affect the temperature rise in the contact zone significantly. In order to use Eq. (7.27) to account for this effect, the thermal–physical properties and thickness of the chemical layer are needed. Because these values are not known, the following thermal–physical properties are used as an approximation:

$$\rho = 3792\,\text{kg/m}^3$$
$$c = 840\,\text{J/(kg} - {}^\circ\text{C)}$$
$$E = 3.45 \times 10^{11}\,\text{Pa}$$
$$K = 0.15\,\text{W/(m} - {}^\circ\text{C)}$$

The $f_t$ values from the above tests are used to find the thickness of the corresponding chemical layers inversely. The result is shown in Fig. 7.30. It is found that the thickness decreases as the load increases as expected because the higher the load, the higher the shear stress in the lubricant, which results in a thinner chemical layer. In Figs 7.31–7.33, the test data are compared with prediction. It can be seen that the chemical layer makes a

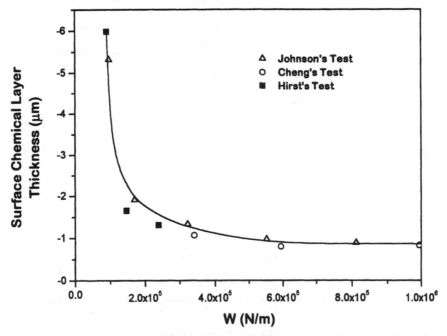

**Figure 7.30** Inversely calculated surface chemical layer thickness vs. load for superfinished surface contacts (roughness ≈ 1 μin. CLA).

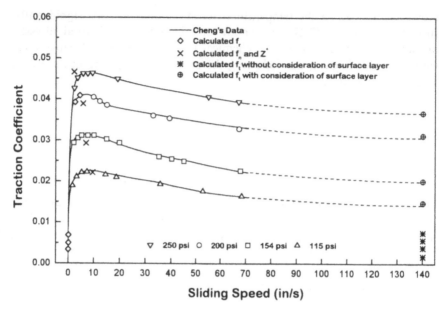

**Figure 7.31**   Comparison of Cheng's experimental data with prediction.

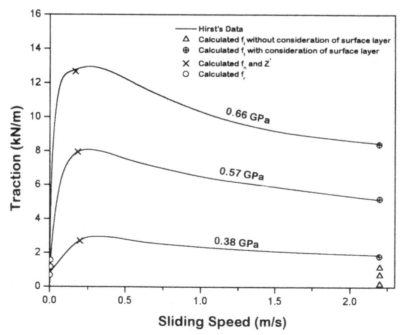

**Figure 7.32**   Comparison of Hirst's test data with prediction.

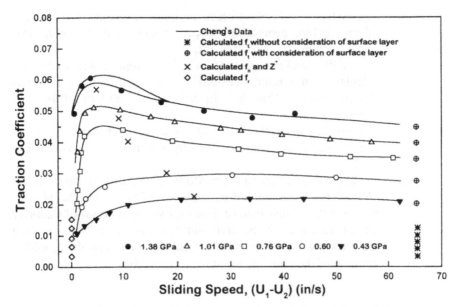

**Figure 7.33** Comparison of Johnson's experimental data with prediction.

great difference in the coefficient of friction for conditions with large $\lambda$ ratios ($>40$) in the thermal regime. Load can have a significant effect on the chemical layer thickness and Fig. 7.30 can be used for evaluating the thickness as a function of normal load.

### 7.7.7 General Observations on the Results

The empirical formulas were checked for different regimes of lubrication, surface roughness, load, speed, and surface coating. The formulas were used for evaluating rolling friction, and traction forces in the isothermal, non-linear, and thermal regimes of elastohydrodynamic lubrication. Because of the current interest in surface coating, the formulas were also applied for determining the coefficient of friction for cylinders with surface layers of any arbitrary thickness and physical and thermal properties.

It can be seen from the empirical formulas that:

1. It appears that in general, the slide/roll ratio has little direct effect on the coefficient of friction in the thermal region (slide/roll $>0.27$).

2.  The surface roughness effect is treated in this study as a function of the surface generating process rather than the traditional surface roughness measurements.
3.  The oil film thickness is found to be better represented for friction calculation in a nondimensional form by normalizing it to the effective radius rather than the commonly used film thickness to roughness ratio $\lambda$.
4.  Coating has a significant effect on the temperature rise in the contact zone. This is represented by a factor $\beta$, as shown in Eq. (7.30).
5.  Coating has an effect on the modulus of elasticity as shown in Eq. (7.35). This is represented by using an effective modulus of elasticity for the coated surface. For tin (whose modulus of elasticity differs from that of the base material most significantly among the three coating materials used), this correction produces a 50% increase in the effective modulus of elasticity.

## 7.8  PROCEDURES FOR CALCULATION OF THE COEFFICIENT OF FRICTION

### 7.8.1  Unlayered Steel–Steel Contact Surfaces

1.  Given: contact surface radii $r_1$, $r_2$ (m, in.)
    Surface velocities, $U_1$, $U_2$ (m/sec, in./sec)
    Dynamic viscosity of lubricant oil at entry condition $\eta_0$ (Pa-sec, reyn)
    Load $F$ (n, lbf)
    Surface roughness $S_1$, $S_2$ (m CLA, in. CLA) or manufacturing processes
    Density of lubricant $\rho$ (kg/m$^3$, (lb/in.$^3$) $\times$ 0.0026)
    Modulus of steel $E$ (Pa, psi)
    Contact width of surfaces $y$ (m, in.)
    Poisson's ratio for steel $v$ (dimensionless)
    Pressure–viscosity coefficient of lubricant $\alpha$ (1/Pa, 1/psi)
2.  Calculate:

Effective modulus of elasticity $E' = \dfrac{E}{1 - v^2}$

Mean rolling velocity $U = \dfrac{U_1 + U_2}{2}$

Effective radius $R = \dfrac{r_1 r_2}{r_1 + r_2}$

Load per unit width $W = \dfrac{F}{y}$

3. Find $S_{e1}$, $S_{e2}$, from Fig. 7.13a by using $S_1$ and $S_2$ (or by manufacturing processes). For $S < 0.05\,\mu\text{m}$, take $S_e = 0.05\,\mu\text{m}$. Then $S_{ec} = \sqrt{S_{e1} + S_{e2}}$.
4. Calculate dimensionless $U, \eta, W, S$ from Eqs. (7.17)–(7.20).
5. Calculate coefficient of rolling friction $f_r$ from Eq. (7.21).
6. Calculate coefficient of friction in the nonlinear region and its location $f_n$ is calculated from Eq. (7.22), $z^*$ is calculated from Eq. (7.23).
7. Calculate minimum oil film thickness $h_0$ from Eq. (7.26), where $G = \alpha E'$.
8. Find $f_0$ from Fig. 7.12. Find $(S_{ec}/R)_e$ from Fig. 7.13b. Calculate coefficient of thermal friction $f_t$ from Eq. (7.24).
9. Use $f_r, f_n, z^*$, and $f_t$ to construct the coefficient of friction curve versus sliding/rolling ratio, as in Fig. 7.11, where sliding speed $= |U_1 - U_2|$, and rolling speed $= U$.

## 7.8.2 Layered Surfaces

1. Given: contact surface radii $r_1$, $r_1$ (m, in.)
surface velocities $U_1$, $U_2$ (m/sec, in./sec)
load $F$ (N, lbf)
surface roughness $S_1$, $S_2$ (m CLA, in. CLA) or manufacturing processes
density of lubricant $\rho$ (kg/m$^3$, (lb/in.$^3$) × 0.0026)
modulus of steel $E$ (Pa, psi)
contact width of surface $y$ (m, in.)
Poisson's ratio for steel $\nu$ (dimensionless)

Lubricant oil properties:
Thermal conductivity $K_0$ (W/m-°C), (BTU/(sec-in.-°F)) × 9338)
Specific heat $c_0$ (J/(kg-°C), (BTU/lb-°F) × 3,604,437)
Pressure–viscosity coefficient $\alpha$ (1/Pa, 1/psi)
Temperature–viscosity coefficient $\beta$ (1/°C, 1/°F)
Dynamic viscosity at entry condition $\eta_0$ (Pa-sec, reyn)

Disk 1 base material properties:
Modulus of elasticity $E_{b1}$ (Pa, psi)
Poisson's ratio $\nu_{b1}$ (dimensionless)

Thermal conductivity $K_{b1}$ (W/(m-°C), (BTU/(sec-in-°F) × 9338)
Specific heat $c_{b1}$ (J/(kg-°C), (BTU/(lb-°F) × 3,604,437)
Density $\rho_{b1}$ (kg/m³, (lb/in.³) × 0.0026)

Disk 1 surface material properties:
Modulus of elasticity $E_{c1}$ (Pa, psi)
Poisson's ratio $\nu_{c1}$ (dimensionless)
Thermal conductivity $K_{c1}$ (W/(m-°C), (BTU/(sec-in.-°F) × 9338)
Specific heat $c_{c1}$ (J/(kg-°C), (BTU/(lb-°F) × 3,604,437)
Density $\rho_{c1}$ (kg/m³, (lb/in.³) × 0.0026)
Thickness $h_{c1}$ (m, in.)

Disk 2 base material properties:
Modulus of elasticity $E_{b2}$ (Pa, psi)
Poisson's ratio $\nu_{b2}$ (dimensionless)
Thermal conductivity $K_{b2}$ (W/(m-°C), (BTU/-sec-in.-°F) × 9338)
Specific heat $c_{b2}$ (J/kg-°C), (BTU/(lb-°F) × 3,604,437)
Density $\rho_{b2}$ (kg/m³, (lb/in.³) × 0.0026)

Disk 2 surface material properties:
Modulus of elasticity $E_{c2}$ (Pa, psi)
Poisson's ratio $\nu_{c2}$ (dimensionless)
Thermal conductivity $K_{c2}$ (W/(m-°C), (BTU/(sec-in.-°F) × 9338)
Specific heat $c_{c2}$ (J/(kg-°C), (BTU/(lb-°F) × 3,604,437)
Density $\rho_{c2}$ (kg/m³, (lb/in.³) × 0.0026)
Thickness $h_{c2}$ (m, in.)

Steel properties:
Modulus of elasticity $E_S$ (Pa, psi)
Poisson's ratio $\nu_S$ (dimensionless)
Thermal conductivity $K_S$ (W/(m-°C), (BTU/(sec-in.-°F) × 9338)
Specific heat $c_S$ (J/(kg-°C), (BTU/(lb-°F) × 3,604,437)
Density $\rho_S$ (kg/m³, (lb/in.³) × 0.0026)
Bulk temperature $T_b$ $(K, R)$ $(K = °C + 273.16)$

2.  Use previous section to calculate $f_t$ for steel–steel contact surfaces. Substitute $\Delta T_S, K_S, c_S, \rho_S$, and $E_S$ for $\Delta T, K_1, c_1, \rho_1, E_1$ and $K_2, c_2, \rho_2, E_2$ in Eqs. (7.27) and (7.28), where $f$ is replaced by $f_t$, $L$ is replaced by $L = 32\sqrt{WR/E_S'}$.

3.  Mean rolling velocity $U = \dfrac{U_1 + U_2}{2}$

    Effective radius $R = \dfrac{r_1 r_2}{r_1 + r_2}$

Load per unit width $W = \dfrac{F}{x}$

Effective modulus of elasticity $\dfrac{1}{E'} = \dfrac{1}{2}\left(\dfrac{1 - v_{c1}^2}{E_{c1}} + \dfrac{1 - v_{c2}^2}{E - c2}\right)$

4.  Half contact width $b = 1.6\sqrt{\dfrac{WR}{E'}}$

5.  Calculate $h_d$ by using Eq. (7.34).
6.  Substitute $E_{b1}, E_{c1}, h_{c1}$ for $E_b, E_c, h_c$ in Eq. (7.35) to calculate $E_{e1}$. Substitute $E_{b2}, E_{c2}, h_{c2}$ for $E_b, E_c, h_c$ in Eq. (7.35) to calculate $E_{e2}$.
7.  Calculate the effective modulus of elasticity of the layered surfaces by:

$$\frac{1}{E_{ec}'} = \frac{1}{2}\left(\frac{1 - v_{b1}^2}{E_{e1}} + \frac{1 - v_{b2}^2}{E_{e2}}\right)$$

8.  Find $S_{e1}, S_{e2}$ from Fig. 7.13a by using $S_1$ and $S_2$ (or by manufacturing processes). For $S < 0.05\,\mu m$ take $S_e = 0.05\,\mu m$. Then $S - ec = \sqrt{S_{e1} + S_{e2}}$.
9.  Calculate dimensionless $U, \eta, W, S$ from Eqs. (7.17)–(7.20) except that $E'$ is replaced by $E_{ec}'$.
10. Calculate coefficient of rolling friction $f_r$ from Eq. (7.21).
11. Calculate coefficient of friction in the nonlinear region and its location $f_n$ is calculated from Eq. (7.22), $z^*$ is calculated from Eq. (7.23).
12. Calculate minimum oil film thickness $h_0$ from Eq. (7.26), where $G = \alpha E_{ec}'$.
13. Calculate $\varepsilon$ by using Eq. (7.28), where $z = 0.27$.
14. Calculate $\Delta T_c$ by Eq. (7.27), where $K_1, c_1, \rho_1, E_1$ are replaced by $K_{c1}, c_{c1}, \rho_{c1}, E_{c1}, K_2, c_2, \rho_2, E_2$ are replaced by $K_{c2}, c_{c2}, \rho_{c2}, E_{c2}$, $L = 2b, f = f_t$ from step 2 for steel–steel contact surfaces.
15. Calculate $D_1$ by using Eq. (7.29) where $K, c, \rho, U$ are substituted by $K_1, c_1, \rho_1, E_1$. Use Eq. (7.30) to calculate $\beta_1$ where $h_c = h_{c1}$, $D = D1$.
16. Calculate $D_2$ by using Eq. (7.29) where $K, c, \rho, U$ are substituted by $K_2, c_2, \rho_2, E_2$. Use Eq. (7.30) to calculate $\beta_2$ where $h_c = h_{c2}$, $D = D2$.
17. Calculate $\Delta T_e$ by using Eq. (7.31) where $\beta = (\beta_1 + \beta_2)/2$.
18. Use Eq. (7.32) to calculate $a$ and $b$ for theparticular lubricant as follows. Suppose that the viscosity is $m1$ at $T1$ and $m2$ at $T2$ (where $T1$ and $T2$ are absolute temperature, say, $K = 273.16 + {}^\circ C$, $m1$ and $m2$ are kinematic viscosity in centi-

stokes), substitute $m1$, $T1$ and $m2$, $T2$ into Eq. (7.32), respectively, and solve these two linear equations simultaneously to get $a$ and $b$. (If lubricant is SAE 10, SAE 20, SAE 30, SAE 40, SAE 50, SAE 60, or SAE 70, use Table 7.3.)

19. Use Eq. (7.32) to calculate the viscosity $\eta$ at temperature $T_e (\eta = \rho_0 v$ where $\rho_0$ is the lubricant density, $v$ is the kinematic viscosity).

20. Use Eq. (7.26) to find $h_0$ where $\eta_0 = \eta$, $G = \alpha E'_{ec}$.

21. Find $f_0$ from Fig. 7.12. Find $(S_{ec}/R)_e$ from Fig. 7.13b. Calculate coefficient of thermal friction $f_t$ from Eq. (7.24).

22. If the difference between $f_t$ value in step 21 and $f_t$ value in step 14 does not satisfy your accuracy requirement, go back to step 14, replace $f_t$ by $f_t$ value in step 21 and iterate until the accuracy requirement is satisfied.

23. Use $f_r, f_n, z^*$, and $f_t$ to construct the coefficient of friction curve versus sliding/rolling ratio as in Fig. 7.11, where sliding speed $= |U_1 - U_2|$, and rolling speed $= U$.

## 7.9  SOME NUMERICAL RESULTS

The following are some illustrative examples for the application of the developed empirical formulas in sample cases.

Figure 7.34 shows calculated coefficient of friction versus sliding/rolling ratio for different rolling speeds, $T = 26°C$ (78.8°F), steel–steel contact, ground surfaces, $S = 0.03 \, \mu m$ (12 μin.), $W = 378,812 \, N/m$ (2160 lbf/in.), 10W30 oil, $R = 0.0234 \, m$ (0.92 in.).

Figure 7.35 shows calculated coefficient of friction versus sliding/rolling ratio for different normal loads, $T = 26°C$ (78.8°F), steel–steel contact, ground surfaces, $S = 0.03 \, \mu m$ (12 μin.), $U1 = 3.2 \, m/sec$ (126 in./sec), 10W30 oil, $R = 0.0234 \, m$ (0.92 in.).

Figure 7.36 shows calculated coefficient of friction versus sliding/rolling ratio for different effective radii, $T = 26°C$ (78.8°F), steel–steel contact ground surfaces, $S = 0.03 \, \mu m$ (12 μin.), $W = 378,812 \, N/m$ (2160 lbf/in.), 10W30 oil, $U1 = 3.2 \, m/sec$ (126 in./sec).

Figure 7.37 shows calculated coefficient of friction versus sliding/ rolling ratio for different viscosity, steel–steel contact, ground surfaces, $S = 0.03 \, \mu m$ (12 μin.), $W = 378,812 \, N/m$ (2160 lbf/in.), 10W30 oil, $U1 = 3.2 \, m/sec$ (126 in./sec), $R = 0.0234 \, m$ (0.92 in.).

Figure 7.38 shows calculated coefficient of friction versus sliding/ rolling ratio for different materials, $T = 26°C$ (78.8°F), ground surfaces,

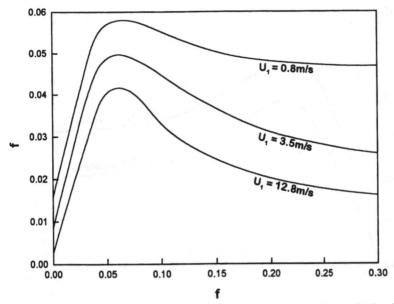

**Figure 7.34** Calculated coefficient of friction vs. sliding/rolling ratio for different rolling speeds, $T = 26°C$ (78.8°F), steel–steel contact, ground surfaces, $S = 0.3\,\mu m$ (12 μin.), $W = 378{,}812\,N/m$ (2160 lbf/in.), 10W30 oil, $R = 0.0234\,m$ (0.92 in.).

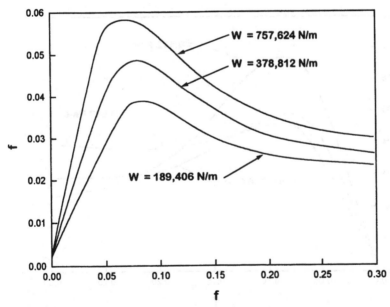

**Figure 7.35** Calculated coefficient of friction vs. sliding/rolling ratio for different normal loads, $T = 26°C$ (78.8°F), steel–steel contact, ground surfaces, $S = 0.3\,\mu m$ (12 μin.), $U = 3.2\,m/sec$ (216 in./sec), 10W30 oil, $R = 0.0234\,m$ (0.92 in.).

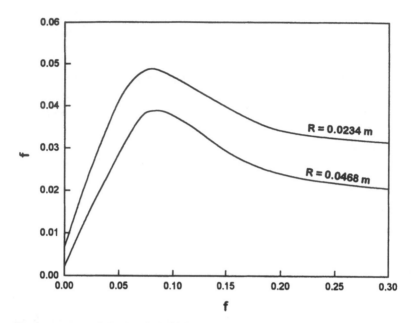

**Figure 7.36** Calculated coefficient of friction vs. sliding/rolling ratio for different effective radii, $T = 26°C$ (78.8°F), steel–steel contact, ground surfaces, $S = 0.3\,\mu m$ (12 µin.), $W = 378{,}812\,N/m$ (2160 lbf/in.), 10W30 oil, $U1 = 3.2\,m/s$ (126 in./sec).

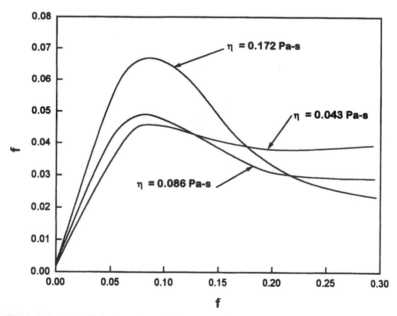

**Figure 7.37** Calculated coefficient of friction vs. sliding/rolling ratio for different viscosity, steel–steel contact, ground surfaces, $S = 0.3\,\mu m$ (12 µin.), $W = 378{,}812\,N/m$ (2160 lbf/in.), 10W30 oil, $U1 = 3.2\,m/s$ (126 in./sec), $R = 0.0234\,m$ (0.92 in.).

306

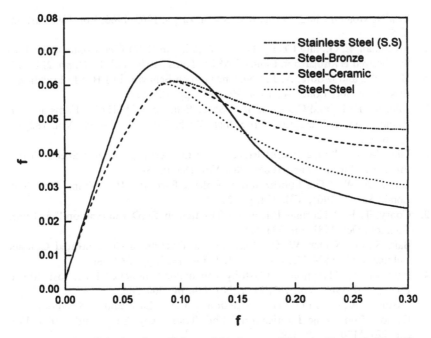

**Figure 7.38** Calculated coefficient of friction vs. sliding/rolling ratio for different materials, $T = 26°C$ (78.8°F), ground surfaces, $S = 0.3\,\mu m$ (12 μin.), $W = 378,812\,N/m$ (2160 lbf/in.), 10W30 oil, $U1 = 3.2\,m/s$ (126 in./sec), $R = 0.0234\,m$ (0.92 in.).

$S = 0.03\,\mu m$ (12 μin.), $W = 378,812\,N/m$ (2160 lbf/in.), 10W30 oil, $U1 = 3.2\,m/sec$ (126 in./sec), $R = 0.0234\,m$ (0.92 in.).

## REFERENCES

1. Palmgren, A., Ball and Roller Bearing Engineering, S. H. Burbank, Philadelphia, 1945.
2. Tabor, D., "The Mechanism of Rolling Friction," Phil. Mag., Vol. 43, 1952, pp. 1055 and Vol. 45, 1954, p. 1081.
3. Rabinowicz, E., Friction and Wear of Materials, John Wiley and Sons, New York, NY, 1965.
4. Dowson, D., and Higginson, G. R., Elastohydrodynamic Lubrication, Pergamon, Oxford, 1977.
5. Grubin, A. N., Book No. 30, English Translation DSIR, 1949.
6. Dowson, D., and Whitaker, A. V., "A Numerical Procedure for the Solution of the Elastohydrodynamic Problem of Rolling and Sliding Contacts Lubricated

by a Newtonian Fluid," Proc. Inst. Mech. Engrs, 1965–1966, Vol. 180, Part 3b, p. 57.

7.  Trachman, E. G., and Cheng, H. S., "Traction in EHD Line Contacts for Two Synthesized Hydrocarbon Fluids," ASLE Trans., 1974, Vol. 17(4), pp. 271–279.

8.  Hirst, W., and Moore, A. J., "Non-newtonian Behavior in EHD Lubrication," Proc. Roy. Soc. Lond. A., 1974, Vol. 337, pp. 101–121.

9.  Johnson, K. L., and Cameron, R., "Shear Behavior of EHD Oil Films at High Rolling Contact Pressures," Proc. Inst. Mech. Engrs, 1967–1968, Vol. 182, Pt. 1, No. 14.

10. Plint, M. A., "Some Recent Research on the Perbury Variable-Speed Gear," Proc. Inst. Mech. Engrs, 1965–1966, Vol. 180, Pt. 3B.

11. Crook, A. W., "The Lubrication of Rollers, Part III," Phil. Trans. Roy. Soc. Lond., Ser. A, 1961, Vol. 254, p. 237.

12. Conry, T. F., "Thermal Effects on Traction in EHD Lubrication," J. Lubr. Technol., Oct. 1981, pp. 533–538.

13. Bair, S., and Winer, W. O., "Regimes of Traction in Concentrated Contact Lubrication," ASME Trans., Vol. 104, July 1982, pp. 382–386.

14. Plint, M. A., "Traction in Elastohydrodynamic Contacts," Proc. Inst. Mech. Engrs, 1967–1968, Vol. 182, Pt. 1, No. 114, pp. 300–306.

15. Dyson, A., "Frictional Traction and Lubricant Rheology in Elastohydrodynamic Lubrication," Phil. Trans. Roy. Soc. Lond., 1970, Vol. 266, No. 1170.

16. Sasaki, T., Okamura, K., and Isogal, R., "Fundamental Research on Gear Lubrication," Bull. JSME, 1961, Vol. 4(14).

17. Drozdov, Y. N., and Gavrikov, Y. A., "Friction and Scoring under the Conditions of Simultaneous Rolling and Sliding of Bodies," Wear, 1968, Vol. 11.

18. O'Donoghue, J. P., and Cameron, A., "Friction and Temperature in Rolling/ Sliding Contacts," ASLE Trans., 1966, Vol. 9, pp. 186–194.

19. Benedict, G. H., and Kelley, B. W., "Instantaneous Coefficients of Gear Tooth Friction," ASLE Trans., 1961, Vol. 4, pp. 59–70.

20. Misharin, J. A., "Influence of the Friction Conditions on the Magnitude of the Friction Coefficient in the Case of Rolling with Sliding," Int. Conf. on Gearing, Proc., Sept. 1958.

21. Ku, P. M., Staph, H. E., and Carper, H. J., "Frictional and Thermal Behavior of the Sliding-Rolling Concentrated Contacts," ASME Trans., J. Lubr. Technol., Jan. 1978, Vol. 100.

22. Li, Y., "An Investigation on the Effects of the Properties of Coating Materials on the Tribology Behavior of Sliding/Rolling Contacts," Ph.D. Thesis, Univ. of Wisconsin, 1987.

23. Rashid, M. K., and Seireg, A., "Heat Partition and Transient Temperature Distribution in Layered Concentrated Contacts," ASME Trans., J. Tribol., July 1987, Vol. 109, pp. 4960–4502.

24. Hsue, E. Y., "Temperature and Surface Damage under Lubricated Sliding/ Rolling Contacts," Ph.D. Thesis, University of Wisconsin-Madison, 1984.

25. Wilson, W. R. D., and Sheu, S., "Effect of Inlet Shear Heating Due to Sliding and EHD Film Thickness," J. Lubr. Technol., April 1983, Vol. 105.
26. Cameron, A., Basic Lubrication Theory, Longman Group, London, England, 1970.
27. Juvinall, R. C., Fundamentals of Machine Component Design, John Wiley & Sons, New York, NY, 1983.

# 8

# Wear

## 8.1  INTRODUCTION

Wear can be defined as the progressive loss of surface material due to normal load and relative motion. This generally leads to degradation of the surface, loss of component functionality, and in many situations, to catastrophic failure.

The wear of mechanical components has been estimated to cost the U.S. economy between 6% and 7% of the gross national product. Understanding the wear process and its control is, therefore, of major practical importance.

The highly complex nature of the wear process has made it difficult to develop generalized procedures for predicting its occurrence and intensity. Even wear tests under seemingly controlled conditions, are not always reproducible. It is not unusual that repeated tests may give wear rates which differ by orders of magnitude.

Surface damage or wear can manifest itself in many forms. Among these are the commonly used terminology: pitting, frosting, surface fatigue, surface cracking, fretting, blistering, plastic deformation, scoring, etc. Wear types include elastic wear, plastic wear, delamination wear, abrasive wear, adhesive wear, corrosive wear, cavitation erosion, etc. The occurrence of a particular type of wear depends on many factors, which include the geometry of the surfaces, the nature of surface roughness, the applied load, the rolling and sliding velocities. Other important factors which influence wear are the environmental temperature, moisture, and chemical conditions, as well as the mechanical, thermal, chemical, and metallurgical properties of the surface layer and bulk material. The microstructure of the surface layer, its ductility, the microhardness distribution in it, and the existence of vacancies and impurities also play critical parts in the wear process. Furthermore, wear is highly influenced by the physical, thermal, and chemical properties of the lubricant, the regime of lubrication, the mutual overlap between the rubbing surfaces, and the potential for removal of the chemical layers and debris generated in the process.

This chapter provides a conceptual evaluation of this extremely complex phenomenon, and presents guidelines for its prediction and control. Although the mechanism of wear is not fully understood, designers of machine components have to rely on judgement and empirical experiences to improve the functional life of their design. The success of their judgement depends on their depth of understanding of which factors are relevant to a particular situation, and which are only accessories.

It is interesting to note that with all the modern tools of experimentation and computation, generalized wear design procedures that would produce practical results are still beyond our reach. We have therefore to rely on thoughtful interpretation of accumulated data and observations. One such poignant observation was documented 2000 years ago by the Roman philosophical poet Titus Caras Lucretius [1]: He said,

> A ring is worn thin next to a finger with continual rubbing. Dripping water hollows a stone, a curved plow share, iron though it is, dwindles imperceptibly in the furrow. We see the cobblestones of the highway worn by the feet of many wayfarers. The bronze statues by the city gates show their right hands worn thin by the touch of all travelers who have greeted them in passing. We shall see that all these are being diminished since they are worn away. But to perceive what particles drop off at any particular time is a power grudged to us by our ungenerous sense of sight.

## 8.2  CLASSIFICATION OF WEAR MECHANISMS

It has not yet been possible to devise a single classification of the different types of wear. Some of the mechanisms by which rubbing surfaces are damaged are [2]:

Mechanical destruction of interlocking asperities;
Surface fatigue due to repeated mechanical interaction between asperities or the variation of pressure developed in the lubrication;
Failure due to work hardening and increasing brittleness caused by deformation;
Flaking away of oxide films;

Mechanical damage due to atomic or molecular interactions;

Mechanical destruction of the surface due to the high temperatures produced by frictional heating;

Adhesion or galling;

Corrosion;

Abrasion due to the presence of loose particles;

Cutting or ploughing of a soft material by a harder rough surface;

Erosion produced by impinging fluid or fluids moving with high rate of shear.

The treatment in this chapter attempts to formulate general concepts about the nature of wear, which can be readily associated with practical experience and to provide equations which can be used for design purposes based on these concepts. The broad categories to be considered are:

Frictional wear

Surface fatigue due to contact pressure

Microcutting

Thermal wear

Delamination wear

Abrasive wear

Corrosion or chemical wear

Erosion wear

## 8.3  FRICTIONAL WEAR

In the broad category of frictional (or adhesive) wear considered in this section, it is assumed that the material removal is the result of the mechanical interaction between the rubbing surfaces at the real area of contact. It has been shown in Chapter 4 that the real area of contact is approximately proportional to the normal load under elastic contact condition. The proportionality constant is a function of the material properties, the asperity density, the radius of the asperities, and the root mean square of the asperity height.

The wear volume per unit sliding distance has been evaluated according to this concept by several investigations. Their results are illustrated in the following.

Archard [3, 4], as well as Burwell and Strang [5], proposed wear equations of the following form:

$$\frac{V}{L} = K \frac{P}{3\sigma_y} = K \frac{P}{H_m} \tag{8.1}$$

where

$V$ = wear volume
$L$ = sliding distance
$P$ = applied load
$\sigma_y$ = yield stress of the softer material
$K$ = proportionality constant depending on the material combination and test conditions (wear coefficient)
$H_m$ = microhardness of the softer material

Results obtained by Archard from dry tests where the end of a cylinder 6 mm diameter was rubbed against a ring of 24 mm diameter under a 400 g load at a speed of 1.8 m/sec are given in Table 8.1.

Rabinowicz [6, 7] gave a similar equation:

$$\frac{V}{L} = \frac{hA}{L} = k\,\frac{P}{9\sigma_y} \qquad (8.2)$$

**Table 8.1** Dry Wear Coefficients for Different Material Pairs

| Sliding against hardened tool steel unless otherwise stated | Wear coefficient, $K$ | Microhardness, $H_m$ ($10^3$ kg/cm$^2$) |
|---|---|---|
| Mild steel on mild steel | $7 \times 10^{-3}$ | 18.6 |
| 60/40 brass | $6 \times 10^{-4}$ | 9.5 |
| Teflon | $2.5 \times 10^{-5}$ | 0.5 |
| 70/30 brass | $1.7 \times 10^{-4}$ | 6.8 |
| Perspex | $7 \times 10^{-6}$ | 2.0 |
| Bakelite (moulded) type 5073 | $7.5 \times 10^{-6}$ | 2.5 |
| Silver steel | $6 \times 10^{-5}$ | 32 |
| Beryllium copper | $3.7 \times 10^{-5}$ | 21 |
| Hardened tool steel | $1.3 \times 10^{-4}$ | 85 |
| Stellite | $5.5 \times 10^{-5}$ | 69 |
| Ferritic stainless steel | $1.7 \times 10^{-5}$ | 25 |
| Laminated bakelite type 292/16 | $1.5 \times 10^{-6}$ | 3.3 |
| Moulded bakelite type 11085/1 | $7.5 \times 10^{-7}$ | 3.0 |
| Tungsten carbide on mild steel | $4 \times 10^{-6}$ | 18.6 |
| Moulded bakelite type 547/1 | $3 \times 10^{-7}$ | 2.9 |
| Polyethylene | $1.3 \times 10^{-7}$ | 0.17 |
| Tungsten carbide on tungsten carbide | $1 \times 10^{-6}$ | 130 |

where

$V$ = wear volume (in.$^3$)

$L$ = sliding distance (in.)

$A$ = surface area (in.$^2$)

$P$ = applied load (lb)

$\sigma_y$ = yield strength of the softer material (psi)

$h$ = depth of wear of the softer material (in.)

$k$ = wear coefficient

Values of $k$ for different material combinations are given in Table 8.2.
The depth of wear of the harder material $h_h$, can be calculated from:

$$\frac{h_h}{h} = \left(\frac{\sigma}{(\sigma_y)_h}\right)^2 \tag{8.3}$$

For conditions where the load and or the surface temperature are high
enough to cause plastic deformation, the wear rate as calculated from Eqs
(8.1) and (8.2) can be several orders of magnitude higher (in the order of

**Table 8.2**  Wear Coefficients, $k$, for Metal Combinations

| Metal combination | $k \times 10^{-4}$ | Metal combination | $k \times 10^{-4}$ |
|---|---|---|---|
| Cu vs. Pb | 0.1 | Zn vs. Zn | 11.6 |
| Ni vs. Pb | 0.2 | Mg vs. Al | 15.6 |
| Fe vs. Ag | 0.7 | Zn vs. cu | 18.5 |
| Ni vs. Ag | 0.7 | Fe vs. Cu | 19.1 |
| Fe vs. Pb | 0.7 | Ag vs. Cu | 19.8 |
| Al vs. Pb | 1.4 | Pb vs. Pb | 23.8 |
| Ag vs. Pb | 2.5 | Ni vs. Mg | 28.6 |
| Mg vs. Pb | 2.6 | Zn vs. Mg | 29.1 |
| Zn vs. Pb | 2.6 | Al vs. Al | 29.8 |
| Ag vs. Ag | 3.4 | Cu vs. Mg | 30.5 |
| Al vs. Zn | 3.9 | Ag vs. Mg | 32.5 |
| Al vs. Ni | 4.7 | Mg vs. Mg | 36.5 |
| Al vs. Cu | 4.8 | Fe vs. Mg | 38.5 |
| Al vs. Ag | 5.3 | Fe vs. Ni | 59.5 |
| Al vs. Fe | 6.0 | Fe vs. Fe | 77.5 |
| Fe vs. Zn | 8.4 | Cu vs. Ni | 81.0 |
| Ag vs. Zn | 8.4 | Cu vs. Cu | 126.0 |
| Ni vs. Zn | 11.0 | Ni vs. Ni | 286.0 |

1000 times). This is generally known as "plastic wear" and often leads to very rapid rate of material removal.

Krushchov and coworkers [8, 9] developed a similar linear relationship between wear resistance and hardness for commercially pure and annealed materials. This relationship is given in Fig. 8.1. A particularly interesting result was obtained by them for heat-treated alloy steels. As shown in Fig. 8.2, the wear resistance for the steels in the annealed condition increased linearly with hardness. However, increasing the hardness of a particular alloy by heat treating produced a smaller rate of increase of the relative wear resistance. This clearly suggests that the relative wear resistance of a material does not only depend on its hardness but is also influenced by the

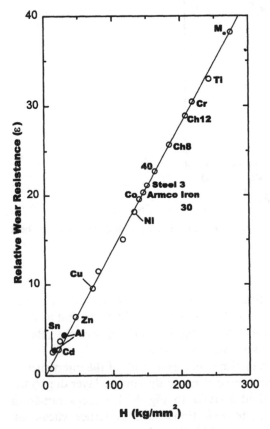

**Figure 8.1** Relationship between relative wear resistance and hardness for some commercially pure metals. (From Ref. 8.)

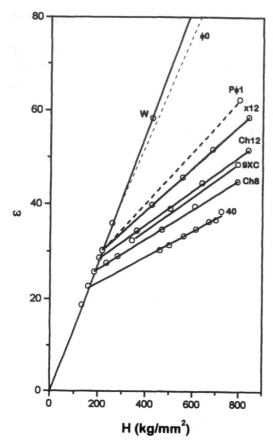

**Figure 8.2** Relationship between relative wear resistance and hardness for heat treated steels. (From Ref. 8.)

presence of microscopic and submicroscopic inhomogeneities in the lattice structure by distortions of the lattice. It was also found by them that increasing the hardness further, by work hardening, did not improve the relative wear resistance and, in some cases, even reduced it.

Frictional surface damage can also occur as a result of the interpenetration of asperities, which produce tensile stress in the surface layer due to the bulge formed ahead of the indentor (refer to Fig. 8.3). Cracks can form perpendicular to the surface at imperfections such as lattice vacancies, grain boundaries and metalurgical defects including pores, gas bubbles, slag inclusions, and marked disparity in grain size.

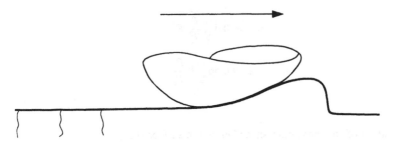

**Figure 8.3** Cracks at surface imperfections due to repeated asperity action.

## 8.4 WEAR DUE TO SURFACE FATIGUE

### 8.4.1 Contact Fatigue

The most common example of the type of surface damage is what is generally known as "pitting" or contact fatigue. It often exists in rolling element bearings and gears and is attributed to the propogation of fatigue cracks originating on or below the surface when the Hertzian pressure exceeds an allowable value. As one element rolls many times over the other element, a subsurface region undergoes cycles of shear ranging from zero to maximum. This situation would be expected to promote fatigue damage when the maximum shearing stress is higher than the fatigue limit for the material in this region. Subsurface cracks may occur and these cracks will propogate to the surface under repeated loading and consequently forming a pit or a spall. The equations for calculating the maximum subsurface shear stress and its location can be written as follows.

For cylindrical contacts:

$$q_0 = \text{maximum contact stress} = 0.418 \sqrt{\frac{PE_e}{LR_e}}$$

$$\tau_{\max} = \text{maximum subsurface shear stress} = 0.304 q_0$$

$$Z \approx 0.84 \sqrt{\frac{PR_e}{LE_e}}$$

For spherical contacts:

$$q_0 = 0.616 \sqrt[3]{\frac{P}{R_e^2} E_e^2}$$

$$\tau_{max} \approx 0.4 \sqrt[3]{\frac{P}{R_e^2} E_e^2}$$

$$Z \approx 0.56 \sqrt[3]{P \frac{R_e}{E_e}}$$

where

$Z$ = location of the maximum shear below the surface (in.)
$P$ = applied load (lb)
$L$ = length of cylinders
$R_e = \dfrac{1}{\dfrac{1}{R_1} + \dfrac{1}{R_2}}$ = effective radius

$E_e = \dfrac{1}{\dfrac{1}{E_1} + \dfrac{1}{E_2}}$ = effective modulus of elasticity

These equations are widely used as the basis for predicting the surface durability of rolling element bearings and gears.

The number of cycles to pitting failure, $N$, generally follows the following fatigue equation:

$$N^{1/n} \tau_{max} = C$$

where

$\tau_{max}$ = maximum shear stress

$C$ and $n$ are constants for each material.

Accordingly, the life ratio depends on maximum shear stress:

$$\frac{N_1}{N_2} = \left(\frac{(\tau_{max})_2}{(\tau_{max})_1}\right)^n$$

The value of $n$ varies between 6 and 18 for most materials.

For cumulative fatigue under different stress cycles, the Miner theory is generally used. It can be expressed as:

$$\sum \frac{N_i}{N_{if}} = 1$$

where

$N_i$ = number of cycles at any stress level
$N_{if}$ = number of cycles to failure at that stress level

## 8.4.2 The IBM Zero Wear Concept

Because of the stringent requirements on the minimization of wear in electronic equipment, IBM conducted extensive wear experiments in order to allow reliable prediction of their useful life [10–12]. The criterion for zero wear is that the depth of the wear scar does not exceed one half of the peak-to-peak value of the surface roughness. This may be a severe requirement for most mechanical equipment, which can tolerate considerably larger amounts of wear without loss of functionality.

The empirical equation developed by IBM is given as follows, based on 2000 cycles as the reference number in their tests:

$$S \leq \left(\frac{2000}{N}\right)^{1/9} GY$$

where

$S_s$ = the maximum shear stress produced by sliding in the vicinity of the contact region

$N$ = number of passes one element undergoes in the relative motion (or number of contact cycles)

$Y$ = yield point in shear (psi) which is a function of the microhardness of the surface as given in Fig. 8.4) and Table 8.3

$G$ = empirical factor determined from the tests. Surprisingly, it was found to take one of the following two values depending on the material pair and the lubrication condition

$G$ = 1.0 for full film lubrication

$G$ = 0.54 for quasihydrodynamic lubrication

For unlubricated or boundary lubrication conditions, $G$, takes one of only two possible values:

$G$ = 0.54 for systems with low susceptibility for transfer

$G$ = 0.20 for systems with high susceptibility for transfer

Table 8.4 gives the values of $G$ for different material combinations tested by IBM.

IBM used the concept of mutual overlap in defining the number of passes. The coefficient of mutual overlap ($K_{mut}$) can be defined as:

$$K_{mut} = \frac{A_a'}{A_a''}$$

**Table 8.3** Values of Yield Point in Shear, $Y$, and Microhardness, $H_m$

| Material | $H_m$ (kg/mm$^2$) | $Y$ (psi × 10$^6$) | Material | $H_m$ (kg/mm$^2$) | $Y$ (psi × 10$^6$) |
|---|---|---|---|---|---|
| **Stainless steels** | | | **Copper alloys** | | |
| 302 | 270 | 58 | Brass | 115 | 17.9 |
| 303EZ | 296 | 63 | Be-Cu | 199 | 31 |
| 321 | 224 | 40 | Cu-Ni | 171 | 35 |
| 347 | 252 | 50 | Phosphor-Bronze | 166 | 27 |
| 410 | 270 | 58 | **Aluminum alloys** | | |
| 416 EZ:$H_m$ | 270 | 58 | 43 aluminum | 60.7 | 8 |
| | 224 | 40 | 112 aluminum | 117 | 15 |
| 440C | 296 | 63 | 195 aluminum | 96.8 | 15 |
| **Steels** | | | 220 aluminum | 124.5 | 18 |
| 1018 | 199 | 33 | 355 aluminum | 90.5 | 14 |
| 1045 | 468 | 106 | 356 aluminum | 62.1 | 8 |
| 1055 | 270 | 58 | **Sintered materials** | | |
| 1060 | 397 | 90 | Sintered brass 1 | | |
| 1085 | 359 | 80 | 7.5 min | 115 | 17.9 |
| 1117 | 160 | 27 | Sintered brass 2 | | |
| 4140 | 180 | 32 | 7.0–7.5 | 96 | 74 |
| 4140LL | 384 | 82.5 | Sintered bronze 1 | | |
| 4150 | 276 | 65 | ASTM B202-58T | 135 | 22.5 |
| 4620 | 242 | 47 | Sintered bronze 2 | | |
| 5130LL | 260 | 55 | ASTM B255-61T | 150 | 25 |
| 8214 | 220 | 40 | Sintered iron 1 | | |
| 8620 | 216 | 40 | 7.5 min | 180 | 31.5 |
| 52100 | 746 | 150 | Sintered iron 2 | | |
| | 220 | 40 | 7.3 min | 150 | 25 |
| Carpenter 11 | | | Sintered iron 3 | | |
| annealed | 226 | 40 | 7.0 min | 110 | 17.9 |
| Hampden steel | | | Sintered iron | | |
| annealed | 262 | 55 | copper 1 | 220 | 40 |
| HYCC(HA) | 340 | 75 | 7.1 copper | 190 | 33 |
| HYCC(PM) | 270 | 58 | infil. – 15% | | |
| Ketos | 296 | 63 | 5.8–6.2 – 20% | | |
| Nitralloy G | 396 | 90 | Sintered steel 1 | 220 | 40 |
| Rexalloy AA | 350 | 80 | 7.0 min | | |
| Star Zenith | | | Stainless 3161 | 150 | 25 |
| annealed | 269 | 58 | Sintered steel 2 | | |
| **Nickel alloys** | | | 7.0 min | | |
| Invar "36" | | | | | |
| annealed | 184 | 30 | | | |
| H$_y$M$_n$80 | | | | | |
| annealed | 270 | 58 | | | |
| Monel C | 184 | 40 | | | |

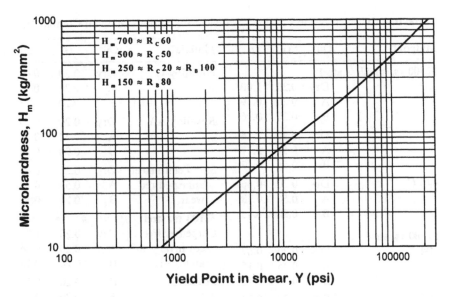

**Figure 8.4** Microhardness, $H_m$, as a function of the yield point in shear.

where $A'_a$, $A''_a$ are the apparent areas subjected to sliding for each of the surfaces. Two extreme examples are illustrated in Fig. 8.5. For the two hollow cylinders condition, $K_{mut} = 1$, and for the pin on disk condition, $K_{mut} \approx 0$.

Several illustrative examples for the method of determining $N$ and calculating $S_s$ used by IBM are given in the following. The coefficient of friction $\mu$ used for calculating $S_s$ for different combinations of materials is given in Table 8.4.

For the cam and follower shown in Fig. 8.6a:

$$N \text{ for the cam} = \text{number of strokes}$$

$$N \text{ for the follower} = \left(\frac{2\pi R}{W}\right) \text{ (number of revolutions)}$$

For the ball reciprocating on a plate shown in Fig. 8.6b:

$$N \text{ for the plate} = \text{number of strokes}$$

$$N \text{ for the ball} = \frac{\text{length of stroke}}{W} \text{ (number of strokes)}$$

For the shaft rotating in a bearing:

$$N \text{ for the shaft} = \text{number of revolutions}$$

$$N \text{ for the journal} \approx 2 \text{ (number of revolutions)}$$

**Table 8.4** *G*-Factors and Friction Coefficient for Various Material Combinations

| Material | Oil[a] | G | μ | Material | Oil[a] | G | μ |
|---|---|---|---|---|---|---|---|
| **52100 vs. stainless steel** | | | | Nitralloy-G | Dry | 0.20 | 0.63 |
| 302 | Dry | 0.20 | 1.00 | | A | 0.20 | 0.15 |
| | A | 0.20 | 0.19 | | B | 0.20 | 0.13 |
| | B | 0.20 | 0.16 | Rexalloy AA | Dry | 0.20 | 0.73 |
| 321 | Dry | 0.20 | 1.16 | | A | 0.54 | 0.13 |
| | A | 0.54 | 0.17 | | B | 0.20 | 0.13 |
| | B | 0.54 | 0.13 | Star Zenith steel | Dry | 0.20 | 0.63 |
| 440 C | Dry | 0.20 | 0.66 | annealed red | A | 0.54 | 0.12 |
| | A | 0.54 | 0.18 | wear | B | 0.20 | 0.12 |
| | B | 0.20 | 0.13 | **52100 vs. steel** | | | |
| **52100 vs. steel** | | | | Carpenter free | Dry | 0.20 | 1.28 |
| 1045 | Dry | 0.20 | 0.67 | cot invar "36" | A | 0.20 | 0.24 |
| | A | 0.45 | 0.15 | annealed | B | 0.20 | 0.18 |
| | B | 0.20 | 0.17 | Monel C | Dry | 0.20 | 0.73 |
| 1060 | Dry | 0.20 | 0.73 | | A | 0.20 | 0.12 |
| | A | 0.20 | 0.14 | | B | 0.54 | 0.14 |
| | B | 0.20 | 0.21 | **52100 vs. copper alloy** | | | |
| 4140 LL | Dry | 0.20 | 0.57 | Cu-Ni | Dry | 0.20 | 1.23 |
| | A | 0.20 | 0.21 | | A | 0.54 | 0.21 |
| | B | 0.20 | 0.17 | | B | 0.54 | 0.15 |
| 52100 | Dry | 0.20 | 0.60 | Phorphorus- | Dry | 0.20 | 0.67 |
| | A | 0.20 | 0.21 | Bronze A | A | 0.20 | 0.19 |
| | B | 0.20 | 0.16 | | B | 0.54 | 0.16 |
| Carpenter 11, | Dry | 0.20 | 0.78 | **52100 vs. aluminum alloy** | | | |
| special steel, | A | 0.45 | 0.18 | 112 Aluminum | Dry | 0.20 | 1.08 |
| annealed | B | 0.45 | 0.16 | | A | 0.54 | 0.25 |
| Hampden steel, | Dry | 0.54 | — | | B | 0.20 | 0.15 |
| annealed, oil | A | 0.54 | 0.13 | 195 Aluminum | Dry | 0.20 | 1.07 |
| wear | B | 0.54 | 0.12 | | A | 0.54 | 0.17 |
| HYCC (HA) | Dry | 0.20 | 0.62 | | B | 0.54 | 0.13 |
| | A | 0.54 | 0.13 | 355 Aluminum | Dry | 0.20 | 1.21 |
| | B | 0.54 | 0.11 | | A | 0.54 | 0.13 |
| HYCC (PM) | Dry | 0.20 | 0.64 | | B | 0.54 | 0.20 |
| | A | 0.20 | 0.16 | **52100 vs. sintered materials** | | | |
| | B | 0.20 | 0.17 | Sintered brass | Dry | 0.20 | 0.32 |
| Ketos | Dry | 0.20 | 0.67 | | A | 0.20 | 0.21 |
| | A | 0.54 | 0.18 | | B | 0.20 | 0.16 |
| | B | 0.54 | 0.15 | | | | |

**Table 8.4** Continued

| Material | Oil[a] | G | μ | Material | Oil[a] | G | μ |
|---|---|---|---|---|---|---|---|
| Sintered bronze | Dry | 0.20 | 0.26 | HYCC (HA) | Dry | 0.54 | 0.89 |
| | A | 0.20 | 0.23 | | A | 0.54 | 0.14 |
| | B | 0.20 | 0.11 | | B | 0.54 | 0.14 |
| Sintered iron | Dry | 0.20 | 0.38 | Nitralloy G | Dry | 0.20 | 0.83 |
| | A | 0.20 | 0.21 | | A | 0.54 | 0.14 |
| | B | 0.54 | 0.23 | | B | 0.54 | 0.14 |
| Sintered iron-copper | Dry | 0.20 | 0.47 | Star Zenith steel | Dry | 0.20 | 0.93 |
| | A | 0.20 | 0.20 | annealed red | A | 0.54 | 0.15 |
| | B | 0.54 | 0.19 | wear | B | 0.54 | 0.14 |
| Sintered steel | Dry | 0.20 | 0.34 | **302 vs. nickel alloy** | | | |
| | A | 0.54 | 0.15 | Carpenter free | Dry | 0.20 | 1.33 |
| | B | 0.54 | 0.15 | cut Invar "36" | A | 0.20 | 0.16 |
| **302 vs. stainless steel** | | | | annealed | B | 0.20 | 0.19 |
| 302 | Dry | 0.20 | 1.02 | **302 vs. nickel alloy** | | | |
| | A | 0.20 | 0.16 | Monel C | Dry | 0.20 | 0.99 |
| | B | 0.20 | 0.15 | | A | 0.20 | 0.15 |
| 321 | Dry | 0.20 | 1.47 | | B | 0.20 | 0.15 |
| | A | 0.54 | 0.15 | **302 vs. aluminum alloy** | | | |
| | B | 0.54 | 0.14 | 112 Aluminum | Dry | 0.20 | 1.16 |
| 440 C | Dry | 0.20 | 0.90 | | A | 0.54 | 0.20 |
| | A | 0.54 | 0.13 | | B | 0.54 | 0.14 |
| | B | 0.20 | 0.15 | 195 Aluminum | Dry | 0.20 | 1.17 |
| **302 vs. steel** | | | | | A | 0.54 | 0.15 |
| 1045 | Dry | 0.20 | 0.71 | | B | 0.54 | 0.14 |
| | A | 0.20 | 0.16 | 355 Aluminum | Dry | 0.20 | 1.11 |
| | B | 0.54 | 0.14 | | A | 0.54 | 0.17 |
| 1060 | Dry | 0.20 | 0.88 | | B | 0.54 | 0.20 |
| | A | 0.54 | 0.16 | **302 vs. plastic** | | | |
| | B | 0.20 | 0.15 | Delrin | Dry | 0.54 | 0.36 |
| 4140 LL | Dry | 0.20 | 0.78 | | A | 0.54 | 0.15 |
| | A | 0.54 | 0.14 | | B | 0.54 | 0.18 |
| | B | 0.54 | 0.14 | Nylatron G | Dry | 0.54 | 0.57 |
| 5130 LL | Dry | 0.20 | 0.84 | | A | 0.54 | 0.22 |
| | A | 0.20 | 0.16 | | B | 0.54 | 0.24 |
| | B | 0.20 | 0.14 | Polyethylene | Dry | 0.54 | 0.26 |
| Carpenter 11 | Dry | 0.20 | 0.84 | | A | 0.54 | 0.17 |
| special steel | A | 0.54 | 0.16 | | B | 0.54 | 0.17 |
| annealed | B | 0.20 | 0.14 | | | | |

**Table 8.4**  Continued

| Material | Oil[a] | G | μ | Material | Oil[a] | G | μ |
|---|---|---|---|---|---|---|---|
| Teflon | Dry | 0.54 | 0.09 | **Brass vs. steel** | | | |
| | A | 0.54 | 0.15 | 1045 | Dry | 0.20 | 0.66 |
| | B | 0.54 | 0.11 | | A | 0.20 | 0.20 |
| Zytel 101 | Dry | 0.54 | 0.60 | | B | 0.20 | 0.12 |
| | A | 0.54 | 0.27 | 4140 LL | Dry | 0.20 | 0.73 |
| | B | 0.54 | 0.27 | | A | 0.20 | 0.22 |
| **Brass vs. stainless steel** | | | | | B | 0.20 | 0.24 |
| 302 | Dry | 0.20 | 0.70 | 52100 | Dry | 0.20 | 0.80 |
| | A | 0.20 | 0.22 | | A | 0.20 | 0.26 |
| | B | 0.54 | 0.19 | | B | 0.54 | 0.20 |
| 321 | Dry | 0.20 | 0.78 | | | | |
| | A | 0.20 | 0.23 | | | | |
| | B | 0.54 | 0.13 | | | | |
| 440 C | Dry | 0.20 | 0.72 | | | | |
| | A | 0.20 | 0.18 | | | | |
| | B | 0.54 | 0.16 | | | | |

[a]Oil A – Socony Vacuum Gargole PE797 (Paraffin type; VI = 105).
[b]Oil B – Esso Standard Millcot K-50 (Naphthenic type; VI = 77).

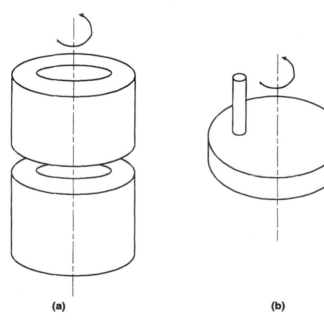

(a)                                        (b)

**Figure 8.5**  Coefficient of mutual overlap. (a) $K_{mut} = 1$; (b) $K_{mut} \approx 0$.

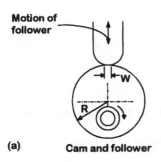

(a)  **Cam and follower**

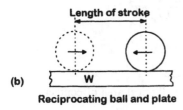

(b)  **Reciprocating ball and plate**

(c)  **Shaft and journal bearing**

**Figure 8.6** Illustration of evaluating number of passes based on the concept of mutual overlap. (a) Cam and follower; (b) reciprocating ball and plate; (c) shaft and journal bearing.

Figure 8.7 shows three groups of different types of contact. The procedure used by IBM for calculating the surface shear stress is as follows:

(a) For area contacts

$$S_s = Kq_0\sqrt{\frac{1}{4} + \mu^2}$$

(b) For line contacts

$$S_s = Kq_0\left[\frac{1+\mu}{2}\right]$$

when sliding in the circumferential direction and

$$S_s = Kq_0\sqrt{\frac{1}{4} + \mu^2}$$

when sliding in the axial direction

where

$q_0$ = the maximum Hertzian contact pressure
$\mu$ = coefficient of friction from Table 8.4
$K$ = stress concentration factor at the edges or corners, which depends on sharpness

(c) For spheres on spheres or crossed cylinders (point contacts)

$$S_s = q_0 \sqrt{\frac{1}{4}\left((1-2v)\,\frac{a}{a+b}\right)^2 + \mu^2}$$

where

$q_0$ = maximum Hertzian contact pressure
$a, b$ = half major and minor axes for the elliptical area of contact
($a = b$ for circular contact)
$v$ = Poisson's ratio

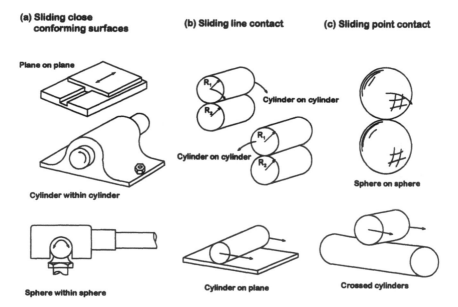

**(a) Sliding close conforming surfaces**

Plane on plane

Cylinder within cylinder

Sphere within sphere

**(b) Sliding line contact**

$R_1$
$R_2$
Cylinder on cylinder

$R_1$
Cylinder on cylinder $R_2$

Cylinder on plane

**(c) Sliding point contact**

Sphere on sphere

Crossed cylinders

**Figure 8.7** Examples of types of contact. (a) Sliding close-conforming surfaces: plane on plane, cylinder within cylinder, sphere within sphere; (b) sliding line contact: cylinder on cylinder, cylinder on cylinder, cylinder on plate; (c) sliding point contact: sphere on sphere, crossed cylinders.

## 8.5   WEAR BY MICROCUTTING

Another mechanism for wear is the penetration of hard asperities into a softer material under conditions, which produce microcutting. An illustrative model for this mechanism is shown in Fig. 8.8, where an asperity with radius $R$ is penetrated a depth $h$ in the softer material and is sliding with respect to it. The equilibrium equations can be written as:

$$\mu N = Q\cos\alpha - P\sin\alpha$$
$$N = Q\sin\alpha + P\cos\alpha$$

from which

$$\tan\alpha = \frac{Q - \mu P}{\mu Q + P}$$

where

$P$ = normal resistance at the contact
$Q$ = shear resistance at the contact
$\mu$ = coefficient of friction

The cutting condition occurs when sliding relative to the bulge is not possible. For this condition:

$$\mu N > Q\cos\alpha - P\sin\alpha$$

because

$$h = R(1 - \cos\alpha)$$
$$\cos\alpha = \frac{1}{\sqrt{1 + \tan^2\alpha}}$$

Therefore, the condition for cutting can be expressed as:

$$\frac{h}{R} \geq 1 - \frac{\mu Q + P}{(Q^2 + P^2)(1 + \mu^2)}$$

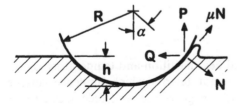

**Figure 8.8**   Model for microcutting.

In general, $P$ and $Q$ are functions of the strength of the surface layer $\sigma_s$. Accordingly,

$$P = C_1 \sigma_s$$
$$Q = C_2 \sigma_s$$

where $C_1$ and $C_2$ are constants. Therefore, the condition for cutting:

$$\frac{h}{R} \geq \left(1 - \frac{\mu C_2 + C_1}{\sqrt{(C_1^2 + C_2^2)(1 + \mu^2)}}\right)$$

and the depth of penetration ratio, which controls sliding or microcutting, depends on the coefficient of friction and the material properties of the surface layer.

Accordingly, by assuming $C_2 = C_1$, the value of $h/R$ for the onset of microcutting can be calculated as:

| $\mu$ | 0.0 | 0.1 | 0.2 | 0.3 | 0.4 | 0.5 | 0.6 |
|---|---|---|---|---|---|---|---|
| $\dfrac{h}{R}$ | 0.293 | 0.226 | 0.168 | 0.12 | 0.081 | 0.05 | 0.0 |

An empirical expression for the relationship between $h/R$ and $\mu$ for the onset of cutting can be written as:

$$\frac{h}{R} = (0.56 - 1.3\mu)^2$$

from which:

| $\mu$ | 0.0 | 0.1 | 0.2 | 0.3 | 0.4 | 0.43 |
|---|---|---|---|---|---|---|
| $\dfrac{h}{R}$ | 0.314 | 0.185 | 0.09 | 0.029 | 0.004 | 0.0 |

## 8.6  THERMAL WEAR

Frictional heating and the associated heat partition and temperature rise in sliding contacts are known to be major factors which influence surface damage. Wear in brakes and scoring in gears are well recognized to be thermally induced surface failures. The former occurs in the unlubricated condition, whereas the latter occurs in the presence of lubrication. It is also well known that in the case of brakes made of hard materials, surface cracks are likely to appear. Similarly in the case of gears made of very hard materials, surface cracks are known to occur and propagate to form surface initiated pits or in certain cases, complete fracture of the tooth.

A network of cracks is frequently observed on surfaces subjected to repeated heating and cooling as a result of the thermal gradients between the surface layers and the bulk material (thermal fatigue). Cracks can also occur if the surface is subjected to one sudden temperature change (thermal shock).

Each thermal cycle creates a microscopic internal change in the material structure. Subsequent cycles cause cumulative change and eventually create microscopic cracks at voids in the material or at the grain boundaries. If these cracks are propagated in the surface layers, they can produce wear or pitting. They can also produce fracture if they propagate deeper in the bulk material, usually under the influence of cyclic mechanical loading.

Microscopic thermal cracks or potential crack sites on or near the surface may occur as a result of the manufacturing or heat treating process, especially for hard materials.

The objective of this section is to illustrate the importance of surface temperature not only on scoring, surface cracking, and pitting, but also on different forms of wear and surface damage.

### 8.6.1 Mechanism of Scoring

Scoring of surfaces is universally recognized as the result of high temperatures at the contacts called the critical temperature. In the case of unlubricated surfaces, this temperature is generally considered to cause softening or melting of the surface layers of the material, the chemical layer, the solid lubricant film or any coated layer which exists on the surface.

In lubricated conditions with nonreactive lubricants, the common hypothesis is that scoring will be initiated when the temperature reaches a critical value beyond which the lubricant loses its adsorption characteristics (desorption) and consequently, fails to wet the surface. This temperature is widely known as the Blok flash temperature $T_f$ (13, 14] and is given for the case of rolling and sliding cylinders as:

$$T_f = K \frac{f W_n (V_1 - V_2)}{(C_1 \sqrt{V_1} + C_2 \sqrt{V_2}) \sqrt{\dfrac{b}{2}}}$$

where

$K$ = constant for the material lubricant

$f$ = coefficient of friction

$W_n$ = normal load per unit length

$b$ = width of the contact band

$V_1, V_2$ = surface velocities

$C_1, C_2$ = constants of materials which are the square root of the product of the thermal conductivity, specific heat, and density

A modification of Blok's formula was proposed by Kelly [15] for similar materials with consideration of surface roughness. The formula is given as:

$$T_T = T_B + K \frac{fW_n(\sqrt{V_1} - \sqrt{V_2})}{\left(1 - \frac{S}{50}\right)\sqrt{\frac{b}{2}}}$$

where

$T_T$ = total surface temperature

$T_B$ = material bulk temperature

$S$ = rms surface roughness (μin.)

$K$ = constant for the material lubricant combination

## 8.6.2 Mechanism for Surface Crack Initiation

It is generally accepted that the penetration of asperities causes plastic deformation in the surface layers where the yield point is exceeded at the real area of contact. Below the plastically compressed layer are layers under elastic compression. As soon as the asperity moves, the elastically compressed layers will exert upon the plastic layer a force, which will create in it a state of tension. Consequently, tensile stresses will appear on the surface in such conditions.

The sliding motion also generates a temperature field, which penetrates the surface layers. The maximum temperature occurs at the contact surface and decreases with increasing distance from the surface as discussed in Chapter 5. Accordingly, the surface layer is thermally elongated more than the subsurface layers and will experience compressive stresses imposed by the bulk material. If this compressive stress exceeds the yield stress, then a tensile residual stress will be induced in the surface after cooling. It should also be noted that the temperature at the real area of contact can be very high at high sliding speeds which results in reducing the yield strength significantly and thus, increasing the stressed zone.

The tensile thermal stress on the surface can be calculated from [16]:

$$\sigma = \alpha E \frac{q}{\pi(1 - \nu)\beta \rho c}$$

where

$q$ = heat flux caused by friction = $\mu P_0 V \alpha_p$

$\mu$ = coefficient of friction

$\alpha$ = coefficient of thermal expansion

$P_0$ = pressure on the real area of contact

$V$ = sliding velocity

$\alpha_p$ = coefficient of heat partition = $\dfrac{\sqrt{K_1 \rho_1 C_1}}{\sqrt{K_1 \rho_1 C_1} + \sqrt{K_2 \rho_2 C_2}}$

$K$ = thermal conductivity

$\rho$ = density

$C$ = thermal capacity

$E$ = modulus of elasticity

$\beta$ = thermal diffusivity = $\sqrt{\dfrac{K}{\rho C}}$

A combination of mechanically induced stresses and thermal stresses in the nominal contact region, or in the real area of contact, generate surface or near surface cracks, which can propogate with repeated asperity action to generate delamination of the surface layer [17] or wear debris from shallow pits. The influence of the thermal effect becomes more significant at high loads, high coefficient of friction, and high sliding speeds.

As illustrated by the parametric analysis in Chapter 5, the physical, chemical, and thermal properties of the lubricant can have significant influence on the maximum surface temperature. These properties control the amount of separation between rubbing surfaces and the thermal properties of the chemical layers generated on them.

## 8.7  DELAMINATION WEAR

Delamination wear denotes the mechanism whereby material loss occurs as a result of the formation of thin sheets (delaminates) with thickness dependent on the normal load and the coefficient of friction. The sequence of events which leads to the delamination can be summarized as follows:

Surface tractions applied repeatedly by asperity action produce subsurface deformation.
Cracks are nucleated below the surface.
Further loading causes the cracks to extend and propagate joining neighboring ones.
The cracks propagate parallel to the surface at a depth governed by the material properties and the coefficient of friction.
After separation from the surface laminates may be rolled due to the sliding action to form wear debris.

A comprehensive analysis of delamination wear can be found in Ref. 17.

## 8.8  ABRASIVE WEAR

Abrasive or cutting wear takes place when hard particles are present between the rubbing surfaces. Such particles include metallic oxides, abrasive dust, and hard debris from the environment. These particles first penetrate the metal and then tear off relatively large particles from the surface. It is one of the most common forms of wear and can be manifested in scratching marks or gouging of the surfaces [18, 19].

The load and the size of the abrasive particles relative to the thickness of the lubricating film are major factors which affect the weight loss by abrasive wear. The equation for abrasive wear can be expressed as:

$$V = k \frac{NL}{3\sigma_s}$$

where

$V$ = wear volume
$N$ = normal load
$L$ = sliding distance
$\sigma_s$ = surface strength
$k$ = wear coefficient

Representative values for $k$ given by Rabinowicz are tabulated below:

It should be noted the abrasive wear may result from, or can be accelerated by, the wear particles themselves. Wear particles for unlubricated steel can be as large as $50\,\mu m$ in size. For well-lubricated steel, they are in the order of $2–3\,\mu m$. Clearance between well-lubricated surfaces should be at least $4\,\mu m$ in order to allow the wear particles to leave the contact region.

## 8.9 CORROSIVE WEAR

Corrosive or chemical wear takes place when the environmental conditions produce a reaction product on one or both of the rubbing surfaces and this chemical product is subsequently removed by the rubbing action. A common example is the corrosive wear of metals in air, which usually contains humidity and other industrial vapors. Oxides or hydroxides of the metals are continuously formed and removed. Carbonates and oxycarbonates may also occur from the normal $CO_2$ present in the air. Chlorides and oxychlorides are known to occur in industrial environments or in near-ocean operations.

The use of an appropriate lubricant can inhibit the corrosion mechanism and provide the necessary protection in a corrosive environment. On the other hand, the lubricant itself may contain chemical elements, which react with the metals. The degree of effectiveness of the lubricant in reducing corrosive wear will depend on its chemical composition and the amount of dissolved water which may naturally exist in it.

An example of intentionally inducing corrosive wear to prevent a more severe condition of surface damage is the use of extreme pressure (EP) additives in the lubricant. This is a common practice when scoring, galling, or scuffing is to be expected. The EP additive reacts with the surface at the locations where high pressures and high speeds create high temperatures and consequently catastrophic galling or seizure are replaced by mild corrosive wear. References 20–26 contain more details and experimental data on the subject for the interested reader.

## 8.10 FRETTING CORROSION

This type of surface damage generally occurs in mechanical assemblies such as press fits and bolted joints due to the combination of high normal pressure and very small cyclic relative motion. It is characterized by discoloration of the mating surfaces and wearing away of the surfaces.

Many examples can be cited in the literature of the existence of fretting corrosion in machine parts and mechanical structures [27–33]. It is reported to be influenced by the hardness of the materials, the surface temperature, the coefficient of friction, humidity, lubrication, and the chemical environment. One of the early empirical formulas is that proposed by Uhlig [30] as:

$$W = (k_0 P^{1/2} - k_1 P)\frac{N}{f} + k_2 aPN$$

where

$W$ = total weight loss (mg)
$P$ = pressure (psi)
$N$ – number of cycles
$f$ = frequency (Hz)
$a$ = slip distance (in.)
$k_0, k_1 k_2$ and constants

The constants for his data are:

$$k_0 = 5.05 \times 10^{-6}, \qquad k_1 = 1.51 \times 10^{-8}, \qquad k_2 = 4.16 \times 10^{-6}$$

Measures, which can be used to reduce fretting include the minimization of the relative movement, reducing friction, use of an appropriate dry or liquid lubricant and increasing the surface resistance to abrasion.

## 8.11 CAVITATION WEAR

Cavitation is defined as the formulation of voids within or around a moving liquid when the particles of the liquid fail to adhere to the boundaries of the passage way. It can produce erosion pitting in the material when these voids collapse. Cavitation was first anticipated by Leonard Euler in 1754 to occur in hydraulic turbines. It is known to occur in ship propellers operating at high speed [34–36].

The mechanism of cavitation wear is generally explained by the formation of bubbles where the absolute pressure drops below the vapor pressure of the surrounding liquid. These bubbles collapse at extremely high velocities producing very high pressures over microscopically small areas. The smaller the size of the bubble, the smaller the velocity of collapse and consequently, the smaller the pressures produced. There appears to be a correlation between the rate of pitting and the vapor pressure and the surface tension of the liquid.

The equilibrium of a vapor bubble can be expressed as:

$$P_i = P_e - \frac{2S}{r}$$

where

$P_i$ = internal pressure
$P_e$ = external pressure
$S$ = surface tension
$r$ = radius of the bubble

and $P_i$ equals the vapor pressure.

The capillary energy $E$ of the bubble can also be expressed as:

$$E = 4\pi r_0^2 S$$

where

$r_0$ = radius of the bubble before collapse

This energy of collapse is generally considered to be the cause of cavitation erosion pitting and wear.


## 8.12  EROSIVE WEAR

Erosive wear occurs due to the change of momentum of a fluid moving at high speed. It has been observed in the wear of turbine blades and in the elbows of high-speed hydraulic piping systems. In its extreme condition, erosive wear is the mechanism utilized in water jet cutting systems. The change in the fluid particle velocity ($\Delta V$) as it impinges on the metal surface can create a high impact pressure which is a function of the density of the fluid and the modulus of elasticity of the impacted material [37, 38]. The effect of the high pressures on wear is partly enhanced by the shearing action of the liquid as it flows across the surface.

The pressure generated due to the change in velocity can be quantified as:

$$P = (\Delta V)\sqrt{E\rho}$$

where

$P$ = impact pressure
$E$ = modulus of elasticity of the material
$\rho$ = density of the material

Surface damage due to erosive wear can be reduced by elastomer coating [39] and cathodic protection [40]. The latter process causes hydrogen to be liberated and to act as a cushion for the impact.

Erosive wear is used to advantage in the cutting, drilling, and polishing of brittle materials such as rocks. The erosive action can be considerably enhanced by mixing abrasive particles in the fluid. Empirical equations for the use of water jets with and without abrasives in cutting and drilling are given later in the book.

## REFERENCES

1.  Hays, D., Wear Life Prediction in Mechanical Components, F. F. Ling Ed., Industrial Research Institute, New York, NY, 1985, p.5.
2.  Kragelski, I. V., Friction and Wear, Butterworths, Washington, D.C., 1965.
3.  Archard, J. F., "Contact and Rubbing of Flat Surfaces," J. Appl. Phys., Vol. 24, 1953.
4.  Archard, J. F., and Hirst, W., "The Wear of Metals Under Lubricated Conditions," Proc. Roy. Soc., 1956, A 236.
5.  Barwell, J. T., and Strang, C. D., "On the Law of Adhesive Wear," J. Appl. Phys., 1952, Vol. 23.
6.  Rabinowicz, E., "Predicting the Wear of Metal Parts," Prod. Eng., 1958, Vol. 29.
7.  Rabinowicz, E., Friction and Wear of Materials, John Wiley & Sons, New York, NY, 1965.
8.  Krushchov, M. M., and Babichev, M. A., Investigation of the Wear of Metals, USSR Acad. Science Publishing House, 1960.
9.  Krushchov, K. K., and Soroko-Navitskaya, A. A., "Investigation of the Wear Resistance of Carbon Steels," Iav. Akad. Nauk, SSSR, Otd. Tekh. Nauk., 1955, Vol. 12.
10. Mechanical Design and Power Transmission Special Report, Prod. Eng., Aug. 15, 1966.
11. Bayer, R. G., Shalkey, A. T., and Wayson, A. R., "Designing for Zero Wear," Mach. Des., Jan. 9, 1969.
12. Bayer, R G., and Wyason, R., "Designing for Measureable Wear," Mach. Des., Aug. 7, 1969.

13. Blok, H., "Les Temperatures de Surfaces dan les Conditions de Craissage sans Pression Extreme," Second World Petroleum Congress, Paris, June 1937.
14. Blok, H., "The Dissipation of Frictional Heat," Appl. Scient. Res., Sec. A, 1955, Vol. 5.
15. Kelly, B. W., "A New Look at Scoring Phenomena of Gears," SAE Trans., 1953, Vol. 61.
16. Barber, J. R., "Thermoplastic Displacement and Stresses Due to a Heat Source Moving over the Surface of a Halfplane," Trans. ASME, J. Eng. Indust., 1984, pp. 636–640.
17. Suh, N. P., and coworkers, The Delamination Theory of Wear, Elsevier, New York, NY, 1977.
18. Haworth, R. D., "The Abrasion Resistance of Metals," Trans. Am. Soc. Metals, 1949, Vol. 41, p. 819.
19. Avery, H. S., and Chapin, H. J., "Hard Facing Alloys of the Chromium Carbide Type," Weld. J., Oct. 1952, Vol. 31(10), pp. 917–930.
20. Uhlig, H. H., Corrosion Handbook, J. Wiley, New York, NY, 1948.
21. Evans, U. R., Corrosion Protection and Passivity, E. Arnold, London, England, 1946.
22. Avery, H. S., Surface Protection Against Wear and Corrosion, American Society for Metals, 1954, Chapter 3.
23. Larsen, R. G., and Perry, G. L., Mechanical Wear, American Society for Metals, 1950, Chapter 5.
24. Godfrey, D., NACA Technical Note No. 2039, 1950.
25. Wright, K. H., Proc. Inst. Mech. Engrs, London, 1B, 1952, p. 556.
26. Row, C. N., "Wear – Corrosion and Erosion, Interdisciplinary Approach to Liquid Lubricant Technology," NASA, SP-318, 1973.
27. Almen, J. O., "Lubricants and False Brinelling of Ball and Roller Bearings," Mech. Eng., 1937, Vol. 59, pp. 415–422.
28. Temlinson, G. A., Thorpe, P. L., and Gough, J. H., "An Investigation of Fretting Corrosion of Closely Fitting Surfaces," Proc. Inst. Mech. Engrs, 1939, Vol. 141, pp. 223–249.
29. Campbell, W. E., "The Current Status of Fretting Corrosion," ASTM Technical Publication, No. 144, June 1952.
30. Uhlig, H. H., "Mechanism of Fretting Corrosion," J. Appl. Mech., 1954, Vol. 21(4), p. 401.
31. Waterhouse, R. B., "Fretting Corrosion," Inst. Mech. Engrs, 1955, Vol. 169(59), pp. 1157–1172.
32. Kennedy, N. G., "Fatigue of Curved Surfaces in Contact Under Repeated Load Cycles," Proc. Int. Conf. on Fatigue of metals, 1956, Inst. Mech. Engrs, Sept. 1956, pp. 282–289.
33. Oding, I. A., and Ivanova, V. S., Fatigue of Metals Under Contact Friction," Proc. of Int. Conf. on Fatigue of Metals, Inst. Mech. Engrs, 1956, pp. 408–413.
34. Poulter, T. C., "Mechanism of Cavitation-Erosion," J. Appl. Mech., March 1942.

35. Nowotny, H., "Destruction of Materials by Cavitation," V.D.I., May 2, 1942, Vol. 86, pp. 269–283.

36. Mousson, J. M., "Pitting Resistance of Metals Under Cavitation Conditions," Trans. ASME, July 1937.

37. Bowden, F. P., and Brunton, J. H., "The Deformation of Solids by Liquid Impact at Supersonic Speeds," Proc. Roy. Soc., 1961, Vol. A263, p. 433.

38. Bowden, F. P., and Field, J. E., "The Brittle Fracture of Solids by Liquid Impact, by Solid Impact, and by Shock," Proc. Roy. Soc., 1964, Vol. A282, p. 331.

39. Kallas, D. H., and Lichtman, J. Z., "Cavitation Erosion," Vol. 1 of Environmental Effects on Polymeric Materials, Chapter 2, Wiley-Interscience, New York, NY, 1968.

40. Plesset, M. S., "On Cathodic Protection in Cavitation Damage," J. Basic Eng., 1960, Vol. 82, p. 808.

# 9

# Case Illustrations of Surface Damage

## 9.1 SURFACE FAILURE IN GEARS

The factors influencing gear surface failures are numerous, and in many cases their interrelationships are not completely defined. However, it can be easily concluded that the gear materials, surface characteristics, and the properties of the lubricant layer are to a great extent responsible for the durability of the surfaces.

It is widely accepted that pitting is a fatigue phenomenon causing cracks to develop at or below the surface. It is also known that lubrication is necessary for the formation of pits [1]. The dependence of pitting on the ratio of total surface roughness to the oil film thickness is suggested by Dawson [2–4].

Wear has been explained as a destruction of the material resulting from repeated disturbances of the frictional bonds [5]. Reduction or prevention or wear may be accomplished by maintaining a lubricant film thickness above a certain critical thickness [6]. Recent work in elastohydrodynamic lubrication [7–19] makes it possible to predict the thickness of the lubricant layer and the pressure distribution within the layer. Scoring is believed to be a burning or tearing of the surfaces. This tearing is caused by metal-to-metal contact at high speed when the lubricant film fails and cannot support the transmitted load. The failure of the lubricant film has been attributed to a "critical temperature" of the lubricant [20]. Experimental evidence shows that the lubricant failure for any particular lubricant–material combination occurs at a constant critical temperature [21, 22].

### 9.1.1  The Significant Parameters for Surface Damage

Surface damage in gear systems is influenced by the following variables: load intensity, geometry of the contacting bodies, physical properties of the surfaces, rolling and sliding velocities, properties of the lubricant, presence of abrasive or corrosive substances, existence of surface layers and their chemical composition, surface finish, and surface temperature. According to the elastohydrodynamic theory [7–19], most of these variables also govern the thickness of the lubricant film, which suggests that the major role is played by the lubricant layer in the control of surface damage.

The first step in structuring a design system is to identify the significant parameters affecting the design. The fundamental parameters for the problem under consideration will be taken as:

Load intensity $W$, normal to the surface;
Oil inlet temperature $T_0$;
Lubricant viscosity at $T_0$ ($\mu_0$);
Effective modulus of elasticity of teeth $E' = 1/[(1/E_1) + (1/E_2)]$;
Effective radius of curvature at contact $R' = 1/[(1/R_1) + (1/R_2)]$;
Rolling and sliding velocity of teeth in contact $U, V$;
Surface finish $S$;
Pressure coefficient of viscosity of the lubricant $\alpha$;
Pressure–temperature coefficient of viscosity $\gamma$;
Thermal properties of the tribological system.

There are certain groups of these parameters, which are believed to collectively affect surface damage. The most important of these groups are the Hertzian contact stress, the lubricant film thickness, and the maximum localized temperature rise in the film. Simplified expressions, which can be used for these groups are:

Maximum Hertzian stress:

$$q \cong 0.59 \left(\frac{E'W}{R'}\right)^{1/2} \tag{9.1}$$

Maximum temperature rise:

$$\Delta T \cong (0.0036) \left(\frac{50}{50-S}\right) W^{3/4} N_p^{1/2} \tag{9.2}$$

Minimum thickness of lubricant film:

$$h = (2.5 \times 10^{-5}) R' (B \times 10^6)^C$$

where

$$B = \frac{2\mu_0 \left( \dfrac{\alpha + \nu}{T_0} \right)}{R'} \left( \frac{W}{E'R'} \right)^{-0.125} \tag{9.3}$$

and

$$C = 1.1 \qquad \text{for } B \leq 10^{-6}$$
$$= 0.64 \qquad \text{for } B \geq 10^{-6}$$

Although there is no uniformity of opinion on the nature of the role played by the Hertzian-type stress field in surface damage, there is general agreement between investigators that the maximum Hertzian stress is a significant parameter, whose value should be kept within certain bounds if damage to the surface is to be avoided. The pressure distribution in the oil film between lubricated rollers is also believed to conform closely to the Hertzian stress distribution. The derivation of the equation for calculating the maximum Hertzian stress, Eq. (9.1), can be found in many texts. The explanation for Eqs. (9.2) and (9.3) is given in the following. It should be noted that the above expressions are intentionally simplified to facilitate the illustration of the design procedure. Among the important factors neglected in these equation are the load distribution across the contact, the errors and the elastic deflection of the teeth, and the effect of the variation of the coefficient of friction on the maximum temperature rise.

### 9.1.2  Maximum Oil Film Temperature

The temperature rise in the oil film is calculated according to AGMA guide for Aerospace Spur and Helical Gears [23]. This gives:

$$\Delta T = W_t^{3/4} N_p^{1/2} \left( \frac{50}{50 - S} \right) \frac{0.0175 \left( \sqrt{\rho_p} - \sqrt{\dfrac{\rho_G}{m_G}} \right)}{(\cos \phi)^{3/4} \left( \dfrac{\rho_p \rho_G}{\rho_p + \rho_G} \right)^{1/4}}$$

where

$W_t$ = tangential load per unit length of contact $= \dfrac{W}{\cos \phi}$

$N_p$ = pinion (rpm)

$S$ = surface finish (rms, $\mu$in.)

$\phi$ = pressure angle

$\rho_p$ = radius of curvature for pinion tooth

$\rho_G$ = radius of curvature for gear tooth

$m_G$ = gear ratio

Analysis of many examples of typical gears showed that the factor:

$$\frac{1}{\cos\phi}\left(\frac{0.0175\left(\sqrt{\rho_P}-\sqrt{\dfrac{\rho_G}{m_G}}\right)}{(\cos\phi)^{3/4}\left(\dfrac{\rho_P\rho_G}{\rho_P+\rho_G}\right)}\right)$$

is approximately equal to 0.0036. Therefore, for convenience, the maximum temperature rise in the oil film can be calculated from Eq. (9.2).

### 9.1.3  Minimum Oil Film Thickness

The analytical, as well as the experimental results on film thickness reported by many investigators [7, 11–14, 16, 18] are plotted in Fig. 9.1 versus the dimensionless parameter:

$$\frac{2\mu_0\left(\alpha+\dfrac{\gamma}{T_0}\right)U}{R'}\left(\frac{W}{E'R'}\right)^{-0.125}$$

Because the oil inlet temperature is conveniently considered as the gear blank temperature and because of the many unknowns in applying a general equation to calculate the minimum film thickness between gear teeth, a conservative design curve can be selected (shown in Fig. 9.1 by the lines a–b–c) which represents a safe lower limit. Since the equation is too conservative, especially for relatively thin films, the following alternative equation suggested by Dowson and Higginson [10] may be used:

$$h = 5 \times 10^{-6}(\mu_0 R' U)^{1/2} \text{ (in.)}$$

where

$\mu_0$ = oil viscosity at the oil temperature (poises)

Some design graphs are given in this section to illustrate the influence of the design and operating parameters on the performance of conventional gears with involute tooth profiles. These graphs can be useful in the initial selection of the main design variables, as well as in gaining qualitative understanding of the effect of the operating variables on the surface durability and the dynamic behavior of cylindrical gears in mesh.

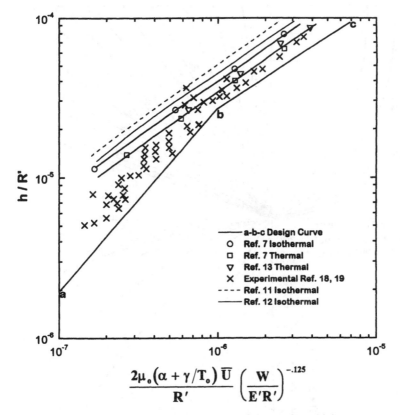

**Figure 9.1**  Selection of design curve for minimum film thickness.

## 9.1.4  Maximum Allowable Oil Sink Temperature for Wear Avoidance

The nomogram given in Fig. 9.2 can be useful in understanding the interaction between the surface roughness and the lubricating oil for wear avoidance in a particular gear pair. Notice the absence of load in the nomogram, as it has little influence on the wear when an adequate lubrication film is achieved. The graph is based on a simplified elastohydrodynamic lubrication analysis where the speed of the gear (the shower element), the shaft center distance, the tooth surface roughness, and the type of oil determine the maximum allowable sink temperature necessary for preventing metal-to-metal contact. The higher the temperature above the allowable sink temperature, the higher the wear rate and the more influence the transmitted load has on it. The nomogram can also guide the selection of an appropriate oil cooling system when necessary.

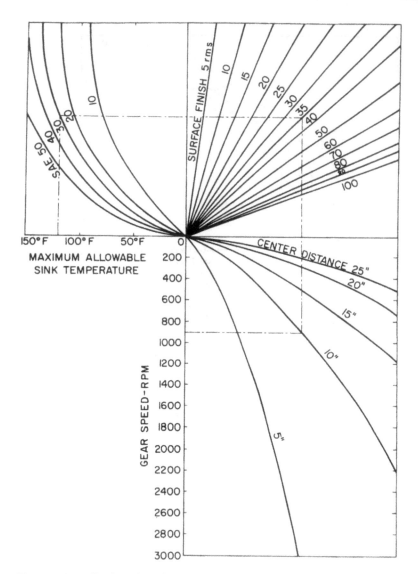

**Figure 9.2**  Design chart for gear lubrication.

### 9.1.5  Lubrication Factor for Surface Durability

Most gear rating and design practices for surface durability are based on the concept of contact fatigue resulting from the Hertzian stress field. This type of approach clearly ignores the effect of lubrication, surface roughness, and the relative sliding between the teeth. Figure 9.3 gives dimensionless relationships that can be used to quantify the reduction in useful life due to wear and pitting which can occur due to inadequate lubrication. The life is normalized with respect to the ideal case of full film lubrication without asperity contacts. This is plotted as a function of the ratio of the elastohydrodynamic film thickness, $h_0$, to the surface roughness, $S$ [24].

### 9.1.6 Dimensionless Maximum Instantaneous Temperature Rise on the Tooth Surface

Figure 9.4 gives a dimensionless plot for the temperature rise at the starting point of contact as a function of the number of teeth, $N_p$, in the pinion and the gear ratio, $m_G$ [25]. The graphs are for standard teeth with 20° pressure angle, where:

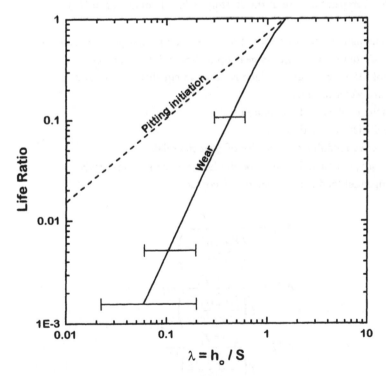

**Figure 9.3** Life ratio for minimum wear with equal load.

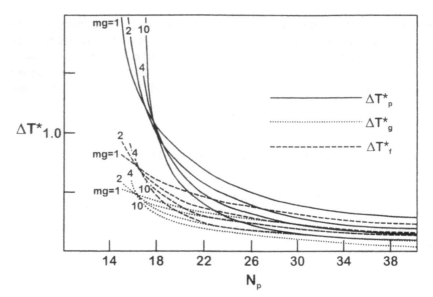

**Figure 9.4** Temperature rise at the starting point of contact ($\phi = 20°$).

$\Delta T_p$ = maximum temperature on pinion dedendum at first point of contact

$\delta T_G$ = maximum temperature on gear tip at first point of contact

$\Delta T_f$ = Blok flash temperature which is based in equality of $\Delta T_p$ and $\Delta T_G$

$C_d$ = shaft center distance

$\omega_p$ = angular velocity of the pinion

$f$ = coefficient of friction

$E_e$ = effective modulus of elasticity of gear materials

$\rho, k, c$ = density, conductivity, and specific heat for the gear materials

$w_t$ = tangential load per unit length of contact

and

$$\Delta T_p^* = \frac{\Delta T_p}{f \left( \dfrac{E_e \omega_p^2 W_t^3 C_d}{\rho^2 k^2 c^2} \right)^{1/4}}$$

$$\Delta T_G^* = \frac{\Delta T_G}{f \left( \dfrac{E_e \omega_p^2 W_t^3 C_d}{\rho^2 k^2 c^2} \right)^{1/4}}$$

$$\Delta T_f^* = \frac{\Delta T_f}{f \left( \dfrac{E_e \omega_p^2 W_t^3 c_d}{\rho^2 k^2 c^2} \right)^{1/4}}$$

### 9.1.7 Qualitative Comparison Between the Nominal Hertz Contact Stress and the Nominal Instantaneous Stress due to Thermal Shock

Here, a set of figures illustrate the conditions where thermal shock due to the transient temperature rise on the surface and subsequent cooling in mixed lubrication becomes significant when compared to the Hertz contact stress [26]. Figures 9.5–9.8 show that surface stress resulting from thermal shock should be given serious consideration for small number of teeth and high pitch line velocity. The figures also show that using the stub teeth or tip relief can considerably reduce the influence of thermal shock. The following parameters are considered in the illustrative example:

Center distance, CD = 10 in. and 60 in.
Gear ratio, GR = 5
Pressure angle = 20°
Coefficient of friction = 0.05
Pinion speed = 1800 rpm

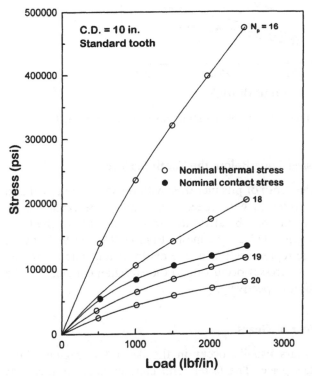

**Figure 9.5** Nominal thermal and contact stress for standard gear teeth.

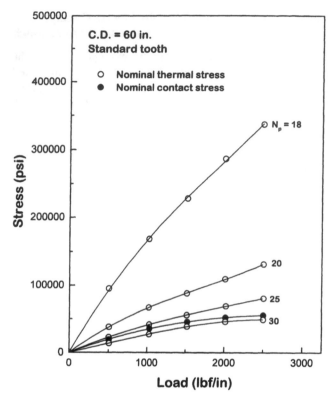

**Figure 9.6**  Nominal thermal and contact stress for standard gear teeth.

### 9.1.8  Depth of Stressed Zone Below the Tooth Surface

It is well known that the depth of pits increases with the increase in load and
size of the gear and decreases with speed. This cannot be explained by
Hertzian stresses alone and may be attributed to the influence of the transient thermal stresses generated at the mesh. Figures 9.9 and 9.10 give a
comparative parametric representation of the depth of the zone below the
surface where significant stresses occur due to the Hertzian contact and the
transient heat generation respectively.

### 9.1.9  Dedendum Wear of Gears

Pitting and wear of gears usually occur in the dedendum region where
"negative sliding" takes place. The latter term characterizes the fact that
the dedendum is always the slowest element of the sliding surfaces. This

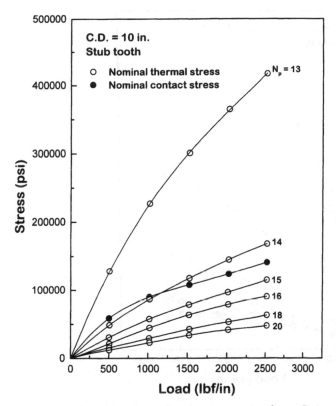

**Figure 9.7** Nominal thermal and contact stress for stub gear teeth.

condition results in higher temperature rise in the dedendum region and consequently higher thermal stresses or thermal shock. The tooth surface in the dedendum region is inherently subjected to cyclic tensile stress due to bending. These two factors to one degree or another can play an important role in initiating and propagating the surface cracks to form wear debris or pits depending on the state of the stress, the microhardness, the metalurgical structure, and the existence of defects of inclusions in the surface region.

## 9.2 ROLLING ELEMENT BEARINGS

Rolling element bearings represents some of the most critical components in rotating machinery. Because of the ever-increasing demands on higher relia-bility and longer life, these bearings are continuously subjected to extensive

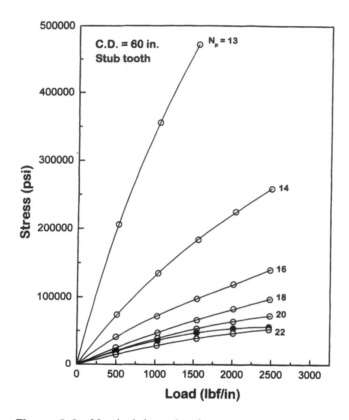

**Figure 9.8** Nominal thermal and contact stress for stub gear teeth.

studies with a view towards improving their design, manufacturing, materials, and lubrication. As in the case of gears, surface damage is the most important factor controlling their useful life.

The desired requisites for steel used for rolling element bearings are:

High fatigue strength
High elastic strength – resistance to plastic indentation
Resistance to sliding or rubbing wear
Structural stability at operating temperatures
A low level of nonmetallic inclusions and alloy or carbide segregation
  which serve as internal stress concentrating factors
Relative insensitivity to internal and external stress concentration
Resistance to environmental chemical corrosion

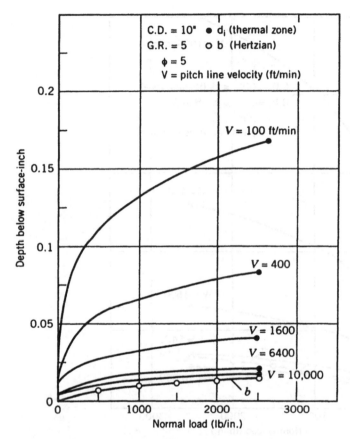

**Figure 9.9** Nominal size of the stressed zones below the surface.

Rolling element bearing performance is strongly dependent upon the degree of separation of the rolling elements and raceways by means of a lubricant film. The ratio of the minimum film thickness under operating conditions should be greater than 1.4, otherwise any skidding which may occur inside the bearing will cause rapid deterioration of its useful life.

### 9.2.1 Contact Stress Calculations

The three types of contacts in rolling element bearings are illustrated in Fig. 9.11. They represent circular, elliptical, and rectangular contacts respectively. The stress distribution in the contact zone can be calculated accord-

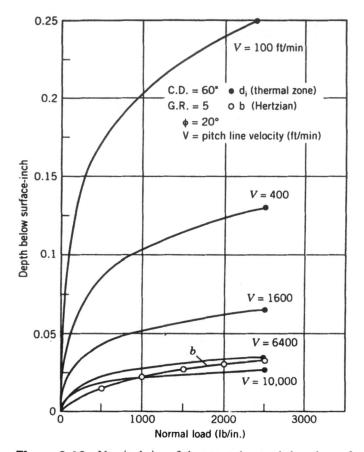

**Figure 9.10**   Nominal size of the stressed zones below the surface.

ing to the Hertz theory discussed in Chapter 2. The maximum compressive stresses for the three conditions are given by:

$$q_0 = \frac{3P_0}{2\pi a^2} \tag{A}$$

$$q_0 = \frac{3P_0}{2\pi ab} \tag{B}$$

$$q_0 = \frac{2P_0}{\pi Lb} \tag{C}$$

where $P_0$ is the total transmitted load. The values of $a$ and $b$ are determined by the radii of curvature of the contacting bodies, as given in Chapter 2.

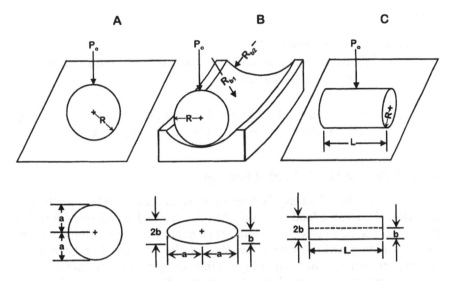

**Figure 9.11** Schematic representation of different types of rolling contacts.

## 9.2.2 Bearing Surface Fatigue Life

Rolling fatigue is a case of fatigue under combined stress where the material in the contact zone is subjected to reversed subsurface shear stress under high triaxial compressive stress [27–30]. Accordingly, it is not surprising that hardened steel can sustain much higher reversed shear stresses in rolling contact than in other loading conditions.

The statistical nature of bearing contact fatigue is evident in the scatter of the life obtained for identical bearings tested under ideal conditions [31–35]. It is not unusual for the longest life in a group of 100 identical bearings to be 50 times the life of the first bearing to fail. The probability of failure in most bearing tests generally follows a Weibull distribution. Consequently, the life for any desired probability of failure can be estimated with a confidence level based on 50% failure tests. The statistical nature of life can be attributed to the fact that most fatigue spalls has been clearly associated with nonmetalltic inclusions. Residual stresses can also influence the scatter in bearing life.

The general equation used for estimating bearing life is:

$$\frac{L_1}{L_2} = \left(\frac{q_2}{q_1}\right)^9 \approx \left(\frac{(P_0)_2}{(P_0)_1}\right)^3$$

where

$L_1, L_2$ = life at load levels 1, 2

$q_1, q_2$ = the corresponding maximum compressive stress on the surface

$(P_0)_1, (P_0)_2$ = applied loads on the bearing

Important factors which influence the life of bearings are the operating temperature and the choice of clearance and lubricant [34, 36, 37].

### 9.2.3 Failure of Lightly Loaded Bearings

It is not uncommon for early failures to occur in bearings which are loaded far below their rated load, because they are selected based on size rather than load. The rolling elements under such conditions usually undergo sliding or skidding action against the race or the retainer. Excessive wear, spalling, or scoring can result which significantly reduce the useful life or lead to premature failure. One such failure is the development of surface thermal cracks in retainers made of steel or aluminum, which propagate into fracture and consequently cause catastrophic failure of the bearing. It is interesting to note that this type of failure does not occur when the retainers are made of bronze. Prevention of skidding under such conditions can be achieved by the appropriate choice of clearance for the purpose of creating an induced load to force a rolling action. Another approach is to use properly designed hollow rolling elements as illustrated by the experimental study discussed in the following section.

### 9.2.4 An Experimental Investigation of Cylindrical Roller Bearings Having Annual Rollers

Hollow roller bearings have long been used for heavy duty applications such as the work rollers of a rolling mill. In such cases, the main purpose of using hollow rollers is to install as many rollers as possible within a limited circumferential space in order to increase the bearing capacity [38].

Annual roller bearings are expected to minimize skidding under low loads and would be useful in marginally lubricated applications where wear can be a problem. They may also improve load distribution between rollers and better thermal characteristics. The expected result would be lower temperature rise, reduced bearing wear, and longer life.

Besides the benefits of reduced skidding between the cage and roller set, an additional benefit occurs when considering the theoretical fatigue life of high speed, radial roller bearings. As outlined by Jones [39], the centrifugal force effects of solid rollers cause an additional loading at the outer race contact (and a second-order, not significant unloading at the inner race contact).

Harris and Aaronson [40] made analytical studies of bearings with annual rollers to investigate the load distribution, fatigue life, and the skidding of rollers. Their work shows that hollow rollers increase the fatigue life of the bearing and decrease the skidding between the cage and roller set. They suggested that attention should be paid, however, to the bending stress of the rollers and to the bearing clearance.

This section describes an experimental study undertaken by Suzuki and Seireg [41] to compare the performance of bearings with uncrowned solid and annular rollers under identical laboratory conditions. Bearing temperature rise and roller wear are investigated in order to demonstrate the advantages of using annual rollers in applications where skidding can be a problem.

## Test Bearings

The two bearings used in the study have the same dimensions and configurations with the exception that one bearing has annular rollers and the other bearing has solid rollers. The details of the bearings are given in Table 9.1.

Brass is selected as the roller material in order to rapidly demonstrate the effect of annular rollers on temperature rise, and roller wear.

The ratio of the inside to the outside diameter of the hollow roller is taken as 0.3. Three sets of inner rings with different outside diameters are used for each bearing in order to produce the different clearances.

Special efforts were undertaken in machining the rollers and rings to approach the dimensional accuracy and surface finish of conventional hardened bearing steels.

## Test Fixture

The experimental arrangement is diagrammatically represented in Fig.9.12. the two test bearings (a) and (b) (one with hollow rollers and another with solid rollers) were placed symmetrically near the middle plane of a shaft (c). The shaft is supported by two self-lubricated ball bearings (d) on both sides. A variable speed drive is used to rotate the shaft through a V-belt (e) and and a pulley (f) at one end of the shaft.

The load is applied radially on the outer rings (g) inside which the bearing is placed by changing the weight (j) suspended at one end of a bar (k). The latter loads a fulcrumed beam-type load divider, which is especially designed to provide identical loads on both bearings. A strain gage ring-type load transducer (i) monitors the load applied on the test bearings to confirm the equality of the load on them at all times. Separate

**Table 9.1**   Test Bearing Specifications

| | | |
|---|---|---|
| Bearing outside diameter | 4.3305 in. | (10.99947 cm) |
| Bearing inside diameter | 1.9682 in. | (4.999228 cm) |
| Bearing width | 1.06 in. | (2.6924 cm) |
| Outer race inside diameter | 3.719 in. | (9.4462 cm) |
| Inner race outside diameter | 2.5658 in. | (6.517132 cm) |
| | 2.5637 in. | (6.511798 cm) |
| | 2.5620 in. | (6.50478 cm) |
| Roller diameter | 0.5766 in. | (1.464564 cm) |
| Roller length | 0.659 in. | (1.67386 cm) |
| Number of rollers | 12 | |
| Roller inside diameter | 0.1719 in. | (0.436626 cm) |
| Diameter ratio | 0.3 | |
| Bearing radial clearance | 0.0021 in. | (0.005334 cm) |
| | 0.0038 in. | (0.009653 cm) |
| Roller material | Brass | |
| Outer and inner race material | Mild steel | |
| Surface finish for rollers and races | 8–10 μin.-rms | |

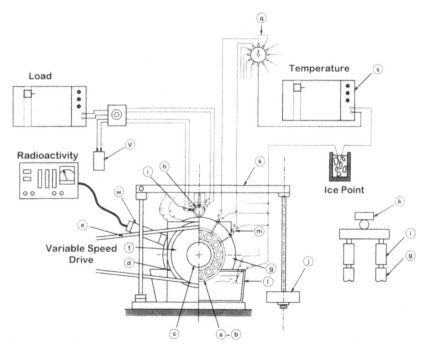

**Figure 9.12**   Diagrammatic representation of experimental setup for dynamic test.

oil pans (l) are placed below each of the test bearings. Oil is filled to the level of the centerline of the lowest roller. Copper–constantan thermocouples are used to measure the bearing temperatures as well as the oil sump temperature. The bearing thermocouples are embedded 30° apart at 0.01 in. (0.25 mm) below the surface of the outer rings where rolling takes place. The thermocouples are connected to a recorder (s) through a rotary selection switch (q), and a cold box (r).

## Results

Figure 9.13a shows the time history of the outer race temperature rise for the bearing with annular rollers. Steady-state temperature conditions are reached after approximately two hours. Figure 9.13b shows the bearing temperature rise as well as oil temperature rise at steady state conditions for a shaft speed of 1000 rpm. The temperature rise for both solid rollers and annular rollers are essentially the same at this speed. At speeds of 2000 rpm and 3000 rpm, on the other hand, the temperature with solid rollers is higher than that with annular rollers. The temperature rise differences are most pronounced at 2000 rpm.

## Wear Measurement

The radioactive tracing technique used in the test is similar to that used by L. Polyakovsky at the Bauman Institute, Moscow for wear measurement in the piston rings of internal combustion engines.

The test specimens (hollow or solid rollers) are bombarded by a high-energy electron beam emitting gamma rays. The strength of the bombardment is governed by the energy of the electron beam, the exposure time, and the material of the specimen. The radioactivity, which naturally decays with time, is also reduced with wear of the bombarded surface. The rate of reduction of the radioactivity is approximately proportional to the depth of wear. The amount of wear can therefore be detected by monitoring the radioactivity of the specimen and using a calibration chart prepared in advance of the test. The main advantage of this method is the ability to detect roller wear without disassembling the bearing. The disassembling process is not only time consuming, but it may also alter the wear pattern of the test specimens.

In this study, one roller in each bearing is bombarded and assembled with the rest of the rollers. A scintillation detector ($\omega$) is placed on the outer surface of the outer ring of the bearing (Fig. 9.12) and a counter is used to monitor the change in radioactivity of the bombarded rollers. The diameter of rollers is periodically measured using an electric height gage to check the accuracy of the radioactive tracing technique.

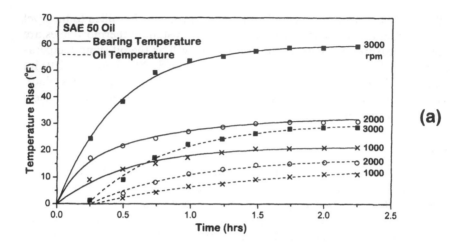

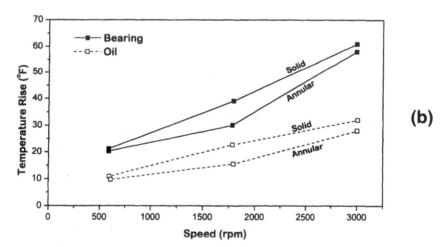

**Figure 9.13** (a) Temperature rise–time history for the bearing with annular rollers. (b) Temperature rise at steady-state conditions.

The shaft speed for the wear test is selected as 3000 rpm and kept unchanged. SAE 10W oil is used as the lubricant for the test bearings to accelerate the roller wear. The bearing outer race temperature and oil temperature are monitored throughout the test.

Figure 9.14a shows a comparison of the wear of the rollers during the test. As can be seen from the figure, the wear of the annular rollers is

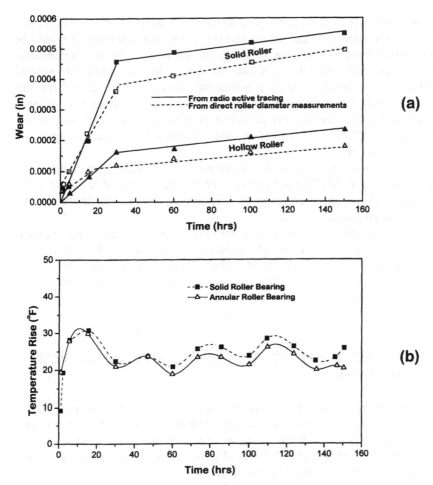

**Figure 9.14** (a) Roller wear. (b) Temperature difference between bearings and oil.

considerably lower than that of the solid rollers. It should be noted that after an initial running period of 30 hours, the oil was changed and a considerably lower rate of wear resulted. The wear rate during this phase of the test is shown as $5.7 \times 10^{-7}$ in./h ($14.5 \times 10^{-6}$ mm/h) for the hollow roller as compared to $8 \times 10^{-7}$ in./h ($20.4 \times 10^{-6}$ mm/h) for the solid roller.

It was observed throughout the test that the wear detected using radio-active tracing technique is slightly higher than that measured directly using the electric height gage. The reason may lie in the fact that the wear detected by the radioactive tracing technique is an average wear, which includes the indentations due to local pittings or flakings. Consequently, if the interest is to study the effect of wear on the change of bearing clearance, it would be more appropriate to use the height gage for measuring the dimensional change. On the other hand, if the interest is to investigate the surface damage, the radioactive tracing technique would be a good tool for this purpose. Better accuracy can be expected with this technique when steel rollers are used. Gamma-ray emission is stronger with steel and consequently the influence of the radioactivity existing in the natural space on the results is reduced.

The temperature rise in the bearings and oil during the wear test is shown in Fig. 9.14b. The temperature of the outer race of the solid roller bearing is shown to be consistently higher than that of the annular roller bearing at all times.

It is interesting to note that the annular roller exhibited a small number of local pits scattered on the rolling surface. In the solid roller, however, a large number of pits were observed in the rolling direction only at the central region of the rolling surface. This may also be due to the cooling effect at the ends of the rollers.

## 9.3 SURFACE TEMPERATURE, THERMAL STRESS, AND WEAR IN BRAKES

The high thermal loads, which are generally induced in friction brakes, can produce surface damage and catastrophic rotor failure due to excessive surface temperatures and thermal fatigue. The temperature gradients and the corresponding stresses are functions of many parameters such as rotor geometry, rotor material, and loading history.

Due to the wide use of frictional brakes, an extensive amount of work has been undertaken to improve the performance and extend the life of their rotors. Some research has been aimed towards studying the effects of rotor geometry on the temperature and stress distribution using classical analytical [42–45] or numerical [46–52] methods. Other studies have concentrated on investigating the effects of rotor materials on the performance of the brake [53–55].

The efforts to improve the automobile braking system performance and meet the ever increasing speed and power requirements had resulted in the introduction of the disk braking system which is considered to be better than the commonly used drum system. A newer system which is claimed to be superior to both of its predecessors is now being introduced. The crown system [56] which can be viewed as a cross brake, with a drum rotor and a disk caliper, combines the advantages of both drum and disk systems. It has the loading symmetry of the disk caliper which results in less mechanical deformation. It also has the larger friction surface areas and heat exchange areas of the drum which result in better thermal performance and lower temperatures. A study by Monza [56], in which the disk and crown are compared, indicated that more weight and cost reduction are attainable by using the crown system. Moreoever, under similar testing conditions, the crown rotor showed 10–20% lower operating tempratures than its counterpart.

This section is aimed at investigating the thermal and thermoelastice performance of rotors subjected to different types of thermal loading. Although there are many procedures in the literature for the analysis of temperature and stress in brake rotors based on the finite element method [1, 3, 8, 9], these procedures would require considerable computing effort. Efficient design algorithms can be developed by placing primary emphasis on the interaction between the design parameters with sufficient or reasonable accuracy. Sophisticated analysis can then be implemented to check the obtained solution and insure that the analytical simplifications are acceptable.

For the thermoelastic analysis in this section, a simplified one-dimensional procedure is used. The rotor is modeled as a series of concentric circular rings of variable axial thickness. Furthermore, it is assumed that the rotor is made of a homogeneous isotropic material and that the axial temperature and stress variations are negligible. The procedure first treats the thermal problem to predict the temperature distribution which is then used to compute the stress distributions.

### 9.3.1 Temperature Rise Due to Frictional Heating

This algorithm used to calculate the temperature rise is a simplified one-dimensional finite difference analysis. The analysis consideres the transient radial temperature variations and neglects both axial and circumferential variations. The rotor, which is subjected to a uniform heat rate, $Q_r$, at its external, internal or both cylindrical surfaces dissipates heat through its exposed surfaces by convention only. The film coefficient depends only on the geometrical parameters.

The proposed analysis is based on the conservation of energy principles for a control volume. This can be stated as:

$$Q_{in} - Q_{out} = Q_{stored} \tag{9.4}$$

where $Q_{in}$ and $Q_{out}$ are the rate of energy entering and leaving the volume, by heat conduction and convection respectively and $Q_{stored}$ is the rate of energy stored in the volume. For the shaded element of Fig. 9.15, Eq. (9.4) with appropriate substitution becomes:

$$Q_{c,n} - Q_{v,n} - Q_{d,n} - Q_{v,n+1} - Q_{c,n+1} = Q_{stored} \tag{9.5}$$

where $Q_{c,n}$ and $Q_{c,n+1}$ are heat quantities entering and leaving the volume by conduction, and $Q_{v,n}$ and $Q_{d,n}$ are geometry dependent convection heat quantities entering and leaving the body depending on the surrounding temperature, $T_s$.

With a current temperature rise above room temperature, $T_{n,t}$ at the interface $n$, one can solve for the future temperature rise, at time $t+1$, for the same location [57]:

$$T_{n,t+1} = \frac{1}{M(A_n + A_{n+1})} \left[ \frac{2\Delta r}{k} (2A_{n-1}T_{n-1,t} + 2A_n T_{n+1,t}) \right.$$
$$+ \frac{2\Delta r}{k} (A_{d,n} + A_{l,n} + A_{u,n+1}) h_n T_s + \left( (M-2)(A_n + A_{n+1}) \right. \tag{9.6}$$
$$\left. - \frac{2\Delta r}{k} (A_{d,n} + A_{l,n} + A_{u,n+1}) h_n \right) T_{n,t} \right]$$

where

$$M = \frac{(\Delta r)^2}{\beta \Delta t}$$

$\beta = \dfrac{k}{\rho c}$ is the thermal diffusivity

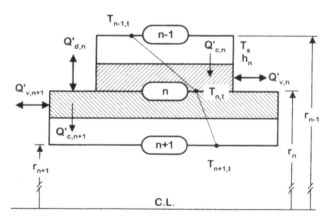

**Figure 9.15** Diagram used for the temperature algorithm.

Similar expressions can be obtained for the temperature at the inner and outer surfaces. The temperature rise in the next time step at the outer radius is:

$$T_{2,t+1} = \frac{1}{MA_2}\left[\frac{2\Delta r}{k}[(A_0 q_0) + (A_0 + A_{u,2})h_2 T_s] + 2A_2 T_{2,t}\right.$$
$$\left. + \left(A_2(M-2) - \frac{2\Delta r}{k}(A_0 + A_{u,2})h_2\right)T_{2,t}\right]$$

(9.7)

and the temperature rise in the next time step at the inner surface is obtained by replacing all the 2,0 and $u$ subscripts in Eq. (9.7) by $m$, $i$, and $l$, respectively.

In the above equations $A_0$, $A_l$, and $A_n$ are the cylindrical areas of the outer, inner, and interface surfaces, respectively. $A_{u,n}$ and $A_{l,n}$ are the ring side areas, upper, and lower halves. $A_{d,n}$ is the area generated by the thickness difference between two adjacent rings (refer to Fig. 9.15). As can be seen, the above algorithm can easily be modified to allow for any variations in heat input, convective film, and surrounding temperatures with location and time.

### 9.3.2 The Stress Analysis Algorithm

The geometrical model of this algorithm is identical to that of the temperature algorithm. For this analysis, both equilibrium and compatibility conditions are satisfied at the rings interfaces. Considering the inner and outer sides of the interface $r_{n+1}$ of Fig. 9.16, the continuity condition (or strain equality) can be expressed as a function of the corresponding stresses as follows [58, 59]:

$$(\sigma_{t,n+1})^0 + (\sigma_{T,n+1})^0 + \nu P_{n+1} = (\sigma_{t,n+1})^i + (\sigma_{T,n+1})^i + \left(\frac{f_n}{f_{n+1}}\right)\nu P_{n+1} \qquad (9.8)$$

where

$(\sigma_{t,n+1})^0, (\sigma_{t,n+1})^i = $ tangential stresses at the outer and inner side of interface $r_{n+1}$, respectively

$(\sigma_{T,n+1})^0, (\sigma_{T,n+1})^i = $ thermal stresses at the corresponding locations

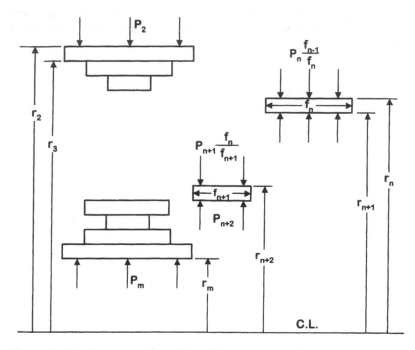

**Figure 9.16** Representation of the disk geometry and the notations used in the stress algorithm.

The radial stress $\sigma_{r,n+1}$ at the radius $r_{n+1}$, which is the average of the pressures on both sides of the interface, can be derived as:

$$\sigma_{r,n+1} = -\frac{1}{2}\left(\frac{1+f_n}{f_{n+1}}\right)P_{n+1} \tag{9.9}$$

The tangential component $\sigma_{t,n+1}$ is calculated by averaging the stress on both sides of the interface as:

$$\sigma_{t,n+1} = \frac{\begin{aligned}O_{n+1}(O_n - 1)P_{n+1} - 2(O_{n+1} - 1)\left(\frac{f_{n-1}}{f_n}\right)O_n P_n + \\ +2(O_n - 1)P_{n+2} - (O_n - 1)(O_{n+1} - 1)\left(\frac{f_n}{f_{n+1}}\right)P_{n+1}\end{aligned}}{2(O_n - 1)(O_{n+1} - 1)} \tag{9.10}$$

$$P_{n+2} = \frac{\beta E(O_{n+1} - 1)(T_n - T_{n+2})}{4(1 - \nu)} - G_n P_n + U_n P_{n+1} \tag{9.11}$$

where

$$G_n = \frac{B_n}{C_n}\left(\frac{f_{n-1}}{f_n}\right)$$

$$U_n = \frac{D_n}{C_n}$$

$$B_n = \frac{2O_n}{O_{n-1}}$$

$$C_n = \frac{2}{O_{n+1} - 1}$$

$$D_n = \frac{\dfrac{(1-v)+(1+v)O_n}{O_n - 1} + \left[[(1+v)+(1-v)O_{n+1}]\left(\dfrac{f_n}{f_{n+1}}\right)\right]}{O_{n+1} - 1}$$

where $O_n$ is a geometry function given by $O_n = (r_n/r_{n+1})^2$. Equations (9.9) and (9.10) are used to determine the radial and tangential stress distributions. Substitution in Eq. (9.11) for each node produces a set of simultaneous equations to be solved for the known boundary pressures $P_2$ and $P_m$ to give the radial distribution in the disk. This set of simultaneous equations is solved by assuming two arbitrary values for $P_3$ and using linear interpolation or extrapolation to satisfy the pressure $P_m$ at the inner boundary [52].

The temperature and stress algorithms are then coupled such that the temperature distribution is automatically used in the stress algorithm. This approach makes it possible to incorporate material properties and heat convectivity that are geometry and temperature dependent [58, 59]. Similar algorithms for disk brakes are given in Refs 60–62.

### 9.3.3 Numerical Examples

The coupled temperature–stress algorithm is used, as a module, to predict the temperature and thermal stresses generated by a given conductive heat flux applied at a given surface or surfaces of a disk of any given material and geometry. Several examples are considered to illustrate the capabilities of the developed algorithm.

The following geometrical, loading and material parameters are used in the considered cases:

Geometry:
Disk outer radius, $r_0 = 12.0$ in.
Disk inner radius, $r_i = 6.0$ in.
Disk thickness, $f_{max} = 12.0$ in.

Material:
Density, $\rho = 0.286\,\text{lb/in.}^3$
Young's modulus, $E = 30 \times 10^6\,\text{psi}$
Coefficient of thermal expansion, $\alpha = 7.3 \times 10^{-6}\,\text{in./(in.°F)}$
Thermal conductivity, $k = 26.0\,\text{BTU/(hr-ft-°F)}$
Specific heat, $c = 0.11\,\text{BTU/(lb-°F)}$

Loading conditions:
Total conductive heat flow rate, $Q_T$ (constant) $= 500{,}000\,\text{BTU/hr}$
Heating time, $t = 180\,\text{sec}$
Average convective heat transfer coefficient at exposed surfaces,
    $h = 5.0\,\text{BTU/(hr-ft}^2\text{-°F)}$

The case of a disk with uniform thickness is considered to investigate the effect of the loading location on the thermal and thermoelastic behavior of the disk by applying the total heating load at the disk outer surface and the inner surfaces respectively. The case where the load is shared equally between the two surfaces is also considered, as well as the case where the thermal load sharing between the surfaces is optimized [59]. The temperatures and tangential stresses for the three loading cases are shown in Tables 9.2–9.4.

The results obtained from the report study illustrate the significant effects of the loading location and load sharing ratio on the thermal and thermoelastic performance of brakes. Tables 9.2–9.4 show that when the thermal load is shared between the internal and external cylindrical surfaces, a considerable reduction can be expected in the temperature and stress magnitudes. It also indicates that the maximum tensile tangential stress is shifted from the inner or outer surface towards the middle where the probability of failure is reduced. The results also show that, for the given case, internal loading produces the highest temperature and stress

**Table 9.2**  The Maximum Temperatures (°F) for the Investigated Cases

| Load condition | $\left(\dfrac{r_i}{r_0}\right) = 0.25$ | $\left(\dfrac{r_i}{r_0}\right) = 0.50$ | $\left(\dfrac{r_i}{r_0}\right) = 0.75$ |
|---|---|---|---|
| 1. Uniform thickness and external loading | 448.9 | 465.2 | 0.75 |
| 2. Uniform thickness and internal loading | 1345.8 | 787.9 | 469.8 |
| 3. Uniform thickness and equal load sharing | 724.3 | 395.2 | 543.8 |
| 4. Uniform thickness and optimal load sharing | 338.9 | 295.9 | 302.8 |

**Table 9.3** The Maximum Tensile Tangential Stresses (psi) for the Investigated Cases

| Load condition | $\left(\dfrac{R_i}{r_0}\right) = 0.25$ | $\left(\dfrac{r_i}{r_0}\right) = 0.50$ | $\left(\dfrac{r_i}{r_0}\right) = 0.75$ |
|---|---|---|---|
| 1. Uniform thickness and external loading | 73,456.4 | 70,378.9 | 50,506.1 |
| 2. Uniform thickness and internal loading | 272,283.7 | 137,739.3 | 67,039.4 |
| 3. Uniform thickness and equal load sharing | 131,164.9 | 54,846.1 | 19,616.8 |
| 4. Uniform thickness and optimal load sharing | 50,283.1 | 33,954.6 | 16,393.5 |

magnitudes. This is due to the fact that the inner surface has a smaller area and consequently for a given heating input the flux is higher. The case of equally shared loading between the inner and outer surfaces allows for a larger area for the heat input, shorter penetration time, lower temperature gradients, and consequently lower thermal stresses. Optimization of the load sharing further improves the design.

### 9.3.4 Wear Equations for Brakes

Wear resistance in brakes is known to increase with increasing thermal conductivity of the material, its density, specific heat, and Poisson's ratio to its ultimate strength and resistance to thermal shock. It is also known to increase with decreasing thermal expansion and elastic modulus of the material.

The following equations can be used for material selection to provide longer wear life.

**Table 9.4** The Maximum Compressive Tangential Stresses (psi) for the Investigated Cases

| Load condition | $\left(\dfrac{R_i}{r_0}\right) = 0.25$ | $\left(\dfrac{r_i}{r_0}\right) = 0.50$ | $\left(\dfrac{r_i}{r_0}\right) = 0.75$ |
|---|---|---|---|
| 1. Uniform thickness and external loading | 73,456.4 | 70,378.9 | 50,506.1 |
| 2. Uniform thickness and internal loading | 272,283.7 | 137,739.3 | 67,039.4 |
| 3. Uniform thickness and equal load sharing | 131,164.9 | 54,846.1 | 19,616.8 |
| 4. Uniform thickness and optimal load sharing | 50,283.1 | 33,954.6 | 16,393.5 |

Brakes without surface coating:

$$\text{Wear resistance} \propto \frac{\sigma_u(1 - \mu)(c\rho k^3)^{1/4}}{\varepsilon E} \tag{9.12a}$$

Brakes with thin coated surface layer:

$$\text{Wear resistance} \propto \frac{\sigma_u(1 - \mu)\sqrt{k\rho c}}{\varepsilon E} \tag{9.12b}$$

Resistance to thermal shock:

$$\text{Resistance to surface crack formation} \propto \frac{k\sigma_0}{\varepsilon E} \tag{9.13}$$

where

$E = $ modulus of elasticity

$\mu = $ Poisson's ratio

$\varepsilon = $ coefficient of thermal expansion

$k = $ thermal conductivity

$c = $ thermal capacity (specific heat)

$\rho = $ density

$\sigma_u = $ ultimate strength

$\sigma_0 = $ resistance to crack formation (ductility)

## 9.4  WATER JET CUTTING AS AN APPLICATION OF EROSION WEAR

One of the beneficial applications of erosion wear is the use of high speed water jets for cutting and polishing. This section presents a review of the literature on the subject and provides dimensionless equations for modeling the erosion process resulting from the momentum change of a high-velocity fluid. The following nomenclature is used in all the equations given in this section.

### 9.4.1  Nomenclature

$c = $ instrinsic speed for rock cutting $= \dfrac{k\tau_0}{\eta_w n \mu_r g_0}$

$c_f, f$ = friction coefficient of rock
$d$ = kerf width = $2.5d_0$
$d, d_j$ = nozzle diameter
$E$ = Young's modulus of rock
$g = 9.81 \, \text{m/s}^2$, gravitational constant
$g_0$ = grain diameter
$h, z$ = depth of cut
$\Delta h$ = increment depth of cut by adding abrasive
$h_w$ = depth of cut by plain water jet
$K_0 = 2100 \, \text{MPa}$, the bulk modulus of the water
$K_1, \alpha$ = experiment constants
$l$ = average grain size of the rock
$n$ = porosity of the rock (%)
$p$ = jet pressure
$p_0$ = jet stagnation pressure
$p_c, p_{th}$ = rock threshold pressure
$Q_a$ = abrasive mass flow rate
$Q_w$ = water mass flow rate based on the measurement
$R$ = drilling rate
$r$ = radius of rotating jet
$s$ = jet standoff distance
$T = \dfrac{d_0}{v}$, the time of exposure
$u$ = feed rate
$v_l, v_j$ = jet velocity at nozzle exit
$v_t$ = jet traverse velocity
$\beta$ = jet inclination angle
$\beta_0$ = experimentally determined constant = 0.025
$\eta$ = damping coefficient
$\eta_w$ = viscosity of the water
$k$ = permeability of the rock
$\mu$ = dynamic viscosity of the water
$\mu_r$ = coefficient of internal friction of rock
$\mu_w$ = coefficient of friction for water
$v = 1.004 \times 10^{-6} \, \text{m}^2/\text{s}$, kinematic viscosity of the water at 20°C
$\rho$ = density of rock
$\rho_0$ = liquid density
$\rho_a$ = density of the abrasives ($\rho_a = 3620 \, \text{kg/m}^3$ for garnet,
$\quad \rho_a = 2540 \, \text{kg/m}^3$ for silica sand)
$\rho_w = 998 \, \text{kg/m}^3$, density of water
$\sigma_c$ = compressive strength of material
$\sigma_y$ = yield strength of target material

$\tau_0$ = force required to shear off one grain per typical grain area
$\omega$ = jet rotating speed

The problem of evaluating the performance of water jet cutting systems has received considerable attention in recent years and some of the many excellent studies are reported in Refs 63–70. Several investigators developed water jet cutting analytical models and several of these studies generated empirical equations based on specific test results. Some of these are briefly reviewed in the following.

Crow [63, 64] investigated the case of a rock feeding at a rate $v$ under a continuous water jet with diameter $d_0$ and pressure $p_0$ cutting a kerf of depth $h$. The jet will fracture the rock because the pressure difference on the exposed grains produce shear stress equal to the shear strength of the rock. During the process, friction along the sides of the kerf causes the jet velocity to decrease and thus the erosive power of the jet decreases.

Crow developed the following predictive equation or the kerf depth $h$:

$$h = 2\mu_w \frac{d_0 p_0}{\tau_0} \int_0^{\theta_0} \frac{e^{\mu_w(\theta-\theta_0)} \sin\theta}{1 + (v/c)\sin\theta} \, d\theta \qquad (9.14)$$

The model was tested by conducting experiments on four different types of rock. The results show that the proposed model gives a reasonable fit for only one rock type, Wilkeson sandstone, over the range $0.1 < v/c < 50$.

Based on a control volume analysis to determine the hydrodynamic forces acting on the solid boundaries in the slot and the Bingham-plastic model, which describes the time-dependent force displacement characteristics of the solid material to be cut, Hashish and duPlessis [65, 66] developed a continuous water jet cutting equation as follows:

$$\frac{z}{d_j} = \frac{1 - \dfrac{\sigma_c}{\rho v_1^2}}{\dfrac{2c_f}{\sqrt{\pi}}} \left[1 - \exp\left(\frac{2c_f}{\sqrt{\pi}} \frac{\rho_0 v_1}{\eta} \frac{v_1}{u}\right)\right] \qquad (9.15)$$

The maximum depth of cut $z_0$ achieved can be expressed as:

$$\frac{z_0}{d_j} = \frac{1 - \dfrac{\sigma_c}{\rho v_1^2}}{\dfrac{2c_f}{\sqrt{\pi}}} \qquad (9.16)$$

Equation (9.16) is used to determine the coefficient of friction, $c_f$, experimentally. Because the damping coefficient $\eta$ in Eq. (9.15) is an unknown material property, the authors had to determine $\eta$ for a particular material using an experimentally measured cutting depth $z$ with known values of $\sigma_c$, $c_f$, $d_j$, $v_1$, and $u$. The authors also pointed out that greater accuracy for a particular material can be achieved by choosing the optimum rheological model for the material.

Hood et al. [67] proposed a physical model of water jet rock cutting based on experimental observations. In this model when the main force of the jet acts on a ledge of rock within the kerf, the ledge is fractured and a new ledge forms against which the jet acts. This process continues until the friction along the wall of the kerf dissipates the jet energy to a value which is insufficient to break off the next ledge, or until the jet moves on to the next portion of the rock surface.

Considering the fact that a large number of variables influence the erosion process, they used the factorial method of experimental design to determine the jet pressure $(p)$ needed to cut a kerf to a specified depth $(h)$ in a given rock type with specified nozzle diameters $(d)$ and traverse velocities $(v)$. The factorial method yields an empirical model for a certain material that identifies and quantifies the relative importance of the variables. The equations are in the form:

$$h = C + K_1 v + K_2 p + K_3 d \tag{9.17}$$

where $C$, $K_1$, $K_2$, and $K_3$ are experimental constants.

Labus [68] developed an empirical water jet cutting equation based on data available in the literature. The general form of his proposed equation is:

$$\frac{h}{d} = K_1 \left[ \left( \frac{d}{s} \right) \left( \frac{p}{\sigma_c} \right) \left( \frac{v_j}{v_t} \right) 0.5 \right] \alpha \tag{9.18}$$

While performing tests in which a plane rock surface was exposed to a vertical stationary jet, Rehbinder [69, 70] observed that the rock immediately beneath the core of the impinging jet was not damaged but that an annular region of rock around this core was fractured. Rehbinder explained that the tensile force exerted in the rock grains by viscous drag of the water flow through the rock around the grains produced this damage.

He assumed that the stagnation pressure $p_0$ at the bottom of a slot drops exponentially (i.e. $p_0 = pe^{-\beta_0 h/D}$), and used Darcy's law to calculate the velocity of the water flow through the rock, and Stokes' law to calculate the velocity of the water flow through the rock, and Stokes' law to compute the force $F$ acting on the grain. He developed the following predictive equation for the kerf depth:

$$h = \frac{D}{\beta_0} \ln\left(1 + \frac{\beta_0 k p_0}{\mu l D} T\right) \quad T < \frac{\mu l D}{\beta_0 k p_0}\left(\frac{p_0}{p_{th}} - 1\right)$$

and

$$h = \frac{D}{\beta_0} \ln\left(\frac{p_0}{p_{th}}\right) \quad T > \frac{\mu l D}{\beta_0 k p_0}\left(\frac{p_0}{p_{th}} - 1\right) \tag{9.19}$$

### 9.4.2 Development of the Generalized Cutting Equations

Some of the significant published experimental data and "test specific" empirical relationships have been studied in the reported investigation with a view towards developing a generalized dimensionless equation for the water jet cutting process. A water jet cutting equation for the three materials considered, namely Barre granite, Berea sandstone, and white marble, has consequently been developed and is expressed as:

$$\left(\frac{h}{d_j}\right)\pi = 1.222 \times 10^{-9}\left(\frac{\rho_w}{\rho}\right)^{0.125}\left(\frac{p_c}{E}\right)^{-3.41}\left[\left(\frac{d_j}{s}\right)\left(\frac{p}{\sigma_c}\right)\left(\frac{p}{\rho v_t^2}\right)^{0.25}\right]^{1.3 f^{0.605} n^{-0.088}} \tag{9.20}$$

In deep kerfing by water jets, a rotary head with a dual nozzle is generally required. Because of the jet inclination angle $\beta$ and the combination of tangential and traverse velocities, Eq. (9.20) is modified and a generalized equation for water jet cutting using a rotary dual jet in deep slotting operations is developed as:

$$\left(\frac{h}{d_j}\right)\pi = 1.222 \times 10^{-9} \cos\beta\left(\frac{\rho_w}{\rho}\right)^{0.125}\left(\frac{p_c}{E}\right)^{-3.41}$$
$$\left[\left(\frac{d_j \cos\beta}{s}\right)\left(\frac{p}{\sigma_c}\right)\left(\frac{p}{\rho v_r^2}\right)^{0.25}\right]^{1.3 f^{0.605} n^{-0.088}} \tag{9.21}$$

where $v_r$ is the resultant traverse velocity which is calculated from $v_r = v_t + r\omega$.

The calculated $(h/d_j)_\pi$ from Eq. 9.20 versus the experimentally based $(h/d_j)_{exp}$ from Eq. (9.18) are plotted for three materials in Figs 9.17a–9.17c, respectively. As can be seen, an almost perfect correlation is achieved in all cases by using Eq. (9.20).

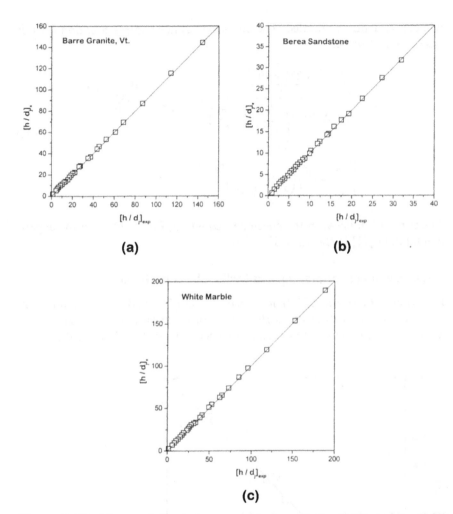

**Figure 9.17** The correlation between empirical model, Eq. (9.18), and Eq. (9.20) for (a) barre granite cutting; (b) berea sandstone cutting; (c) white marble cutting.

### 9.4.3 The Generalized Equation for Drilling

Equation (9.20) is extended for water jet drilling by taking into account the effect of the submerged cutting on the jet performance. Because in the vertical drilling process, the consumed water and the material fractured have to be squeezed out the hole and the jets have to penetrate the cut material before they can reach the target. A considerable amount of energy will consequently be needed. The developed dimensionless equation for this case is found to be:

$$\left(\frac{R}{d_j\omega}\right)_\pi = \frac{\dfrac{1.222 \times 10^{-9}}{2\pi}}{1 + 0.00527 \left(\dfrac{d_j}{\sqrt[3]{\dfrac{v^2}{g}}}\right)^2 \left(\sqrt{\dfrac{p}{K_0}}\right)} \cos\beta \left(\frac{\rho_w}{\rho}\right)^{0.125} \left(\frac{p_c}{E}\right)^{-3.41}$$

$$\left[\left(\frac{d_j \cos\beta}{s}\right)\left(\frac{p}{\sigma_c}\right)\left(\frac{p}{\rho(r\omega)^2}\right)^{0.25}\right]^{1.3 f^{0.605} n^{-0.088}}$$

(9.22)

The correlation between the developed equation, Eq. (9.22), and the experimental data [71] is shown in Fig. 9.18.

### 9.4.4 Equations for Slotting and Drilling by Abrasive Jets

Equations (9.20) and (9.22) for deep slot cutting and hole drilling were extended for predicting the performance of abrasive jets. A dimensionless modifying factor has been developed to account for the effect of the added

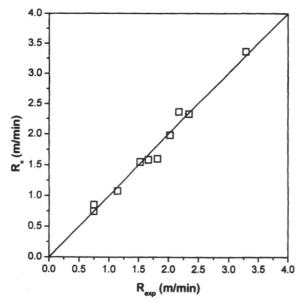

**Figure 9.18** Comparison between the model and the experimental data for drilling.

abrasives to the plain water jets, which gives excellent correlation with the published experimental data [72] for two types of abrasives, namely, silica sand and garnet, as shown in Fig. 9.19. This factor is expressed as:

$$\left(\frac{\Delta h}{h_w}\right)_\pi = 1.278\left(\frac{Q_a}{Q_w}\right)^{0.763}\left(\frac{\rho_a}{\rho_w}\right)^{1.769}$$  (9.23)

The deep slot cutting equation by abrasive jets can be readily generated by combining Eqs (9.21) and (9.23) as:

$$\left(\frac{h}{d_j}\right) = 1.222 \times 10^{-9} \cos\beta\left(\frac{\rho_f}{\rho}\right)^{0.125}\left(\frac{p_c}{E}\right)^{-3.41}\left[\left(\frac{d_j\cos\beta}{s}\right)\left(\frac{p}{\sigma_c}\right)\left(\frac{p}{\rho v_r^2}\right)0.25\right]^{1.3f^{0.605}n^{-0.088}}$$
$$\left[1 + 1.278\left(\frac{Q_a}{Q_w}\right)^{0.763}\left(\frac{\rho_a}{\rho_w}\right)^{1.769}\right]$$  (9.24)

Similarly, the dimensionless drilling rate equation by abrasive water jets can be readily obtained by multiplying the plain water drilling equation, Eq. (9.22), with the abrasive factor:

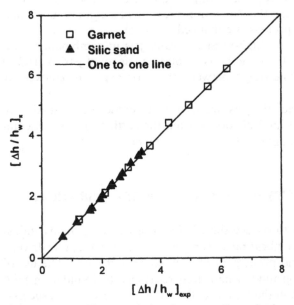

**Figure 9.19** The correlation between dimensionless analysis and the empirical model.

$$\left(\frac{R}{d_j\omega}\right) = \frac{\dfrac{1.222 \times 10^{-9}}{2\pi}}{1 + 0.00527\left(1 + \dfrac{Q_a}{q_w}\right)\left(\dfrac{d_j}{\sqrt[3]{\dfrac{v^2}{g}}}\right)^2\left(\sqrt{\dfrac{p}{K_0}}\right)} \cos\beta\left(\frac{\rho_f}{\rho}\right)^{0.125}\left(\frac{p_c}{E}\right)^{-3.41}$$

$$\times \left[\left(\frac{d_j\cos\beta}{s}\right)\left(\frac{p}{\sigma_c}\right)\left(\frac{p}{\rho(r\omega)^2}\right)^{0.25}\right]^{1.3 f^{0.650} n^{-0.088}}$$

$$\times \left[1 + 1.278\left(\frac{Q_a}{Q_w}\right)^{0.763}\left(\frac{\rho_a}{\rho_w}\right)^{1.769}\right]$$

(9.25)

Details of the determination of these equations are given in Ref. 73.

### 9.4.5 Jet-Assisted Rock Cutting

Various methods have been developed to improve the erosion process using water jets. One of these methods involves introducing cavitation bubbles into the jet stream. The rate of erosion is greatly enhanced when the bubbles collapse on the rock surface. Another method is to break up the continuous jet stream into packets of water that impact the surface. The stresses generally are much greater than the stagnation pressure of a continuous jet and consequently the rosion process is enhanced.

Two other methods use high-pressure water jets in combination with mechanical tools. One of these methods employs an array of jets to erode a series of parallel kerfs. The ridges between the kerfs are then removed by mechanical tools.

In the second method, the jets are utilized to erode the crushed rock debris formed by the mechanical tools during the cutting process. A comprehensive review of this approach is given by Hood et al. [74].

## 9.5 FRICTIONAL RESISTANCE IN SOIL UNDER VIBRATION

Vibration is widely used to reduce the frictional resistance in many industrial applications, such as vibrating screens, feeders, conveyors, pile drivers, agricultural machines, and processors of bulk solids and fluids. The use of vibration to reduce the ground penetration resistance to foundation piles was first reported in 1935 in the U.S.S.R. Resonant pile driving was successfully developed in the U.K. in 1965 and proved to be a relatively fast and quiet method.

Since the early 1950s, there has been increasing interest in the application of vibration to soil cutting and tillage machinery. Research has been carried out in many countries on different soil cutting applications. Successful implementations include vibratory cable plows for the direct burial of telephone and power cables in residential areas, and oscillatory plows and tillages.

Some of the published studies on the effect of vibrations on the frictional resistance in soils are briefly reviewed in the following.

Mogami and Kubo [75] investigated the effect of vibration on soil resistance. They related the reduction of strength in the presence of vibration to what they called "liquefication".

Savchenko [76] reported that the coefficient of internal friction of sand decreases with the increase in the amplitude and/or frequency of vibration. His tests on clay soils indicated similar reduction in shear strength by increasing the frequency and amplitude, but very little reduction at amplitudes greater than 0.6 mm.

Shkurenko [77] studied the effect of oscillation on the cutting resistance of soil. His result showed that at fairly high oscillation velocities, there is a considerable reduction in the cutting resistance in the range 50–60%.

Mackson [78] attempted to reduce the soil to metal friction by utilizing electro-osmosis lubrication. Mink et al. [78] experimented with an air lubrication method. The electric potential was large and the power for air compression was too high. Both methods have been shown to be uneconomical.

Choa and Chancellor [79] introduced combined Coulomb friction and viscous damping to represent the soil resistance to blade penetration. They determined the coefficient of viscous damping of soil by drawing a sub soiler into the soil several times at a depth of 10 in. without vibration and at varying forward speed $V$. The soil resistance $R$ was found as $R = 71.17V + 2250$ where 71.17 lb-sec/ft is the equivalent viscous coefficient and 2250 lb is the Coulomb friction.

All reported studies show that soil frictional resistance is greatly reduced under the influence of vibration. This is illustrated in the dimensionless plots given in Fig. 9.20, which are derived from the published experimental data [80].

## 9.6   WEAR IN ANIMAL JOINTS

Degradation of the cartilage in human and animal joints by mechanical means may be one of the significant causes of joint disease. It can influence almost all different types of degenerative arthritis or "osteoarthrosis". Although biochemical, enzymatic, hereditary, and age factors are important

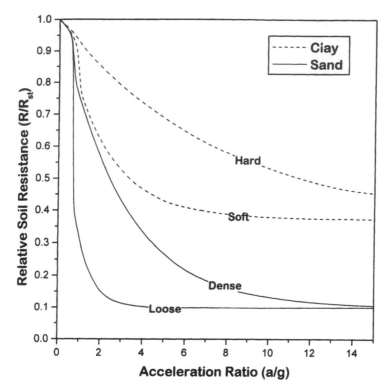

**Figure 9.20** Effect of vibration of soil resistance.

in controlling the structure and the characteristics of the cartilage and syno-
vial fluid, the joint is primarily a mechanical load-bearing element where the
magnitude and the nature of the applied stress is expected to be a major
contributor to any damage to the joint. There is no exclusive evidence in
the literature that joint degradation is simply a wear-and-tear phenomenon
related to lubrication failure [81–84]. However, many investigations give
strong indications that primary joint degeneration is not simply a process
of aging [85–87]. Also cartilage destruction resembling the changes seen
clinically can be created in the knees of adult rabbits by subjecting them to
daily intervals of physiologically reasonable impulsive loading [88]. The joint
degeneration by mechanical means may result from two types of forces:

    1.   A suddenly applied normal force with large magnitude and short
           duration that produces immediate destruction of the tissues as in
           severe crushing injuries or initiate damage in the form of micro-
           fractures.

2. A continually degrading force that leads to gradual destruction of the joint.

Radin et al. [88–90] and Simon et al. [91] have extensively investigated the effects of suddenly applied normal loads. The experiment reported in this section [92] investigates the effects of continuous high-speed rubbing of the joint *in vivo* when subjected to a static compressive load which is maintained constant during the rubbing. The patella joint of the laboratory rat was tested in a specially modified version of the apparatus developed by Seireg and Kempke [93] for studying the behavior of *in vivo* bone under cyclic loading. The load is applied to one joint while the other remains at rest. The factors investigated include changes in surface temperature at the joint, surface damage, cellular structure, and mineral content in the cartilage and bone.

### 9.6.1 The Experimental Apparatus and Procedure

The apparatus used in this investigation is shown diagrammatically in Fig. 9.21. It has a slider crank mechanism to produce a small reciprocating sliding motion at the rat joint. The amplitude of motion is controlled by adjusting the crank length on an eccentric wheel mounted on the shaft of a variable speed motor. A soft fabric strap transmits the cyclic motion to the leg. The leg is cantilevered to the mounting jig through a specially designed attachment, which provides a firm fixation of the leg at the distal end of the tibia with minimum ill effects. The tight clamp may restrict the blood circulation to the foot but not the leg. The blood supply to the joint would

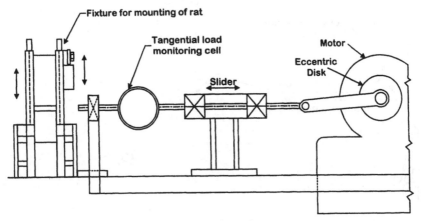

**Figure 9.21** Diagrammatic representation of test apparatus.

remain near normal. The configuration of the joint is shown on Fig. 9.22A. A schematic diagram is given in Fig. 9.22B to show the animal's leg in place with the static load and cyclic rubbing motion identified. A special fixture is designed for applying constant compressive loads to the joint Fig. 9.23. It has a spring-actuated clamp which can be adjusted to apply static

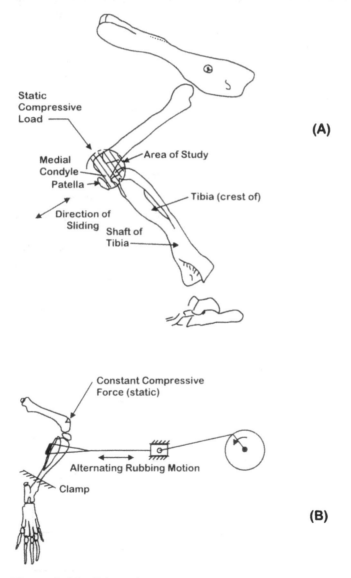

**Figure 9.22** Schematic representation of applied load and rubbing motion.

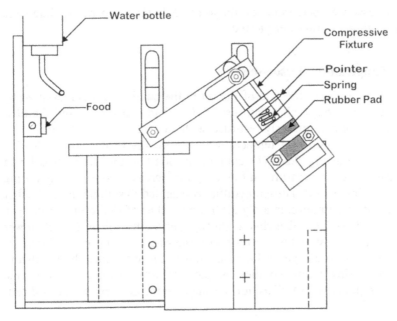

**Figure 9.23** Constraining rig showing the fixture for application of compressive load.

loads between 450 g and 3.6 kg. The spring is calibrated for continuous monitoring of the normal load which is applied to the rat joint through a soft rubber pad.

*Test Specimens*

The test specimens were all male white albino rats. Their weight varies from 300 to 350 g. The rats were maintained on mouse breeder blox and water. The room temperature was kept between 80 and 84°F. Nine rats were tested in this study with each three specimens subjected to identical load levels.

*Test Plan*

The right tibia of each rat was subjected to an alternating pull force between 0.0 and 90 g at a rate of 1500 cycles/min. All the tests were conducted at this value of the cyclic load with the compressive normal load fixed at 0.45, 0.9, and 1.8 kg, respectively. The duration of the testing was 2–3 hr every day for a period of 14 days. After that period the rats were sacrificed and the different tests were performed on the joint. Only the temperature data were obtained while the rats were tested.

### 9.6.2 Temperature Measurements

The temperature over the skin of the rat at the patella joint is measured by thermocouples. The combination used is iron and constantan and the temperature can be continuously recorded with an accuracy of ±0.1°F.

A typical variation of temperature on both the loaded and unloaded joints is shown in Fig. 9.24. The compressive load on the test joint is 1.8 kg in this case. The temperature on the test joint increased considerably during the first loading period. The temperature rise tended to stabilize after the first week of test to an approximately 2.5°F above that of the joint at rest. The latter showed no detectable change throughout the test. Progressively lower temperature rise resulted in the tests with the smaller compressive forces.

These results are in general agreement with those obtained by Smith and Kreith [94] using thermocouples on patients with acute gouty arthritis, rheumatoid patients, as well as normal subjects during exercise and bed rest.

### 9.6.3 Measurement of Changes in Mineral Content

The mineral content of the bone and the cartilage can be determined through the absorption by bone of monochromatic low-energy photon

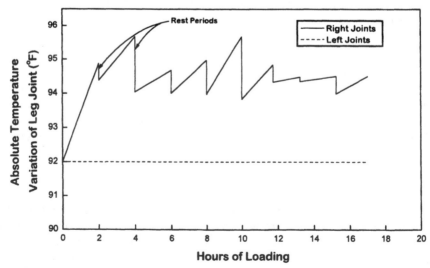

**Figure 9.24** Sample of skin temperature data near the joints.

beam which originates in a radioactive source (iodine 125 at 27.3 keV). The technique has been developed by Cameron and Sorenson [95].

The source and the detector system are rigidly coupled by mechanical means and are driven simultaneously in 0.025 in. steps in a direction transverse to the bone by a milling head attachment. Measurements of the transmitted photon beam through the bone are made for a 10 sec interval after each stop and are automatically used to calculate the mineral content.

A typical summary result is shown in Fig. 9.25 where the change in bone mineral ratio between the test joint and the one at rest are plotted as measured at different locations below the surface. In this case, the rat joint was subjected to a 1.8 kg compressive load for approximately three hours daily for a period of 14 days. It can be seen from the figure that the tested joint showed a significantly higher mineral content ratio at 0.025 in. below the surface which gradually reaches 1 at a distance of approx. 0.075 in. below the surface. Progressively smaller increases in the mineral content ratio resulted from the lower compressive loads. This result is interesting in view of the finding of Radin et al. [88] that increased calcification and stiffening of the rabbit joints occurred as a result of repeated high impact load. It shows that increased calcification can occur as well due to rubbing of the joint under static compression.

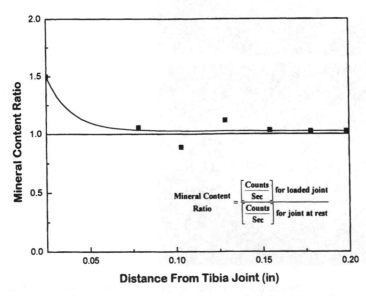

**Figure 9.25** Change in mineral content ratio below the surface.

### 9.6.4   Investigation of Surface Characteristics and Cellular Structure

The surface texture and condition of the loaded and the intact joints for each rat are investigated by means of the biological microscope for general observation, histological slides for the cellular structure, and the scanning electron microscope for close investigation of the load-bearing areas.

At the end of each test, the rat is sacrificed. The joint is then dissected and put in fixative so that the cells retain their shape. The fixative used is 0.1% glutra-aldehyde. When viewed under a biological microscope to a magnification of 25–40×, considerable wear of the smooth surfaces can be observed in the loaded joint as shown in Fig. 9.26 for a static compressive load of 1.8 kg.

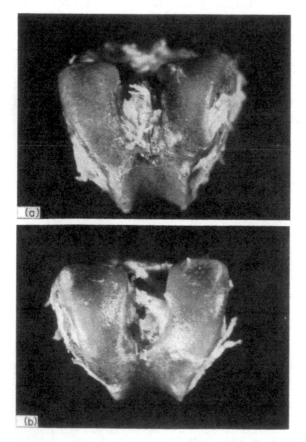

**Figure 9.26**   General views of (a) the surface of the rat joint tested under a 1.8 kg compressive load; (b) the joint at rest.

The slides of the histology studies are prepared at four different sections of the joint in both the tested and immobilized joints. The cellular structure is compared as shown in Fig. 9.27 for a compressive load of 1.8 kg and the following differences are observed:

1.    The surface is significantly rougher in the loaded joint as compared to the one at rest.
2.    The surface structure is compressed at some locations causing an increase in the mineral content. This observation is supported by the results of the photon absorption technique.

The procedure used for the electron microscope study of the structure of the cartilage is explained in detail by Redhler and Zimmy [96]. The specimens from the cartilage are fixed in 0.1% glutra–aldehyde in Ringers solution. The fixation takes approximately 4 hr. They are then passed through graded acetone. The concentration of the acetone is changed from 50, 70, 90, and 100% for a duration of 0.5 hr each. This is done to ensure that no moisture exists which may cause cracking when coated with gold and palladium alloys. The magnification used is 1000–3000× and the areas seen are primarily load bearing areas. The differences observed among the loaded, Fig. 9.28, and the intact, Fig. 9.29, joints can be summrized in the following:

1.    In the loaded specimens, the zoning which predominates in the normal cartilage disappears. The upper surface is eroded, and the radial pattern predominates throughout. The relatively open mesh underneath the surface is replaced by a closely packed

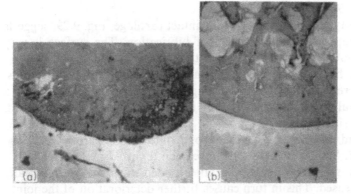

**Figure 9.27**   Section of rat joint tested (a) under a 1.8 kg compressive load; (b) at rest.

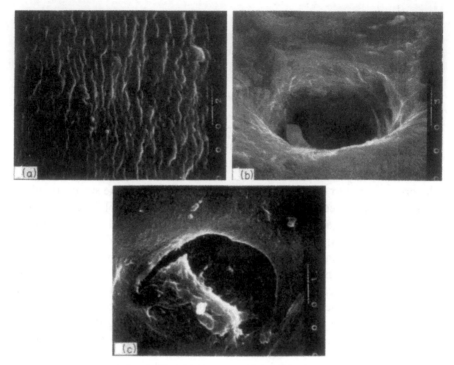

**Figure 9.28** Electron microscope results showing (a) surface roughness for joint subjected to 0.9 kg normal load; (b) surface pits for joint subjected to 1.8 kg compressive load; (c) surface tear for joint subjected to 0.9 kg compressive load.

network of thick coarse fibers, all radial in direction as shown in Fig. 9.28a.

2. The type of the surface of the intact cartilage, Fig. 9.29, suggests a trapped pool mechanism of lubrication. The surface is very smooth without serious asperities.

3. The fibers in the intact cartilage are oriented in all directions, whereas in the loaded cartilage they reorient themselves in a radial form, Fig. 9.28a. This is known to be common in old, arthritic joints [97].

4. Dead cells can be seen under some of the load-bearing areas, Fig. 9.28. Similar observations have been reported by McCall [97].

5. The surface roughness of the loaded cartilage is drastically increased. This in turn causes further deterioration of the joint.

6. Pits and tears appear in the loaded cartilage as illustrated in Figs 9.28b and c.

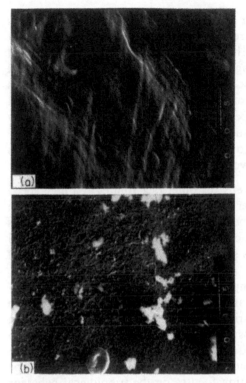

**Figure 9.29** (a), (b) Electron microscope results for the joint at rest for the rats of Figs. 9.28a and c.

## 9.7 HEAT GENERATION AND SURFACE DURABILITY OF RAMP–BALL CLUTCHES

### 9.7.1 Introduction

This section deals with the thermal-related problems and surface durability of ramp–ball clutches, which are generally used for one-directional load transmission and can be utilized in developing mechanical function generators. The surface temperature rise under fluctuating load conditions is predicted by using a simplified one-dimensional transient heat transfer model that is found to be in good agreement with finite element analysis. The depth of fretting wear due to repeated high-frequency operation is evaluated from the viewpoint of frictional energy density. A simplified model for fretting wear due to fluctuation of load without gross slip in the wedging condition is proposed by qualitatively guiding the design of the clutch.

It is well known that during sliding contact, the frictional energy is transformed to thermal energy, resulting in high surface temperature at the contact [98, 99]. If the high heat flux is periodic, the sharp thermal gradient might cause severe damage such as thermal cracking and thermal fatigue. High temperature can also cause change of the material properties of the surface layer, acceleration of oxidation, poor absorption of oil, and material degradation. High temperature may also occur at the asperity contacts due to cyclic microslip, such as in fretting corrosion [100–104].

In a ramp–ball clutch (refer to Fig. 930), the heat generated during its operation can be classified into two categories:

1.  *Overrunning mode.* Usually the outer race rotates at a high speed with respect to the inner race during the overrunning mode. The balls, under the influence of the energizing spring, will always contact both races and consequently produce a sliding frictional force. This condition is similar to the case of lightly loaded ball bearing.
2.  *Wedging mode.* The ramp–ball clutch utilized in mechanical function generators [105] can be ideally designed to operate on the principle of wedge. During the wedging mode, the combination of high oscillating pressure and microslip at the contact due to load fluctuation generates frictional heat on the surfaces of the balls and both races and, consequently may cause fretting-type damage. This investigation focuses on the tribological behavior in the wedging mode only because of its importance to the function generator application.

### 9.7.2   Analysis of the Wedging Condition

Many studies have been conducted on the temperature rise on the asperities during sliding and in fretting contacts [100–104].

Due to the nature of the contact and variation of Hertzian contact stress, the magnitude and the extent of the microslip area is a function of time. However, because of the high stiffness of the clutch system, the windup angle is very small; consequently the center of the contact area does not move appreciably. In order to simplify the analysis, the following assumptions are made:

1.  The contact area is a Hertzian circle area.
2.  The center of the contact area remains unchanged.
3.  Frictional heat is equally partitioned between the contacting surfaces due to the existence of thin, chemical, surface layers with low conductivity.

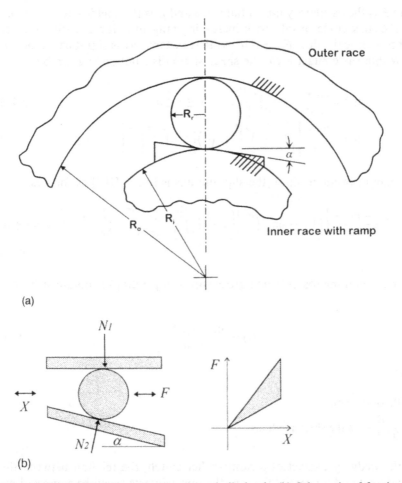

(a)

(b)

**Figure 9.30** (a) Schematic of a ramp–ball clutch. (b) Schematic of fretting contact with wedging condition and corresponding hysteresis.

4. All surfaces not in contact are adiabatic.

According to Mindlin's stick–slip model [106], the contact area of sphere on a flat subjected to a tangential force is a mixed stick–slip circle. The boundary between the slip and stick regime is a circle with radius:

$$c = a\left(1 - \frac{f}{\mu N}\right)^{1/3} \qquad (9.26)$$

where $F$ is the oscillatory tangential force and $\mu$ is the coefficient of friction.

The stick circle shrinks with increasing tangential force, until the force reaches a critical value, $F_{cr} = N\mu$. At that instant, gross slip starts to occur. Within the contact area, the shear stress distribution is given by:

$$\tau(r) = \frac{3\mu N}{2\pi a^2}\left[1 - \left(\frac{r}{a}\right)^2\right]^{1/2}, \qquad c \leq r \leq a \tag{9.27}$$

$$\tau(r) = \frac{3\mu N}{2\pi a^2}\left\{\left[1 - \left(\frac{r}{a}\right)^2\right]^{1/2} - \left(\frac{c}{a}\right)\left[1 - \left(\frac{r}{c}\right)^2\right]^{1/2}\right\}, \qquad r \leq c \tag{9.28}$$

The amount of microslip in the slip annulus is found [107] as follows:

$$\delta(r) = \frac{3(2 - v)\mu N}{16Ga}\left\{\left[1 - \frac{2}{\pi}\sin^{-1}\left(\frac{c}{r}\right)\right]\left[1 - 2\left(\frac{c}{r}\right)^2\right] + \frac{2c}{\pi r}\sqrt{1 - \left(\frac{c}{r}\right)^2}\right\}, \qquad c \leq r \leq a \tag{9.29}$$

The maximum microslip (when gross slip is impending) is obtained by setting $c = 0$:

$$\delta(r) = \frac{3(2 - v)\mu N}{16Ga} \tag{9.30}$$

where

$v = $ Poisson's ratio

$G = \dfrac{E}{2(1 + v)}$  is the shear modulus

For the wedging contact of a ramp–roller clutch, the relation between the normal load and tangential force at the upper interface can be expressed as:

$$N = F\left(\frac{1 + \cos\alpha}{\tan\alpha}\right) \tag{9.31}$$

Substituting Eq. (9.31) into Eq. (9.26) yields:

$$\frac{c}{a} = \left(1 - \frac{\tan\alpha}{\mu(1 + \cos\alpha)}\right)^{1/3} \tag{9.32}$$

Equation (9.32) shows that the ratio of the radius of the stick circle to that of the contact area is constant. If the ramp angle is properly chosen, the sphere will never slip, no matter how large the tangential force is.

Because no surface is perfectly smooth, the contact occurs only at discrete asperities and the real contact area is so small that it leads to extremely high local stress and high temperature rise under sliding condition. The real contact area is approximately proportional to the normal load under elastic contact condition. According to the Greenwood–Williamson elastic microcontact model [108], the average real contact pressure can be an order of magnitude higher than the nominal contact pressure. Accordingly, if the normal load is concentrated on the real contact area, the resulting stress and heat flux can be very high.

### 9.7.3 Frictional Energy and Average Heat Flux

The frictional energy generated per unit time during fretting contact is the product of the interface shear stress (surface traction) and the amount of microslip per unit time on each point within the slip annulus:

$$\dot{E}(t) = \int_0^{2\pi} \int_c^a \tau(r, t) |\dot{\delta}(r, t)| r \, dr \, d\theta \tag{9.33}$$

where

$D$ = roller diameter = $2R_r$

$$a(t) = 0.881 \sqrt[3]{\frac{ND}{E}}, \qquad \text{for steel with Poisson's ratio } \nu = 0.3 \tag{9.34}$$

$$c(t) = 0.881 \sqrt[3]{\frac{ND}{E} \left(1 - \frac{\tan \alpha}{\mu(1 + \cos \alpha)}\right)} \tag{9.35}$$

$$\tau(r, t) = \frac{3\mu N}{2\pi a^2} \left[1 - \left(\frac{r}{a}\right)^2\right]^{1/2} \tag{9.36}$$

$$\delta(r, t) = \frac{3(2 - \nu)\mu N}{16Ga} \left\{\left[1 - \frac{2}{\pi} \sin^{-1}\left(\frac{c}{r}\right)\right]\left[1 - 2\left(\frac{c}{r}\right)^2\right] + \frac{2c}{\pi r} \sqrt{1 - \left(\frac{c}{r}\right)^2}\right\} \tag{9.37}$$

Equations (9.36) and (9.37) in the wedging condition are plotted in normalized form as shown in Figs 9.31 and 9.32, respectively. The change of the $c$ value with increasing ramp angle can be readily seen in Fig. 9.32.

For the impending gross slip conditon, $c = 0$, Eq. (9.32) gives:

$$\frac{\tan \alpha}{\mu(1 + \cos \alpha)} = 1 \tag{9.38}$$

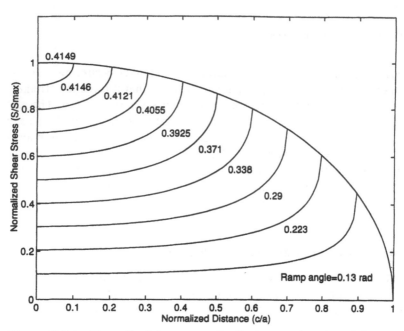

**Figure 9.31** Normalized shear stress distribution as a function of the ramp angle.

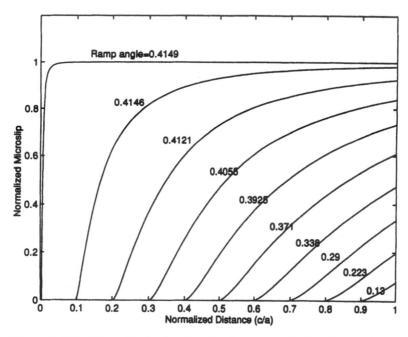

**Figure 9.32** Normalized microslip distribution within the slip annulus as a function of the ramp angle.

In this case, microslip occurs over the entire contact area and the corresponding deflection can be expressed as:

$$\delta(r, t) = \frac{3(2 - v)\mu N}{16 G a_{max}} \tag{9.39}$$

Therefore, Eq. (9.33) can be rewritten as:

$$\dot{E}(t) = |\dot{\delta}(r, t)| \int\limits_{0}^{2\pi} \int\limits_{c}^{a} \tau(r, t) r \, dr \, d\theta \tag{9.40}$$

In Eq. (9.40), the double integral is the total tangential force, $\mu N(t)$, applied on the ball. Therefore, the energy rate can be expressed as:

$$\dot{E} = \frac{3(2 - v)\mu^2}{16 G a_{max}} N(t)|\dot{N}(t)| \tag{9.41}$$

For the clutch, the normal force can be represented as a function of the applied torque:

$$N(t) = \frac{T(t)}{n(b + 1)R_r} \left( \frac{1 + \cos\alpha}{\tan\alpha} \right) \tag{9.42}$$

where $n$ is the number of balls; $b$ is the ratio of the radii $(R_i/R_r)$; $R_i$ and $R_r$ are the radii of an inner race and the balls, respectively.

Substituting Eq. (9.38) into Eq. (9.42) yields:

$$N(t) = \frac{T(t)}{N\mu(b + 2)R_r} \tag{9.43}$$

If the applied torque can be expressed as the product of its magnitude and a normalized continuous function of time as follows:

$$T(t) = T_{max} f(t) \tag{9.44}$$

then, the frictional energy generated per unit time in the contact under wedging conditions can be obtained by substituting Eq. (9.44) into Eq. (9.41):

$$\dot{E}(t) = 0.169 \frac{(2 - v)E^{1/3}\mu^{1/3}}{GR_r^2} \left( \frac{T_{max}}{n(b + 2)} \right)^{5/3} f(t)|\dot{f}(t)| \tag{9.45}$$

or in normalized form:

$$\dot{E}\left(\frac{t}{\tau}\right) = 0.169 \frac{(2-v)E^{1/3}\mu^{1/3}}{GR_r^2}\left(\frac{T_{max}}{n(b+2)}\right)^{5.3} f\left(\frac{t}{\tau}\right)\left|\dot{f}\left(\frac{t}{\tau}\right)\right| \quad (9.46)$$

We can also obtain the friction energy generated per cycle by integrating Eq. (9.45) over one period, $\tau$:

$$E_{cycle} = 0.169 \frac{(2-v)E^{1/3}\mu^{1/3}}{GR_r^2}\left(\frac{T_{max}}{n(b+2)}\right)^{5/3}\int_0^{2\pi/w} f(t)|\dot{f}(t)|\, dt \quad (9.47)$$

The average heat flux is found from:

$$\dot{Q} = \frac{\dot{E}}{\pi a_{max}^2} \quad (9.48)$$

where the radius of the maximum contact area is:

$$a_{max} = 1.11\left(\frac{T_{max}}{n\mu E(b+2)}\right)^{1/3} \quad (9.49)$$

The average heat flux can be found by substituting Eq. (9.45) and (9.49) into Eq. (9.48):

$$\dot{Q}(t) = 0.0437 \frac{(2-v)E\mu T_{max}}{n(b+2)GR_r^2} f(t)|\dot{f}(t)| \quad (9.50)$$

All the equations derived above are based on the assumption of fretting contact, that is, without gross slip. For the case of gross slip, the frictional energy generated per unit time is:

$$\dot{E} = N(t)\mu R_0|\dot{\theta}(t)| \quad (9.51)$$

where $R_0$ is the radius of an outer race and $\dot{\theta}(t)$ is the angular velocity of the outer race relative to the inner race. Accordingly, the average heat flux for gross slip condition can be obtained by substituting Eq. (9.51) into Eq. (9.48):

$$\dot{Q}(t) = 0.258\left(\frac{T_{max}}{n}\right)^{1/3}[\mu(b+2)E]^{2/3}f(t)|\dot{\theta}(t)| \quad (9.52)$$

An illustrative example is considered by using the design parameters, listed in Table 9.5, for a steel clutch subjected to a 22.6 N/m (200 lbf-in.) peak torque.

**Table 9.5** Parameters Used in the Illustrations

| | |
|---|---|
| Material | 4340 steel |
| Hardness | 352 BHN |
| Ramp angle ($\alpha$) | 0.4149 rad |
| Radius ratio ($b$) | 4 |
| Radius of the roller ($R_r$) | 3.8 mm (0.15 in.) |
| Young's modulus ($E$) | $2 \times 10^{11}$ N/m$^2$ ($30 \times 10^6$ psi) |
| Poisson's ratio ($v$) | 0.3 |
| Coefficient of friction ($\mu$) | 0.23 |
| Radius of the outer race ($R_0$) | $(b+2)R_r$ |
| Thermal conductivity ($k$) | 45 W/(m-°C) (26 BTU/(hr-ft-°F)) |
| Density ($\rho$) | 7850 kg/m$^3$ (490 lb/ft$^3$) |
| Specific density ($c$) | 0.42 kJ/(kg-°C) (0.1 BTU/(lb-°F)) |

Assuming that the system is subjected to a periodic versed sine load:

$$T_0 = 22.6 \left( \frac{1 - \cos 2\pi f t}{2} \right) \text{(N-m)}$$

the corresponding average heat flux during the microslip condition is plotted in Fig. 9.33.

### 9.7.4 Estimation of the Temperature Rise

*One-Dimensional Model*

The temperature rise of a semi-infinite solid subjected to stationary uniform heat supply over a circular area was first investigated by Blok [109]. In the same report, Blok also shows that if the same amount of heat flux has a parabolic distribution the maximum temperature rise is 4/3 times as high as the uniform distribution. Jaeger [110] also gives:

$$\Delta T(t) = \frac{1.128 \dot{Q} \sqrt{t}}{\sqrt{k \rho c}} \qquad (9.53)$$

where

$\dot{Q}$ = steady heat flux
$k$ = thermal conductivity
$\rho$ = density
$c$ = specific heat

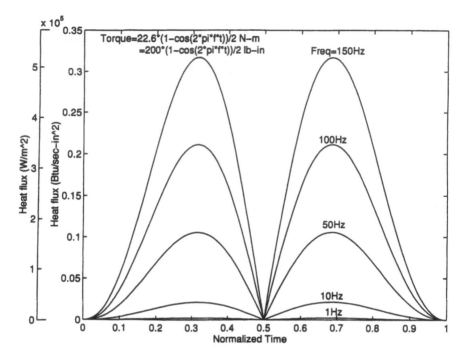

**Figure 9.33** The average heat flux generated during the microslip condition due to versed sine load. (Time normalized to the period τ of the torque cycle.)

The temperature rise of a ramp–ball clutch subjected to a versed sine load can therefore be calculated by integrating Eq. (9.53) after substituting Eq. (9.50) as shown in Fig. 9.34 and 9.35. From Fig. 9.34, we can observe the effect of cooling due to heat convection to the surrounding lubricant as the contact area is reduced by microslip. This cooling effect causes the temperature rise to approach an asymptotic limit. The thickness of the lines in Fig. 9.35 represents the range of temperature fluctuation due to cooling.

Due to the extensive computation necessary over long periods, extrapolation is undertaken by curve fitting of the results from a limited number of cycles. The temperature rise within a limited time can be approximately predicted by the following form:

$$\Delta T(t) = \beta \sqrt{t} \tag{9.54}$$

where constant β can be obtained by curve-fitting the envelope of the results given in Fig. 9.34. The curves in Fig. 9.35 show the extrapolated results.

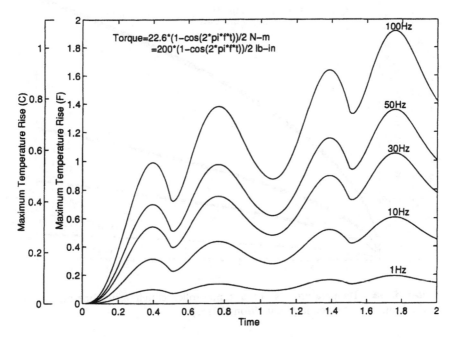

**Figure 9.34** Temperature rise of a ramp–ball clutch using one-dimensional method under fretting contact condition (two cycles). (Time normalized to the period of torque cycle.)

## Finite Element Analysis

The finite element method, based on ANSYS code, is used to verify the accuracy of the result from the one-dimensional theory. Being axisymmetric, a sphere can be modeled by using a 90° segment, with 10-point thermal element (SOLID87).

Instead of the average heat flux derived in the previous section, multiple heat flux is used in the analysis for better results. The maximum contact area is divided into four rings, as shown in Fig. 9.36, and each has a key point. The heat flux history at each key point represents the local average heat flux on corresponding annulus. Convective cooling is taken into consideration outside the contact region in the finite element analysis in order to simulate the effect of the lubricant.

The heat flux history of each key point can be found in differential form:

$$\dot{Q}(r, t) = \tau(r, t) \cdot \left| \frac{\delta(r, t) - \delta(r, t - \Delta t)}{\Delta t} \right|, \qquad c \leq r \leq a \qquad (9.55a)$$

$$\dot{Q}(r, t) = 0, \qquad r \leq c \qquad \text{or} \qquad r \geq a \qquad (9.55b)$$

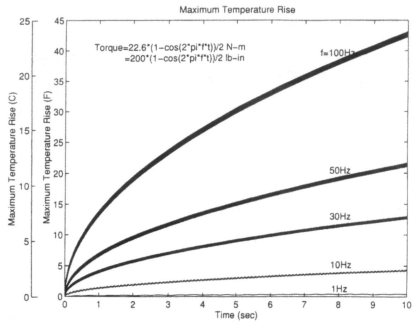

**Figure 9.35** Temperature rise of a ramp–roller clutch using one-dimensional method under fretting contact condition (10 sec).

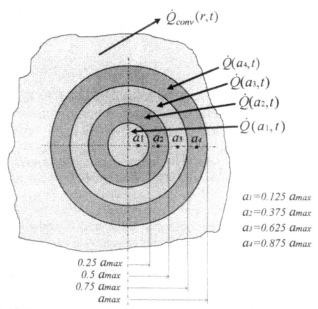

**Figure 9.36** Four heat sources along with heat convection are used to calculate the temperature rise.

where $r = a1, a2, a3, a4$ (referring to Fig. 9.36), and $\tau$ and $\delta$ are described by Eqs (9.36) and (9.37), respectively.

Figure 9.37 shows a typical heat flux history at every key point for one cycle for the microslip condition. The input load is a versed sine load as considered before.

Transient analysis is used throughout the finite element calculations. The insulated surface and symmetric surfaces are, by definition, adiabatic, therefore no other boundary conditions are necessary. Figures 9.38 and 9.39 show the solutions at $a1$ (the central region of the contact) for load frequency of 50 and 100 Hz, respectively. By comparing the results with those from the one-dimensional theory, the difference is found to be relatively small in the first two cycles. However, a stronger cooling effect results for longer load duration in the finite element method due to convection to the lubricant.

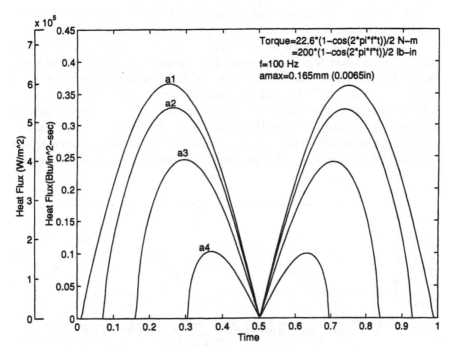

**Figure 9.37** The heat flux at four key points under the fretting contact condition (versed time pulse). (Time normalized to a period of the torque cycle.)

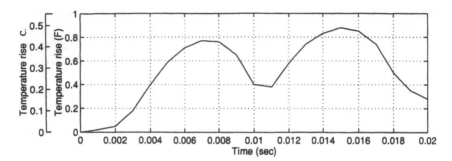

**Figure 9.38** Temperature rise of a ramp–ball clutch using finite element method (versed sine pulse, 50 Hz).

### 9.7.5 Wear Depth Prediction

Due to the complex nature of the wear behavior in this case, a precise prediction of the amount of wear is by no means an easy task. The great majority of published wear equations apply to fixed sliding conditions, usually without any measurement of temperature or energy produced [111].

The situation for fretting wear is even more ambiguous, because no single equation is available. Most of the investigations dealing with fretting wear are case-study-type experiments under specific conditions. Kayaba and Iwabuchi [112] report that fretting wear decreases with increasing temperature up to 300°C (570°F), and the trapped debris is $Fe_3O_4$, which has a lubricating effect. Fretting wear at high temperatures has been receiving particular attention [113–116]. However, the results are inconsistent because of different materials, experimental conditions, and estimates of wear. The influence of hardness and slip amplitude on the fretting wear are investi-

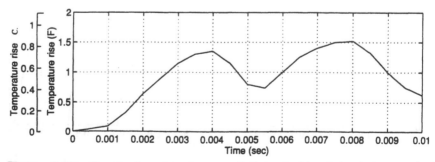

**Figure 9.39** Temperature rise of a ramp–ball clutch using finite element method (versed sine pulse, 100 Hz)

gated by Kayaba and Iwabuchi [117]. It is also reported [118] that the wear rate and the form of fretting damage depend on the chemical nature of the environment and on whether the debris, mostly oxide, can escape. If the ddbris, oxide or chemical compound, is trapped and acts as a buffer or lubricant, then the wear rate may slow down considerably when the temperature builds up to 200°C (400°F) [112].

The concept of frictional energy has been used to deal with adhesive and abrasive wear in some models, including the Archard's equation, in which the debris is assumed to be hemispherical in shape. Rabinowicz [119] concludes that the ratio of frictional energy to material hardness is an important factor in wear and may have some effect on debris size. Although the hypothesis of hemispherical debris is questionable, the concept of energy needed to generate debris makes the Archard's model [120] a viable approach for predicting wear depth.

Seireg and Hsue [121] indicate that the wear depth is dependent on the temperature rise and the heat input at the contacting surfaces. Suzuki and Seireg [122] also provide evidence for the correlation between wear and energy input. Due to the nature of fretting contact, the frictional energy can be accumulated within a limited area with minimum convection to the surroundings. Therefore, the energy accumulation can be used as a potential tool to predict the fretting wear depth.

Archard's and Rabinowicz's equations can be rewritten as follows, respectively:

$$V = \left(\frac{KA}{3\mu}\right)\left(\frac{Q}{H_v}\right) \tag{9.56}$$

and

$$h = \left(\frac{K}{9\mu}\right)\left(\frac{Q}{S}\right) \tag{9.57}$$

In Eqs (9.56) and (9.57), $Q = F\mu L/A$ denotes the frictional energy density. The yield strength, $S$, is about 1/3 of its Vickers hardness, $H_v$. Therefore, from the energy standpoint, both abrasive and adhesive wear depth share a common expression as follows:

$$h \propto \frac{Q}{H_v} \tag{9.58}$$

In the case of a clutch switching between engagement and disengagement, the debris is not completely trapped as in the conventional fretting case of fastened or press-fitted assemblies. Accordingly, this fretting process can be assumed to be of the same type expressed in Eq. (9.58).

In order to quantify the relation between accumulated energy and fretting wear depth, the work by Sato [123, 124] has been adapted. In his series of experiments, carried out on a glass plate in contact with a steel ball of 5 mm (0.2 in.) diameter, Sato obtains good agreement with other researchers and suggests that the coefficient of friction increases steadily up to 0.5 as the microslip annulus grows. After gross slip, amplitude $= 3\,\mu m$ (120 µin.) at 9.8 N (2.2 lbf) normal load, the coefficient of friction remains unchanged and is considered to be the sliding coefficient of friction.

Figure 9.40 shows the wear depth after 50,000 cycles plotted against the oscillation amplitude [124]. At small amplitudes, the amount of wear is found to be negative, because the debris is unable to escape from the stick area and accumulates and wedges up within the contact area. For the clutch case, the debris is not easily trapped; therefore, only the data corresponding to the slip region are considered.

Based on Sato's experimental data, the frictional energy density can be calculated by the following equation:

$$Q = 35.4 N \mu a l \qquad (9.59)$$

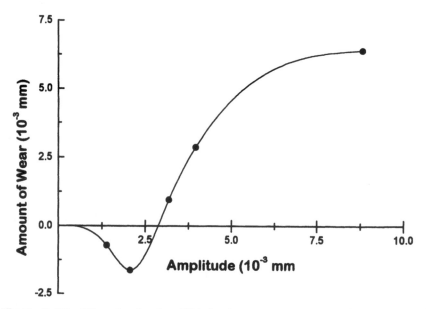

**Figure 9.40**  Wear depth after 5000 fretting cycles (steel on glass). (From Ref. 124.)

where

$Q$ = frictional energy density
$N$ = normal load
$\mu$ = coefficient of friction
$a$ = amplitude (center-to-peak)
$l$ = total number of cycles

The relation between wear depth and the energy input can then be obtained by curve fitting. A linear fuction relating frictional energy density and Vickers hardness is found as:

$$h = 0.147 \frac{Q}{H_v} \qquad (9.60)$$

where

$h$ = fretting wear depth
$H_v$ = Vickers hardness

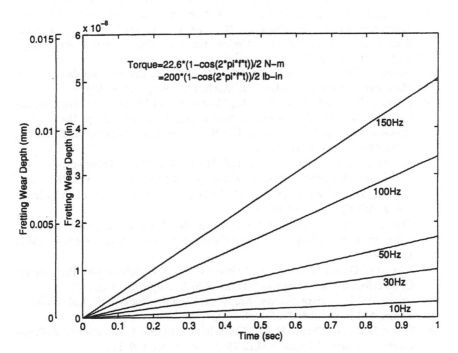

**Figure 9.41** Fretting wear prediction using Eq. (9.60).

Equation (9.60) can be used for approximate prediction of the wear depth of a none-directional clutch. Considering the thermal loading conditions given in Fig. 9.33 and using Eq. (9.60), an illustrative example of the qualitative prediction of clutch wear for different engagement frequencies is shown in Fig. 9.41.

## REFERENCES

1. Way, S., "Pitting Due to Rolling Contact," ASME Trans., December 1934.
2. Dawson, P. H., "The Pitting of Lubricated Gear Teeth and Rollers," Power Transmiss., April–May 1961.
3. Dawson, P. H., "Effect of Metallic Contact on the Pitting of Lubricated Rolling Surfaces," J. Mech. Eng. Sci., 1962, vol. 4(1).
4. Dawson, P. H., "Contact Fatigue in Hard Steel Specimens with Point and Line Contacts," J. Mech. Eng. Sci., 1967, Vol. 9(1).
5. Kragelskii, I. V., Friction and Wear, Butterworths, 1965.
6. Feng, I-Ming, and Chang, C. M., "Critical Thickness of Surface Film in Boundary Lubrication," ASME J. Appl. Mech., September 1956, Vol. 23(3).
7. Cheng, H. S., and Sternlicht, B., "A Numerical Solution for the Pressure, Temperature and Film Thickness Between Two Infinitely Long, Lubricated Rolling and Sliding Cylinders Under Heavy Loads," ASME J. Basic Eng., September 1965, p. 695.
8. Dowson, D., and Higginson, G. R., "A Numerical Solution to the Elastohydrodynamic Problem," J. Mech. Eng. Sci., Vol. 6(1), p. 6.
9. Dowson, D., and Higginson, G. R., "The Effect of Material Properties on the Lubrication of Elastic Rollers," J. Mech. Eng. Sci., Vol. 2(3), p. 188.
10. Dowson, D., and Higginson, G. R., Elastohydrodynamic Lubrication, Pergamon Press, Oxford, 1966.
11. Archard, G. D., Gair, F. C., and Hirst, W., "The Elastohydrodynamic Lubrication of Rollers," Proc. Roy. Soc. (Lond.) 1961, Ser. A., Vol. 262, p. 51.
12. Dowson, D., and Whitaker, A. V., "The Isothermal Lubrication of Cylinders," Trans. ASLE, 1965, Vol. 8(3).
13. Cheng, H. S., "A Refined Solution to the Thermal-Elastohydrodynamic Lubrication of Rolling and Sliding Cylinders," Trans. ASLE, October 1965, Vol. 8(4).
14. Niemann, G., and Gartner, F., "Distribution of Hydrodynamic Pressure on Counterformal Line Contacts," Trans. ASLE, July 1965, Vol. 8(3).
15. Orcutt, F. K., "Experimental Study of Elastohydrodynamic Lubrication," Trans. ASLE, October 1965, Vol. 8(4).
16. Sibley, L. B., and Orcott, F. K., "Elastohydrodynamic Lubrication of Rolling–Contact Surfaces," Trans. ASLE, 1962, Vol. 5, pp. 160–171.

17. Kannel, J. W., Bell, J. C., and Allen, C. M., "Methods for Determining Pressure Distributions in Lubricated Rolling Contact," Trans. ASLE, 1965, Vol. 8(3).
18. Crook, A. W., "The Lubrication of Rollers," Phil. Trans. Roy. Soc. (Lond.), 1958, Ser. A, Vol. 250 (981), pp. 387–409.
19. Crook, A. W., "The Lubrication of Rollers," Part II, Phil. Trans. Roy. Soc. (Lond.), 1961, Ser. A, Vol. 254, p. 223.
20. Blok, H., "The Flash Temperature Concept," Wear, 1963, Vol. 6.
21. Kelley, B. W., and Leach, E. F., "Temperature – The Key to Lubricant Capacity," Trans. ASLE, July 1965, Vol. 8(3).
22. Niemann, G., Rettig, H., Lechner, "Scuffing Tests on Gear Oils in the FZG Apparatus," ASLE Trans., 1961, Vol. 4, pp. 71–86.
23. "Gear Scoring Design Guide for Aerospace Spur and Helical Power Gears," AGMA Information Sheet, 217.01, October 1965.
24. Seireg, A., and Conry, T., "Optimum Design of Gear Teeth for Surface Durability," Trans. ASLE, 1968, Vol. 11.
25. Taylor, T. C., and Seireg, A., "Optimum Design Algorithm for Gear Systems Incorporating Surface Temperature," ASME Trans., J. Mech, Transmiss. Autom. Des., July 1985.
26. Dooner, D., and Seireg, A., The Kinetic Geometry of Gears: A Concurrent Engineering Approach, Wiley Interscience, New York, NY, 1995.
27. Styri, H., "Fatigue Strength of Ball Bearing Races and Heat Treated 52100 Steel Specimens," Proc. ASTM, 1951, Vol. 51.
28. Fessler, H., and Ollerton, E., "Contact Stresses in Toroids Under Radial Loads," Br. J. Appl. Phys., October, 1957, Vol. 8, pp. 387–393.
29. Radzimovsky, E. I., "Stress Distribution and Strength Condition of Two Rolling Cylinders Pressed Together," Univ. Illinois Eng. Exper. Stat. Bull. Ser. No. 408, Vol. 50(44), 1953.
30. Rowland, E. S., "Resistance of Materials to Rolling Loads, an Engineering Approach to Surface Damage," C. Lipson, and L. V. Colwell (Eds), Univ. of Michigan, 1958 Summer Conference on Wear of Metals.
31. Johnson, L. G., Ball Bearings Engineers Statistical Guide Book, New Departure Division, General Motors Corporation, Bristol, Connecticut, April, 1957.
32. Lieblein, J., and Zelen, M., "Statistical Investigation of the Fatigue Life of Deep Groove Ball Bearings," J. Res. Nat. Bur. Stand., November 1956, Vol. 57(5), Res. Pap. 2719, p. 273.
33. Macks, E. F., "The Fatigue Spin Rig – a New Apparatus for Rapidly Evaluating Materials and Lubricants for Rolling Contact," Lubr. Eng., October 1953, Vol. 9(5), p. 254.
34. Butler, R. H., and Carter, T. L., "Stress Life Relation of the Rolling Contact Fatigue Spin Rig," NACA Technical Note 3930, March 1957.
35. Butler, R. H., Bear, H. R., and Carter, T. L., "Effect of Fiber Orientation on Ball Failures Under Rolling Contact Conditions," NACA Technical Note 3933, February 1957.

36. Carter, T. L., "Effect of Temperature on Rolling Contact Fatigue Life with Liquid and Dry Powder Lubricants," NACA Technical Note 4163, January 1958.
37. Barwell, F. T., and Scott, D., "Effect of Lubricant on Pitting Failure of Ball Bearings," Engineering, July 6, 1956, p. 9.
38. Harris, T. A., "Optimizing the Design of Cluster Mill Rolling Bearings," ASLE Trans., April 1964, Vol. 7, pp. 127–132.
39. Jones, A. B., "The Dynamic Capacity of High Speed Roller Bearings," MRC Corp., Engineering Devel. Report #7 (Circa 1947).
40. Harris, T. A., and Aaronson, S. F., "An Analytical Investigation of Cylindrical Roller Bearings having Annular Rollers, ASLE Trans., 1967, Vol. 10, pp. 235–242.
41. Suzuki, A., and Seireg, A., "An Experimental Investigation of Cylindrical Roller Bearings Having Annular Rollers," ASME Trans., J. Lubr. Technol., October 1976, pp. 538–546.
42. Chichinadze, A. V., "Temperature Distribution in Disk Brakes," Frict. Wear Mach. (Translat. ASME), 1962, vol. 15, pp. 259–275.
43. Fazekas, G. A. G., "Temperature Gradients and Heat Stresses in Brake Drums," SAE Trans., 1953, Vol. 61, pp. 279–308.
44. Limpert, R., "Cooling Analysis of Disk Brake Rotors," SAE Paper 750104.
45. Timoshenko, S. P., Strength of Material, Van Nostrand Company, Inc., Princeton, NJ, 1955.
46. Ashworth, R. J., El-Sherbiny, M., and Newcomb, T. P., "Temperature Distributions and Thermal Distortions of Brake Drums," Proc. Inst. Mech. Engrs, 1977, Vol. 191, pp. 169–176.
47. Day, A. J., Harding, P. R. J., and Newcomb, T. P., "A Finite Element Apporach to Drum Brake Analysis," Proc. Inst. Mech. Engrs, 1979, Vol. 193, pp. 401–406.
48. Evans, D. J., and Newcomb, T. P., "Temperatures Reached in Braking when the Thermal Properties of Drum or Disk Vary with Temperature," J. Mech. Eng. Sci., 1961, Vol. 3(4), pp. 315–317.
49. Fensel, P. A., "An Axisymmetric Finite Element Analysis of Mechanical and Thermal Stresses in Brake Drums," SAE Paper 740321.
50. Johnson, M. R., Welch, R. E., and Yeung, R. S., "Analysis of Thermal Stresses and Residual Stress Change in Railroad Wheels Caused by Severe Drag Braking," Trans. ASME, 1977, Ser. B, Vol. 99(1), pp. 18–23.
51. Ozisik, M. N., Heat Conduction, John Wiley & Sons, New York, NY, 1980.
52. Seireg, A. A., "A Method for Numerical Calculation of Stresses in Rotating Disks with Variable Thickness," University of Wisconsin Report, 1965.
53. Rainbolt, J. D., "Effect of Disk Material Selection on Disk Brake Rotor Configuration," SAE Paper 750733.
54. Rhee, S. K., and Byer, J. E., "A Comparative Study by Vehicle Testing of Copper Alloy and Gray Iron Brake Discs," SAE Paper 720930.
55. Rhee, S. K., Rusnak, R. M., and Spurgeon, W. M., "A Comparative Study of Four Alloys for Automotive Brake Drums," SAE Paper 690443.

56. Monza, J. C., "Valeo Crown Brake," SAE Paper 820027.
57. Kreith, F., and Bohn, M. S., Principles of Heat Transfer, Harper & Row, New York, NY, 1986.
58. Elbella, A. M., "Optimum Design of Axisymmetric Structures Subjected to Thermal Loading," Ph.D. Dissertation, University of Wisconsin-Madison, August 1984.
59. Reigel, M. S., Levy, S., and Sliter, J. A., "A Computer Program for Determining the Effect of Design Variation on Service Stresses in Railcar Wheels," Trans. ASME, November 1966, Ser. B, Vol. 88(4), pp. 352–362.
60. Timtner, K. H., "Calculation of Disk Brakes Components using the Finite Element Method with Emphasis on Weight Reduction," SAE Paper 790396.
61. Takeuti, Y., and Noda, N., "Thermal Stress Problems in Industry 2: Transient Thermal Stresses in a Disk Brake," J. Therm. Stresses, 1979, Vol. 2, pp. 61–72.
62. Dike, G., "An Optimum Design of Disk Brake," Trans. ASME, 1974, Ser. B, Vol. 96(3), pp. 863–869.
63. Crow, S. C., "A Theory of Hydraulic Rock Cutting," Int. J. Rock Mech. Miner. Sci. Geomech. Abst., 1973, Vol. 10, pp. 567–584.
64. Crow, S. C., "Experiments in Hydraulic Rock Cutting," Int. J. Rock Mech. Miner. Sci. Geomech. Abst., 1975, Vol. 12, pp. 203–212.
65. Hashish, M., and duPlessis, M. P., "Theoretical and Experimental Investigation of Continuous Jet Penetration of Solid," Trans. ASME, J. Eng. Indust., February 1978, Vol. 100, pp. 88–94.
66. Hashish, M., and duPlessis, M. P., "Prediction Equations Relating High Velocity Jet Cutting Performance to Stand Off Distance and Multipasses," Trans. ASME, August 1979, Vol. 101, pp. 311–318.
67. Hood, M., Nordlund, R., and Thimons, E., "A Study of Rock Erosion Using High-Pressure Water Jets," Int. J. Rock Mech. Sci. Miner. Geomech. Abst., 1990, Vol. 27(2), pp. 77–86.
68. Labus, T. J., "Material Excavation Using Rotating Water Jets," Proc. 7th Int. Sym. on Jet Cutting Tech., BHRA Fluid Engr., June 1984, Paper No. P3.
69. Rehbinder, G., "Some Aspects on the Mechanism of Erosion of Rock with a High Speed Water Jet," Proc. 3rd Int. Sym. on Jet Cutting Tech., BHRA Fluid Engr., May 1976, Paper No. EI, Chicago.
70. Rehbinder, G., "Slot Cutting in Rock with a High Speed Water Jet," Int. J. Rock Mech. Miner. Sci. Geomech. Abst., 1977, Vol. 14, pp. 229–234.
71. Veenhuizen, S. D., Cheung, J. B., and Hill, J. R. M., "Waterjet Drilling of Small Diameter Holes," 4th Int. Sym. on Jet Cutting Tech., England, April 1978, Paper No. C3, pp. C3-39–C3-40.
72. Iihoshi, S., Nakao, K., Torii, K., and Ishii, T., "Preliminary Study on Abrasive Waterjet Assist Roadheader," 8th Int. Symposium on Jet Cutting Tech., Durham, England, Sept., 1986, Paper #7, pp. 71–77.
73. Yu, S., "Dimensionless Modeling and Optimum Design on Water Jet Cutting Systems," Ph.D. Thesis, University of Wisconsin-Madison, 1992.
74. Hood, M., Knight, G. C., and Thimos, E. D., "A Review of Jet Assisted Rock Cutting," ASME Trans., J. Eng. Indust., May 1992, Vol. 114, pp. 196–206.

75. Mogami, T., and Kubo, K., "The Behavior of Soil during Vibration," Proc. of 3rd Int. Conf. of Soil Mech. Foundation Engineering, 1953, Vol. 1, pp. 152–155.

76. Savchenko, I., "The Effect of Vibration of Internal Friction in Sand," Soil Dynamics Collection of Papers, No. 32, State Publishing House on Construction and Construction Materials, Moscow, NTML Translations, 1958.

77. Shkurenko, N. s., "Experimental Data on the Effect of Oscillation on Cutting Resistance of Soil," J. Agric. Eng. Res., 1960, Vol. 5(2), pp. 226–232.

78. Verma, B., "Oscillating Soil Tools – A Review," Trans. ASAE, 1971, pp. 1107–1115.

79. Choa, S., and Chanceller, W., "Optimum Design and Operation Parameters for a Resonant Oscillating Subsoiler," Trans. ASAE, 1973, pp. 1200–1208.

80. Kotb, A. M., and Seireg, A., "On the Optimization of Soil Excavators with Oscillating Cutters and Conveying Systems," Mach. Vibr., 1992, Vol. 1, pp. 64–70.

81. Hohl, M., and Luck, J. V., "Fractures of the Tibial Condyle: A Clinical and Experimental Study," J. Bone Joint Surg. (A), 1956, Vol. 38, pp. 1001–1018.

82. Lack, C. H., and Ali, S. Y., "Cartilage Degradation and Repair," Nat. Acad. Sci., Nat. Res. Council, Washington, D.C., 1967.

83. Palazzi, A. S., "On the Operative Treatment of Arthritis Deformation of the Joint," Acta Orthop. Scand., 1958, Vol. 27, pp. 291–301.

84. Weiss, C., Rosenberg, L., and Helfet, A. J., "Bone Surgery," (A), 1968, Vol. 50.

85. Trias, A., "Cartilage, Degeneration and Repair," Nat. Acad. Sci., Nat. Res. Council, Washington, D.C., 1967.

86. Luck, J. V., "Cartilage Degradation and Repair," Nat. Acad. Sci., Nat. Res. Council, Washington, D.C., 1967.

87. Sokoloff, L., The Biology of Degenerative Joint Disease, University of Chicago Press, Chicago, IL, 1969.

88. Radin, E. L., et al., "Response of Joints to Impact Loading – III," J. Biomech., 1973, Vol. 6, pp. 51–57.

89. Radin, E. L., and Paul, I. L., "Does Cartilage Compliance Reduce Skeletal Impact Loads?" Arth. Rheum., 1970, Vol. 13, p. 139.

90. Radin, E. L., Paul, I. L., and Tolkoff, M. J., "Subchondral Bone Changes in Patients with Early Degenerative Joint Disease," Arth. Rheum., 1970, Vol. 14, p. 400.

91. Simon, S. R., Radin, E. L., and Paul, I. L., "The Response of Joint to Impact Loading – II. In vivo Behavior of Subchondral Bone," J. Biomech., 1972, Vol. 5, p. 267.

92. Seireg, A., and Gerath, M., "An in vivo Investigation of Wear in Animal Joints," J. Biomech., 1975, Vol. 8, pp. 169–172.

93. Seireg, A., and Kempke, W., J. Biomech., 1969, Vol. 2.

94. Smith, J., and Kreith, F., Arth. Rheum., 1970, Vol. 13.

95. Cameron, J. R., and Sorenson, J., "Cameron Photon Absorption Technique of Bone Mineral Analysis," Science, 1963, Vol. 142.
96. Redhler, I., and Zimmy, L., Arth. Rheum., 1972, Vol. 15.
97. McCall, J., Lubrication and Wear of Joints, J. B. Lippincott Company, Philadelpha, PA, 1969, pp. 30–39.
98. Blok, H., "The Flash Temperature Concept," Wear, 1963, Vol. 6, pp. 483–494.
99. Seif, M. A., and Abdel-Aal, H. A., "Temperature Fields in Sliding Contact by a Hybrid Laser Speckle-Strain Analysis Technique," Wear, 1995, Vol. 181–183, pp. 723–729.
100. Attia, M. H., and D'Silva, N. S., "Effect of Motion and Process Parameters on the Prediction of Temperature Rise in Fretting Wear," Wear, 1985, Vol. 106, pp. 203–224.
101. Bowden, F. P., and Tabor, D., "Friction and Lubrication," John Wiley, New York, NY, 1956.
102. Gecim, B., and Winer, W. O., "Transient Temperature in the Vicinity of an Asperity Contact," J. Tribol., July 1985, Vol. 107, pp. 333–342.
103. Tian, X., and Kennedy, F. E., "Contact Surface Temperature Models for Finite Bodies in Dry and Boundary Lubricated Sliding," J. Tribol., July 1993, Vol. 115, pp. 411–418.
104. Greenwood, J. A., and Alliston-Greiner, A. F., "Surface Temperature in a Fretting Contact," Wear, 1992, Vol. 155, pp. 269–275.
105. Chang, C. T., and Seireg, A., "Dynamic Analysis of a Ramp-Roller Clutch," ASME paper No. DETC '97/VIB-4043, 1997.
106. Mindlin, R. D., "Compliance of Elastic Bodies in Contact," J. Appl. Mech., 1949, Vol. 71, pp. 259–268.
107. Johnson, K. L., "Surface Interaction Between Elastically Loaded Bodies Under Tangential Forces," Proc. Roy. Soc. (Lond.), 1955, A, Vol. 230, p. 531.
108. Greenwood, J. A., Williamson, J. B. P., "Contact of Nominally Flat Surface," Proc. Roy. Soc. (Lond.), Ser. A, 1966, Vol. 295, pp. 300–319.
109. Blok, H., "Theoretical Study of Temperature Rise at Surfaces of Actual Contact under Oilness Lubricating Conditions," Proc. General Discussion on Lubrication and Lubricants, Inst. Mech. Engrs, London, 1937, Vol. 2, pp. 222–235.
110. Jaeger, J. C., "Moving Sources of Heat and the Temperature at Sliding Contacts," Proc. Roy. Soc. (N.S.W.), 1942, Vol. 56, p. 203.
111. Meng, H. C., and Ludema, K. C., "Wear Models and Predictive Equations: Their Form and Content," Wear, 1995, Vol. 181–183, pp. 443–457.
112. Kayaba, T., and Iwabuchi, A., "The Fretting Wear of 0.45%C Steel and Austenitic Stainless Steel from 20 to 650°C in Air," Wear, 1981–1982, Vol. 74, pp. 229–245.
113. Hurricks, P. L., and Ashford, K. S., "The Effects of Temperature on the Fretting Wear of Mild Steel," Proc. Inst. Mech. Engrs, 1969–1970, Vol. 184, p. 165.
114. Feng, I. M., and Uhlig, H. H., "Fretting Corrosion of Mild Steel in Air and in Nitrogen," J. Appl. Mech., 1954, Vol. 21, p. 395.

115. Hurricks, P. L., "The Fretting Wear of Mild Steel from Room Temperature to 200°C," Wear, 1972, Vol. 19, p. 207.

116. Bill, R. C., "Fretting Wear of Iron, Nickel and Titanium under Varied Environmental Condition," ASME, Proc. Int. Conf. on Wear of Materials, Dearborn, MI, 1979, p. 356.

117. Kayaba, T., and Iwabuchi, A., "Influence of Hardness on Fretting Wear," ASME, Proc. Int. Conf. on Wear of Materials, Dearborn, MI, 1979, p. 371.

118. Jones, M. H., and Scott, D., "Industrial Tribology," Elsevier, New York, 1983.

119. Rabinowicz, E., Friction and Wear of Materials, John Wiley and Sons, New York, NY, 1965.

120. Archard, J. F., "Contact and Rubbing of Flat Surfaces," J. Appl. Phys., 1953, Vol. 24, pp. 981–988.

121. Seireg, A., and Hsue, E., "An Experimental Investigation of the Effect of Lubricant Properties on Temperature and Wear in Sliding Concentrated Contacts," ASME-ASLE Int. Lubrication Conf., San Francisco, CA, Aug. 18–21, 1980.

122. Suzuki, A., and Seireg, A., "An Experimental Investigation of Cylindrical Roller Bearings Having Annular Rollers," ASME, J. Tribol., Oct. 1976, pp. 538–546.

123. Sato, J., "Fundamental Problems of Fretting Wear," Proc. JSLE Int. Tribology Conf., July 8–10, 1985, Tokyo, Japan, pp. 635–640.

124. Sato, J., "Damage Formation During Fretting Fatigue," Wear, 1988, Vol. 125, pp. 163–174.

# 10

# Friction in Micromechanisms

## 10.1 INTRODUCTION

The emerging technology of micromechanisms and microelectromechanical systems (MEMS) is integrating mechanical, material, and electronic sciences with precision manufacturing, packaging, and control techniques to create products as diverse as microminiaturized robots, sensors, and devices for the mechanical, medical, and biotechnology industries. New types of micromechanisms can now be built to measure very small movements and produce extremely low forces. Such devices can even differentiate between hard and soft objects [1–3].

Although many of the advanced and still experimental processes which are currently being investigated for the microelectronic devices can be applied to the manufacturing of micromechanical components, the conventional semiconductor processing based on lithography and etching still is the predominant method. Other techniques include beam-induced etching and deposition as well as the LIGA process which can be used for metal, polymer, and ceramic parts.

The method of fabrication known as the sacrificial layer technique can be employed to manufacture complex structures such as micromotors by successive deposition and etching of thin films [4–7].

The Wobble motor manufactured of silicon at the University of Utah is driven by electrostatic forces generated by applying a voltage to the motor walls. The micromotor developed at the University of California at Berkeley is only 60 μm in diameter. Although some silicons have proven to be almost as strong as steels, researchers in microfabrication technology are experimenting with the mass production of metallic components. Examples of this are gears made of nickel and gold which are approximately 50 μm thick and can be made even smaller.

Microscopic parts and precise structural components are now being created on silicon chips by depositing ultrathin layers of materials in some areas and etching material away from others. Templates for batches of tiny machines can be positioned using high-powered microscopes.

Scaling laws dictate that the ratio of surface area to volume ratio increases inversely with size.

Because of their very large surface-area-to-volume ratios, adhesion, friction, drag, viscous resistance, surface tension, and other boundary forces dominate the behavior of these systems as they continue to decrease in size. The surface frictional forces in MEMS may be so large as to prevent relative motion. Understanding frictional resistance on a microscale is essential to the proper design and operation of such systems.

Some important factors which influence frictional resistance, besides surface geometry and contamination, are other surface forces such as electrostatic, chemical, and physical forces which are expected to be significant for microcomponents. The influence of capillary action and adsorbed gas films, environmental temperature and humidity is also expected to be considerably greater in MEMS.

Although the frictional resistance and wear phenomena in MEMS are far from being fully understood, this chapter presents illustrative examples of frictional forces from measurements on sliding as well as rolling contacts between materials of interest to this field.

## 10.2 STATIC FRICTION

A number of researchers have examined the frictional forces in microelectromechanical systems. In recent experiments, the frictional properties of different materials were examined by sliding components made of different materials under the same loading conditions.

Tai and Muller [8] studied the dynamic coefficient of friction in a variable capacitance IC processed micromotor. Friction coefficients in the range 0.21–0.38 for silicon nitride–polysilicon surfaces were reported. Lim et al. [9] used a polysilicon microstructure to characterize static friction. They reported friction coefficients of $4.9 \pm 1.0$ for coarse-grained polysilicon–polysilicon interfaces and $2.5 \pm 0.5$ for silicon nitride–polysilicon surfaces. Mehregany et al. [10] measured both friction and wear using a polysilicon variable-capacitance rotary harmonic side-drive micromotor. They report a frictional force of 0.15 mN at the bushings and 0.04 mN in the bearing of the micromotor. Both the bushings and bearing surfaces were made of heavily phosphorus-doped polysilicon. Noguchi et al. [11] examined the coefficient of maximum static friction for various materials by sliding millimeter-sized movers electrostatically. The value obtained (0.32) for the static friction coefficient of silicon nitride and silicon surfaces in contact is smaller by a factor of 8 that the one reported by Lim et al. [9]. However, the measured values for the dynamic coefficient of friction are close to those reported in Ref. 8.

Suzuki et al. [12] compared the friction and wear of different solid lubricant films by applying them to riders and disks of macroscopic scale and sliding them under the same loading conditions. Larger values of the dynamic coefficient of friction (0.7–0.9) were obtained for silicon nitride and polysilicon surfaces than the ones reported by Tai and Muller.

A comprehensive investigation of the static friction between silicon and silicon compounds has been reported by Deng and Ko [13]. The materials studied include silicon, silicon dioxide, and silicon nitride. The objectives of their study are to examine different static friction measurement techniques and to explore the effects of environmental factors such as humidity, nitrogen, oxygen, and argon exposure at various pressures on the frictional resistance.

Two types of tribological pairs were used. In the first group of experiments, flat components of size 2 mm were considered. In the second group of experiments, a 3 mm radius aluminum bullet-shaped pin with spherical end coated with the test material is forced to slide on a flat silicon substrate. The apparent area of contact in the second group was measured by a scanning electron microscope and estimated to be in the order of $0.03–0.04 \, mm^2$.

The tests were performed in a vacuum chamber where the different gases can be introduced. The effect of humidity was determined by testing the specimens before and after baking them. The normal force was applied electrostatically and was in the range of $10^{-3} \, N$. The tangential force was applied by a polyvinylide difluoride bimorph cantilever, which was calibrated to generate a repeatable tangential force from 0 to $8 \times 10^{-4} \, N$.

Excellent correlation was obtained between the normal force and the tangential force necessary to initiate slip. The slope of the line obtained by linear regression of the data represents the coefficient of friction.

Their results are summarized in Tables 10.1 and 10.2 for the different test groups.

Several significant conclusions were drawn from the study, which are stated as:

Humidity in air was found to increase the coefficient of friction from 55% to 157%.

**Table 10.1** Measurement Results from Experiment A (SiN$_x$: PECVD Silicon Nitride)

| | Air (before baking) | Air (after braking) | 10$^{-5}$ Torr (after baking) |
|---|---|---|---|
| SiN$_x$ on SiN$_x$[a] | 0.62–0.84 | 0.62–0.84 | 0.53–0.71 |
| SiO$_2$ on SiO$_2$ | 0.54 ± 0.03 | 0.21 ± 0.03 | 0.36 ± 0.02 |
| SiO$_2$ on Si | 0.48 ± 0.02 | 0.31 ± 0.03 | 0.33 ± 0.03 |

[a]Measured at different locations with maximum deviation ±0.03.
*Source*: Ref. 13.

Exposure to argon produced no change in friction.
Exposure to nitrogen resulted in either no change or a decrease in the coefficient of friction.
Exposure to oxygen increased the frictional resistance.

## 10.3 ROLLING FRICTION

Rolling element bearings are known to exhibit considerably lower frictional resistance than other types of bearings. They are therefore expected to be extensively used in MEMS because of their lower frictional properties, improved life, and higher stability in carrying loads.

Microroller bearings can therefore play an important role in improving the performance and reducing the actuation power of micromechanisms. This section presents a review of the fabrication processes for such bearings. Results are also given from tests on the frictional resistance at the onset of motion in bearings utilizing stainless steel microballs in contact with silicon micromachined v-grooves with and without coated layers [14]. A macro-model is also described based on the concept of using the width of the hysteresis loop in a full motion cycle of spring-loaded bearings to evaluate the rolling friction and the effect of sliding on it. A test method is presented for utilizing the same basic concept for test rolling friction in very small microbearings [15].

### 10.3.1 Fabrication Processes

The silicon micromachined v-grooves are made using 3 in., 0.1 Ω-cm (100) p-type silicon wafers 508 μm thick. The wafers were cleaned using a standard RCA procedure. A thin layer (700 Å) of thermal oxide was grown at 925°C. A 3000 Å LPCVD silicon nitride was deposited on the thermal oxide. The

**Table 10.2** Measurement Results from Experiment B ($SiN_x$: PECVD Silicon Nitride)

| | Air (before baking) | UHV ($\sim 5 \times 10^{-10}$ Torr) | Ar ($<10^{-6}$ Torr) | $N_2$ ($<10^{-6}$ Torr) | $O_2$ ($<10^{-6}$ Torr) | R-$N_2$[c] | R-$O_2$[c] | R-($O_2/N_2$)[d] |
|---|---|---|---|---|---|---|---|---|
| $SiN_x$ on $SiN_x$ | 0.55–0.85 | 0.40–0.70[a] | 0.40–0.70[a] | Decrease from 0.58 to 0.35[b] | Increase from 0.44 to 0.68[b] | ~0.6 | ~1.6 | ~1.9 |
| $SiN_x$ on Si | 0.40–0.55[a] | 0.35±0.05 | 0.35±0.05 | 0.35±0.05 | Increase to 0.45±0.05 | ~1.0 | ~1.3 | ~1.3 |
| $SiO_2$ on $SiO_2$ | 0.43±0.05 | 0.20±0.02 | 0.20±0.02 | Decrease to 0.15±0.02 | Increase to 0.75±0.05 | ~0.8 | ~3.8 | ~5.0 |
| $SiO_2$ on Si | 0.55±0.05 | 0.39±0.04 | | Decrease to 0.20±0.02 | Increase to 0.55±0.04 | ~0.5 | ~1.4 | ~2.7 |

[a]Measured at different locations with maximum deviation ±0.05.
[b]Measured at the same location with maximum deviation ±0.05.
[c]R-$N_2$ and R-$O_2$ are ratios of the coefficients of friction measured in nitrogen and oxygen to those measured in UHV, respectively.
[d]R-($O_2/N_2$) is the ratio of the coefficients of friction measured in oxygen to those measured in nitrogen.
*Source:* Ref. 13.

samples were patterned photolithographically. A plasma etch ($CF_4/O_2$) was used to etch the silicon nitride and thermal oxide to form the anisotropic etch mask. The photoresist was removed using a chemical resist remover. The samples were then cleaned in a solution of $NH_4OH:H_2O_2:H_2O$ 1:1:6 in an ultrasonic bath for 5 min. Prior to micromachining, the samples were put in a dilute HF bath for 10 sec to remove the native oxide. The patterned samples were immersed in a quartz reflux system containing an anisotropic etchant solution of $KOH:H_2O$ (40% by weight) at 60°C constant temperature for 12 hr. The micromachined samples were then immersed in a reflux system containing concentrated phosphoric acid at 140°C for 2 hr in order to remove the silicon nitride and then in a buffered-oxide etch (BOE 1:20) bath for 10 min to remove the thermal oxide. The samples were rinsed with deionized $H_2O$ and blow-dried with nitrogen gas [14].

### 10.3.2 Rolling Friction at the Onset of Motion

A recent investigation by Ghodssi et al. [14] utilized a tilting table with 0.01° incremental movement to study the tangential forces necessary to initiate rolling motion of stainless steel microballs (285 µm in diameter) in micromachined v-grooves (310 µm wide, 163 µm deep, 10,000 µm long and 14,000 µm edge to edge) with and without the deposited thin films. A schematic representation of the bearing is given in Fig. 10.1. The average values

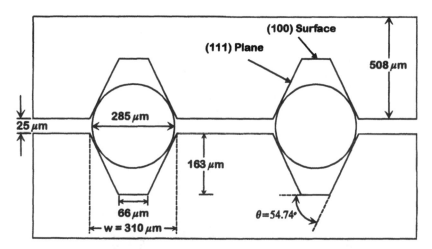

**Figure 10.1**  Schematic representation of the cross-sectional view of the test specimen. Dashed lines show the width of the etched v-groove (*w*) and the angle $\theta$ between the (100) surface and (111) plane. (From Ref. 14.)

of the frictional resistance at the onset of rolling friction obtained from 20 measurements in both directions of motion were found to be as follows for the three test materials used for the grooves:

$$F_T = 0.046 + 0.0076F_N$$

for the silicon grooves,

$$F_T = 0.059 + 0.0083F_N$$

when a 0.3 μm silicon nitride thin film was deposited on the surface of the grooves,

$$F_T = 0.036 + 0.0076F_N$$

when a 0.5 μm sputtered-chromium thin film was deposited on the surface of the grooves, where

$F_T$ = frictional force (mg) at the onset of rolling
$F_N$ = normal force (mg)

### 10.3.3 Rolling Friction During Motion

The frictional resistance in rolling element bearings in micromechanical systems has not yet been thoroughly investigated. The previous investigation [14] dealt only with the resistance at the onset of motion but not during the rolling motion. In the study reported in [15], a macro (scaled-up) model is used to investigate the feasibility of measuring rolling friction on a microscale. Such investigations can provide useful information on important factors which have to be taken into consideration in the design of an experiment for reliable measurements on a microscale because the forces required to sustain the rolling motion after the start are expected to be extremely small.

### 10.3.4 The Macroscale Test

A setup was designed as shown in Fig. 10.2 for the feasibility study. It represents a scaled-up model utilizing v-grooves (4 in. long, 0.5 in. wide, and 1.3 in. thick) in steel blocks and stainless steel balls (0.375 in. in diameter). A soft spring is attached to the top v-block or slider, at one end. A string is attached to the opposite end to apply the tangential force and is

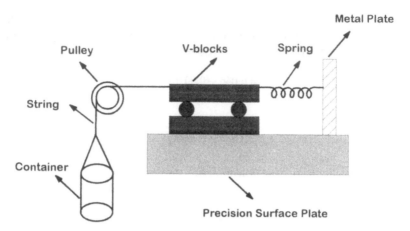

**Figure 10.2** An experimental setup for characterizing the rolling friction on a macroscale. This concept can be implemented for measuring rolling friction on a microscale.

supported by a pulley with low friction. The normal load as well as the tangential loads are applied by placing weights of known magnitude on the top v-block and pouring sand in the container attached to the string respectively.

The hysteresis in the setup is measured with and without the slider in place. Figure 10.3 shows the measured applied force versus displacement for the spring case and the spring with the slider case. First the string is attached directly to the spring and is poured into the container and the displacement is measured. Additional amounts of sand are added to yield an increased applied force up to about 70 gm. Then sand is removed to reduce the applied force and complete the hysteresis loop as shown in Fig. 10.3. In the second part of the experiment, the set of large model metal v-grooves and stainless ball bearings are used. Two ball bearings are positioned on the front and rear of a v-groove, respectively. The other v-groove is put on top of the ball bearings and used as a slider. The same procedure is performed as before with increasing and decreasing applied normal loads. In this case the hysteresis is larger. The arrows in the figure show the difference between the hysteresis loops which represent the rolling friction between the balls and grooves. The normal load in this case is equal to 500 gm. It can be deduced from the figure that the rolling friction in this case is equal to:

$$\mu = \frac{4.33}{400} = 0.00866$$

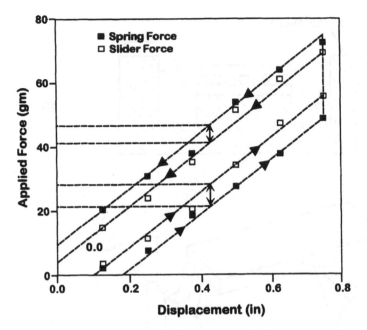

**Figure 10.3** The measured force versus displacement for the system with and without the bearing. The difference in hysteresis is due to the rolling friction in the bearing.

The macroscale test serves a very useful function in quantifying the effect of normal load on the relative sliding which takes place between the balls and groove during the rolling action. This is monitored during the tests by tracing the ball movement on the upper and lower v-grooves. The slide-to-roll ratio is found to be significant and can be as high as 30% in the performed test.

### 10.3.5 The Microscale Test

A microscale test setup is described in this section which can be utilized for testing micromachined bearings. It is based on the same concept as the macroscale test described in the previous section.

A schematic representation of the setup is shown in Fig. 10.4. Three U springs made of thin Ti–Ni wire are attached to each end of the top v-block. The motive force can be gradually applied by activating the springs on one end of the block by passing an electric current in the wire. The force can also be applied by using polyvinylide difluoride bimorph cantilevers [13].

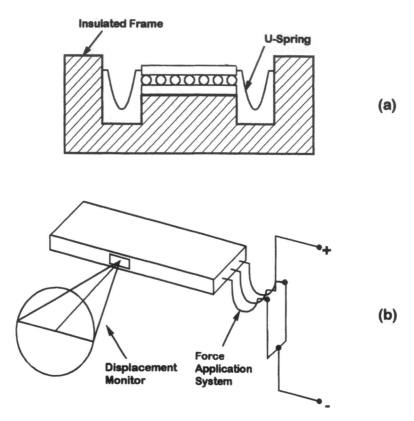

**Figure 10.4** (a) Mechanical setup. (b) Force application and displacement monitoring systems.

The movement of the block can be monitored by optical encoders and interferometers or by using a calibrated cathode follower [16]. The system is self-contrained and can be conveniently calibrated using a traveling microscope. The hysteresis can be displayed on the screen of a cathode ray tube. The springs can be designed to generate tangential forces in the microgram range for any desired range of micromovements.

The effective use of microroller bearing in micromechanisms is highly dependent on the accurate prediction of their frictional resistance. The macormodel used in the reported study shows that the frictional resistance during movement can be evaluated from the hysteresis loop obtained from the spring supported upper block of the bearing. The friction in the bearing is measured from the differential change of the width of the loop with normal load. The observation of the behavior of the scaled-up model including the observed slip was very helpful in the planning of the proposed microscale test setup. The frictional resistance measured in all the performed tests were found to be considerably lower (by orders of magnitude) than those reported in the literature for microsliding bearings.

## REFERENCES

1. Hazelrigg, G. A., "Microelectromechanical Devices, an Overview," SPIE, Vol. 1, Precision Engineering and Optomechanics, 1989, p. 114.
2. Hayashi, T., "Micro Mechanisms," J. Robot. Mechatron., Vol. 3(1).
3. Seireg, A., "Micromechanisms: Future Expectations and Design Methodologies," 1st IFTOMM, Int. Micromechanism Symposium, Japan, June 1–3, 1993, pp. 1–6.
4. Csepregi, L., "Micromechanics: A Silicon Microfabrication Technology," Microelect. Eng., 1985, No. 3, p. 221.
5. Peterson, K. E., "Silicon as a Mechanical Material," Proc. IEEE, 1982, Vol. 70, p. 420.
6. Benecke, W., "Silicon Micromachining for Microsensors and Microactuators," Microelect. Eng., 1990, No. 11, p. 73.
7. Mehregany, M., Senturia, S. D., Lang, J. H., and Nagarkar, P., "Micromotor Fabrication," IEEE Trans. Electron Dev., September 1992, Vol. 38(9).
8. Tai, Y. C., and Muller, R. S., "Frictional Study of IC-Processed Micromotors," Sens. Actuat., 1990, A21–A23, pp. 180–183.
9. Lim, M. G., Chang, J. C., Schultz, D. P., Howe, R. T., and White, R. M., "Polysilicon Microstructures to Characterize Static Friction," Proc. of IEEE Workshop on Micro Electro Mechanical Systems (MEMS), Napa Valley, CA, Feburary 1990, pp. 82–88.
10. Mehregany, M., Senturia, S. D., and Lang, J. H., in "Technical Digest of IEEE Solid State Sensors and Actuators Workshop," Hilton Head Island, South Carolina, June 1990, p. 17.
11. Noguchi, K., Fujita, H., Suzuki, M., and Yoshimura, N., "The Measurements of Friction on Micromechatoronics Elements," Proc. of the IEEE Workshop on Micro Electro Mechanical Systems (MEMS), Nara, Japan, February 1991, pp. 148–153.
12. Suzuki, S., Matsuura, T., Uchizawa, M., Yura, S., Shibata, H., and Fujita, H., "Friction and Wear Studies on Lubricants and Materials Applicable to MEMS," Proc. of the IEEE Workshop on Micro Electro Mechanical Systems (MEMS), Nara, Japan, February 1991, pp. 143–147.
13. Deng, K., and Ko, W. H., "A Study of Static Friction between Silicon and Silicon Compounds," J. Micromech. Microeng., 1992, Vol. 2, pp. 14–20.
14. Ghodssi, R., Denton, D. D., Seireg, A. A., and Howland, B., "Rolling Friction in a Linear Microactuator," JVST A, August 1993, Vol. 11, No. 4, pp. 803–807.

15. Ghodssi, R., Seireg, A., and Denton, D., "An Experimental Technique for Measuring Rolling Friction in Micro-Ball Bearings," Proc. First IFTOMM Int. Micromechanism Symp., Japan, June 1–3, 1993, pp. 144–149.
16. Seireg, A., Mechanical System Analysis, International Textbook Co., Scranton, PA., 1969.

# 11

# Friction-Induced Sound and Vibration

## 11.1 INTRODUCTION

The phenomenon of sound and vibration generation by rubbing action has been known since ancient times. Its undesirable manifestation as in the case of the squeal of chariot wheels has been remedied by the use of wax or fatty lubricants. Friction-induced sound phenomena have been used to advantage in developing musical instruments where rubbing strings causes them to vibrate at their natural frequency and generate sound with predictable tones.

Modern advances in sound monitoring instrumentation are now making it possible for the formation of cracks due to material fatigue to be readily detected at an early stage by the acoustic emission caused by rubbing at the crack site.

This chapter gives a brief introduction to the mechanism of sound generation. Two aspects of the phenomena will be considered. The first is the rubbing noise due to asperity interaction and the resulting surface waves. The second is the sound generated due to the vibration of a mechanical element or structure, which is self-excited with its intensity controlled and sustained by the rubbing action.

## 11.2 FRICTIONAL NOISE DUE TO RUBBING

One of the major sources of noise in machines and moving bodies is friction. Examples of the numerous studies of the noise generated by relative displacements between moving parts of machines and equipment are reported in

Refs 1–6. Only a few studies have been carried out to investigate the distinctive properties of such noise. In 1979 Yokoi and Nakai [7] concluded, based upon experimental studies, that frictional noise could be classified into two categories: rubbing noise which is generated when the frictional forces between sliding surfaces are relatively small, and squeal noise which occurs when those forces are high. In 1986, Symmons and McNulty [8] investigated the acoustic signals due to stick–slip friction by comparing the vibration and noise emission from perspex–steel junction with those of cast iron–steel and steel-steel junctions. The results indicated the presence of acoustic signals in some sliding contact cases and not in others. An important consideration in frictional noise is how sound due to sliding is influenced by surface roughness and material properties.

An experimental investigation into the nature of the noise generated, when a stylus travels over a frictional surface, has been carried out by Othman et al. [9], using several engineering materials. The relation between the sound pressure level (SPL) and surface roughness under various contact loads was established. An acoustic device was designed and constructed to be used as a reliable tool for measuring roughness. For each tested material, it has been found that the filtered noise signal within a certain spectrum bandwidth contains a specified frequency at which the amplitude is maximum. This frequency, called the dominating frequency, was found to be a material constant independent of surface roughness and contact load. It was also found that the dominating frequency for a given material is proportional to the sonic speed in that material.

## 11.2.1  Experimental Setup

The device shown in Fig. 11.1 was constructed to study the relation between frictional noise properties and surface roughness of the material. The main features of the transducer shown schematically in Fig. 11.1 are a spring-loaded stylus (1) (numbers refer to the components) attached to a rotating disk (2) which is driven by a DC servomotor (3). The end of the spring has a tungsten carbide tip (4) which constitutes the sliding element. The rotating disk is dynamically balanced by a small mass (5) to minimize disk rotational vibration. As the tip slides over the specimen surface (6), a frictional noise is generated. The noise intensity depends on surface material, roughness, sliding velocity, and spring load. The load can be increased incrementally by raising the moving plate (7) with the hydraulic jack (8) in order to compress the spring. The movement is monitored by the dial gage (9). The load may also be decreased by lowering the jack. The disk and spring rotate inside a chamber (10). The chamber is internally covered by a foamy substance (11) which acts as a sound-insulating material that eliminates the surrounding

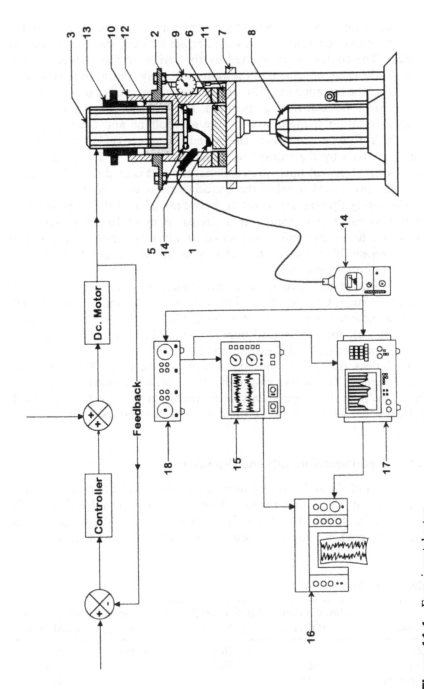

**Figure 11.1** Experimental setup.

noise. The chamber, which houses the DC motor, is lined with an additional sound insulating material (12) at the interface between the motor and the chamber. The contact load, exerted by the spring on the surface, is controlled by the axial movement of the motor assembly relative to the chamber by means of the threaded nut (13). The motor was selected to produce as low a noise level as possible during operation. The frictional sound generated by the stylus rotation is monitored by the microphone and the sound level meter, B & K type 2209 (14). The sound pressure signal is recorded by spectral analysis by the storage oscilloscope (15) and is displayed on the strip chart (16). A real-time spectrum analyzer (17) is used as well. A band-pass filter (18) is used to select the frequency range of interest.

The spring tip was set to rotate by means of a 12 V DC motor at a constant speed of 1000 rpm over a circular path of 10 mm radius. This results in a linear circumferential speed of 1.05 m/s, which was found to produce repeatable noise spectra. The motor speed was checked regularly by means of a stroboscope.

The faces of test specimens, approximately 80 mm in diameter, were turned to obtain a range of roughness from 1 to 20 μm. Three different specimen materials were used: steel SAE 1040, annealed yellow α brass (65 Cu–35 Zn), and commercial pure aluminum 1100 (99.9 + % Al). Table 11.1 lists the properties of these materials.

In all tests, the spring stylus axis of rotation was offset 20 mm from the specimen center. This was to ensure that the stylus tip traveled across the lay most of the time. The experiments were carried out in a $2 \times 3 \times 2$ m sound-insulated room where the background noise did not exceed 6 dB.

### 11.2.2  Experimental Results and Discussion

The stiffness of the stylus spring used in the device was 510 N/m. For each material tested, the sound pressure level (SPL) was recorded for the tip circumferential speed of 1.05 m/s. In order to compare the results, which were obtained when using the transducer with conventional direct measure-

**Table 11.1**  Material Properties

| Material | Elastic modulus (GPa) | Specific weight (kN/m³) | Sonic speed (m/s) | Surface wave speed (m/s) |
|---|---|---|---|---|
| Steel | 207 | 76.5 | 5196 | 3080 |
| Brass | 106 | 83.8 | 3415 | 1950 |
| Aluminum | 71 | 26.6 | 5156 | 2971 |

ments of surface roughness, a commercial roughness meter (Talysurf 10; Taylor and Robson Ltd.) was used. The SPL signals and the average roughness readings that were obtained from both instruments are shown in Fig. 11.2. The contact loads at the stylus tip were 0.25, 0.50, 0.75, and 1.00 N, as indicated in the figure. The relationship between the generated sound pres-

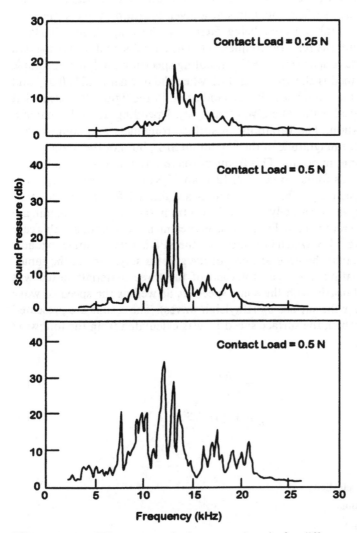

**Figure 11.2**   SPL spectrum in frequency domain for different materials (contact load = 0.50 N, all cases).

sure levels and average roughness was found to be a straight line on the log–log scale, and thus could be expressed as follows:

$$R_n = \left(\frac{\text{SPL}}{B + CF}\right)^n \tag{11.1}$$

where $F$ is the contact force and $B$, $C$, and $n$ are experimental parameters. This indicates that SPL can be used as a reliable alternative means of quantifying the average surface roughness at a given location on the surface.

The SPL is analyzed after being filtered in the range from 400 Hz to 20 kHz to capture the relevant frequencies. The sound signals are converted to one-third octave spectra by FFT computing spectrum analyzer. A sample of spectra obtained is shown in Fig. 11.2 when the normal load is 0.5 N and the average surface roughness $R_n$ is as indicated in the figure. It is clear that the SPL has a peak value at a given frequency depending upon the material under investigation. The variations in surface roughness and contact load will alter only the magnitude of the maximum SPL, but not the frequency at which this maximum occurs. This frequency is referred to as the dominating frequency and was found to be 12.2, 12.1, and 7.8 kHz for steel, aluminum, and brass, respectively. The sound signals are filtered for those frequency bands and analyzed separately. The SPL spectra in the frequency domain are observed at different contact loads for steel specimens. The results in this case are shown in Fig. 11.3, which indicates that the SPL is very sensitive to loads, despite the fact that the general trend of the spectra stays almost the same.

The dominating frequency for each of the three materials tested was found to vary linearly with the sonic speed, $v$, as well as the speed of wave propagation over the surface $v_R$ (Rayleigh waves). The results are presented in Fig. 11.4 in which the surface speed $v_R$ was calculated from the following expression [10]:

$$v = \sqrt{\frac{Eg}{\omega}} \tag{11.2}$$

$$v_R = 0.9194\sqrt{\frac{Gg}{\omega}} \tag{11.3}$$

where

$E$ = modulus of elasticity

$G$ = shear modulus

$\omega$ = specific weight

$g$ = gravitational acceleration

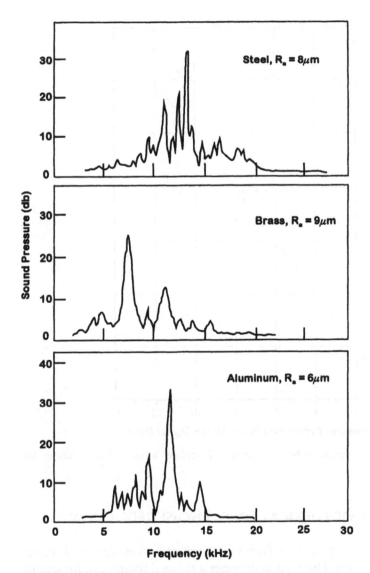

**Figure 11.3** SPL spectrum in frequency domain for steel, brass, and aluminum.

It is interesting to note that the SPL increases with increasing the contact load when the sound signal is filtered at the dominating frequency. The results also show that the filtered SPL increases linearly with roughness.

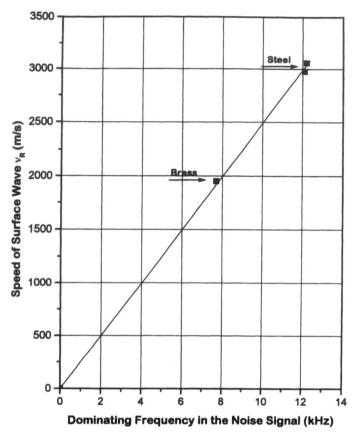

**Figure 11.4** Correlation between speed of surface wave and dominating frequency.

## 11.3 EFFECT OF LUBRICATION ON NOISE REDUCTION

It is generally accepted that frictional noise reduction can be achieved through lubrication. This section provides a rational framework for quantifying the role played by the lubricating film between the rubbing surfaces in reducing the intensity of sound generated by relative motion.

The hypothesis considered in this section is that frictional rubbing noise is the result of asperity penetration into the surface. The movement of the asperity therefore disturbs the surface layer and generates surface waves. The intensity of the sound can be assumed to be dependent on the depth of penetration which can in turn be assumed to be proportional to the real area of contact. As discussed in Chapter 4, the real area, as well as the frictional resistance, change in an approximately linear function with the normal load. It can therefore be assumed that the real area of contact between the lubricated solids can be used as a quantitative indicator of the intensity of the sound generated during sliding.

The roughness profile data given by McCool [11] for five different surface finishing processes are used to determine the input parameters for the Greenwood–Williamson microcontact model [12]. The model is then used to compute the ratio of the real area of contact to the nominal area for the given normal load and the thickness of the lubricant film separating the surfaces.

The model used for the illustration (Fig. 11.5) is represented by two rollers with radii $R_1 = R_2 = 10$ in. subjected to a load of 2000 lb/in. The lubricant viscosity and speed are changed to produce different ratios of film thickness to surface roughness ranging from 0 to 3.0. The considered surface roughness conditions are given in Table 11.2.

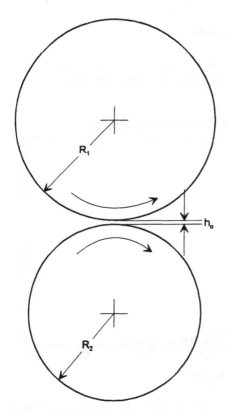

**Figure 11.5**   Contact model.

**Table 11.2**  Roughness Conditions

|   | Surface finishing process | rms roughness $\sigma$ (μin.) | Slope of the roughness profile |
|---|---|---|---|
| 1 | Fine grinding | 2.74 | 0.01471 |
| 2 | Rough grinding | 21.5 | 0.09017 |
| 3 | Lapping | 3.92 | 0.05254 |
| 4 | Polishing | 1.70 | 0.01157 |
| 5 | Shot peening | 45.9 | 0.07925 |

This ratio of the real area of contact to the nominal area, assuming smooth surfaces, is given in Fig. 11.6. This ratio can be used to represent the change in the relative sound intensity (dB) with lubrication for the different surface finishing processes under the given load.

## 11.4  FRICTIONAL NOISE IN GEARS

Gears are a well-known, major source of noise in machinery and equipment, and it is no surprise that gear noise has been the subject of extensive inves-

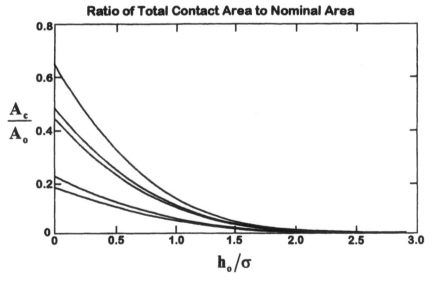

**Figure 11.6**  Ratio between the real area $A_c/A_0$ and nominal area of contact $A_0$ for different roughness and lubricant film conditions.

tigations. The nature of gear noise is quite complex because of the multitude of factors that contribute to it. The survey conducted by Welbourn [13] and the comparative study presented by Attia [14] on gear noise revealed that the published literature on the subject did not show how friction during the mesh of the rough contacting teeth influences noise generation.

A procedure for determining the effect of the different design and operating parameters on frictional noise in the gear mesh is presented by Aziz and Seirig [15]. The Greenwood asperity-based model [12] with Gaussian distribution of heights is utilized to evaluate the penetration of asperities in the contacting surface of the teeth. A parametric relationship is developed for relating the interpenetration of the asperities to the relative noise pressure level (NPL) in the lubricated and dry regimes. Numerical results are given in the following example to illustrate the effect of gear ratio, roughness, load, speed, and lubricant viscosity on noise.

The frictional noise generated from a pair of helical involute steel gears of 5 in. (127 mm) center distance was calculated in both dry and lubricated regimes for different design and operating parameters. The teeth are standard with a normal pressure angle equal to 25° and a helical angle equal to 31°. The relation between the relative NPL and different gear ratios was determined. Figures 11.7 and 11.8 present the effect of change of load for surface contact stresses 68.9, 689, 1378, and 1722.5 MPa (10,000, 100,000, 200,000, and 250,000 psi) on the relative NPL with an average surface roughness of 0.005 mm (CLA), gear oil viscosity of 0.075 N-sec/m$^2$ (0.075 Pa-sec), and pinion speeds of 1800 and 500 rpm, respectively. The results show that the relative NPL increases with the increase of load for all gear ratios. The rate of the relative NPL decreases as gear ratio increases. When reducing the pinion speed from 1800 to 500 rpm the same trend occurs but with higher noise levels, as shown in Fig. 11.8. This could be attributed to the change in the film thickness and consequently the amount of penetration. This effect is clearly shown in Fig. 11.9 for surface contact stress 689 MPa (100,000 psi), viscosity 0.075 N-sec/m$^2$ (0.075 Pa-sec) and surface roughness 0.005 mm.

The effect of change of surface roughness at different gear ratios on the relative NPL is shown in Fig. 11.10 for contact stress 689 MPa (100,000 psi), pinion speed 1800 rpm, viscosity 0.075 N-sec/m$^2$ (0.075 Pa-sec), and surface roughness 0.0015, 0.002, 0.003, and 0.005 mm, respectively. As can be seen in this figure, the relative NPL increases as would be expected with the increase of surface roughness for every gear ratio since the number of asperities subject to deformation is high for higher roughness and consequently the associated NPL becomes higher.

The effect of change of lubricant viscosity on the relative NPL at different gear ratios is presented in Fig. 11.11 for surface contact stress 689

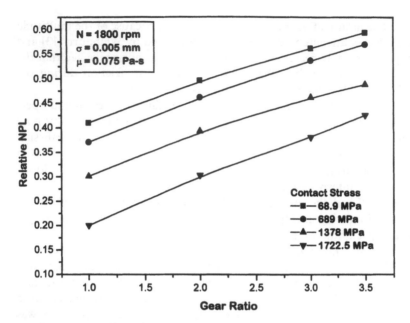

**Figure 11.7**  Effect of load change on NPL.

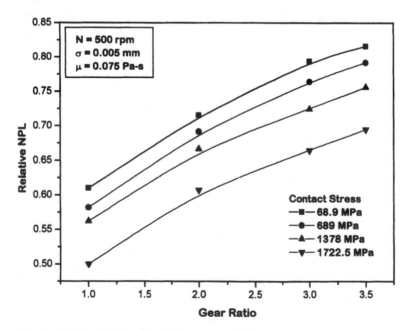

**Figure 11.8**  Effect of load change on NPL.

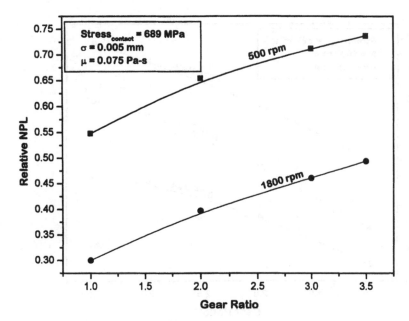

**Figure 11.9** Effect of speed change on NPL.

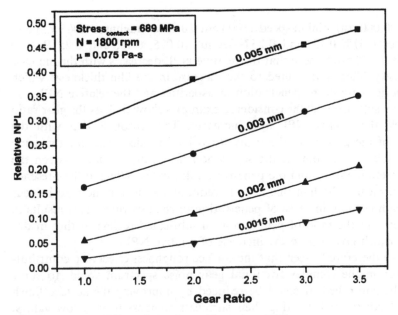

**Figure 11.10** Effect of roughness change on NPL.

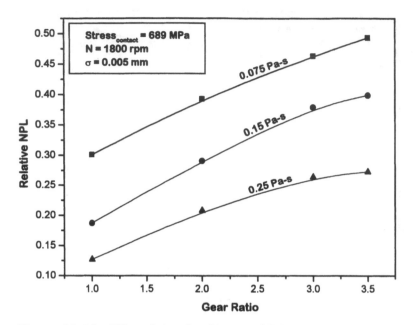

**Figure 11.11**   Effect of viscosity change on NPL.

MPa (100,000 psi), pinion speed 1800 rpm, surface roughness 0.005 mm and viscosities 0.075, 0.15, and 0.25 N-sec/m$^2$ (0.075, 0.15, and 0.25 Pa-sec) respectively. It can be seen that the relative NPL decreases with the increase of viscosity. This is attributed to the increase in the film thickness, which causes a decrease in the penetration of asperities and the relative NPL.

The results from all the considered examples show that, as the gear ratio increases, the relative NPL also increases. The reason is that with the increase of the gear ratio, the transmitted load is decreased for the same contact stress. Accordingly, the separation in the dry regime between the mating teeth increases and the penetration decreases. While in the presence of the lubricant, the film thickness is reduced due to the increase in gear ratio, leading to an increase of penetration of the asperities. This results in an increase in the ratio of penetration in lubricated regime to that in dry regime, which gives rise to the increase of relative NPL.

It can therefore be seen that the surface roughness effect as a contributing factor in the complex frictional gear noise spectrum could, to some extent, be controlled. It could be reduced by improving the surface finish of gear teeth through limiting their surface roughness to very low values, and by using lubricating oils of high viscosities.

Gear ratios of values greater than unity can have a significant effect on increasing the NPL even with lower values of roughness. The developed procedure can be used to guide the designer in selecting the appropriate parameters for minimizing the frictional noise for any particular application.

## 11.5  FRICTION-INDUCED VIBRATION AND NOISE

There are numerous cases in physical systems where sound due to vibrations is developed and sustained by friction. Such cases are generally known as *self-excited vibrations*. They are described as such because the vibration of the system itself causes the frictional resistance to provide the necessary energy for sustaining the motion. The frequency of the vibration is therefore equal to (or close to) the natural frequency of the system. Some of the common examples of self-excited (or self-sustained) vibrations are the chatter vibration in machine tools and brakes, the vibration of the violin strings due to the motion of the bow and numerous other examples of mechanical systems subjected to kinetic friction.

We shall now consider as an illustration the well-known case of a single-mass system vibrating in a self-excited manner under the influence of kinetic friction, Fig. 11.12. Assuming that $\mu$ is the coefficient of kinetic friction, and $N$ is the normal force between the mass $m$ and the frictional wheel, the unidirectional frictional force acting on the mass will therefore be equal to $\mu N$. It is well known that the coefficient of kinetic friction is not a constant value but diminishes slightly as the velocity of relative sliding increases (see Fig. 11.13). If, due to some slight disturbance, the mass starts to vibrate, the frictional force $\mu N$ will not remain constant but will be larger when the mass moves in the direction of the tangential velocity $V_0$ of the wheel than when it moves opposite to it.

Assuming that the velocity of the oscillation $\dot{x}$ is much smaller than the tangential velocity of the frictional wheel, the frictional force $\mu N$, which is a function of the relative velocity $(V_0 - \dot{x})$, will therefore always be in the direction of $V_0$. Over a complete cycle of vibration, the frictional force will therefore produce net positive work on the mass and the amplitude of its vibration will build up.

In order to study this vibration, the equation of motion can be written as:

$$m\ddot{x} + c\dot{x} + kx = \mu N = (\mu_0 + \alpha\dot{x})N \tag{11.4}$$

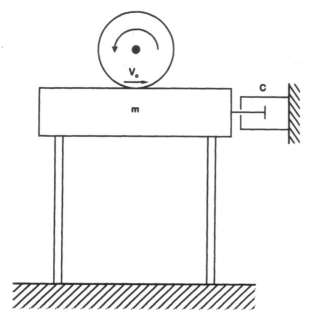

**Figure 11.12**   Frictional drive for self-excited vibration.

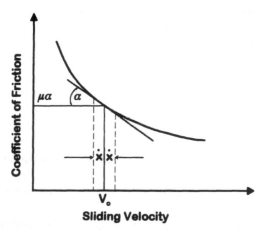

**Figure 11.13**   Coefficient of kinetic friction.

where

$c$ = coefficient of viscous damping

$\mu_0$ = coefficient of kinetic friction at $V_0$ relative velocity

$\alpha$ = slope of the friction curve at $V_0$ and can be considered constant
  = for a small $\dot{x}$

This equation can be rearranged as:

$$m\ddot{x} + (c - \alpha N)\dot{x} + kx = \mu_0 N \qquad (11.5)$$

The term $(c - \alpha N)$, which represents a net damping coefficient, will determine whether the vibration will be stopped or built up in a self-excited manner. If $\alpha N < c$, the resultant damping term will be positive and the vibration will decay signifying the stability of the system. On the other hand, if $\alpha N > c$, a negative damping term will exist and the vibration will build up as shown in Fig. 11.14a. The system in this case is unstable.

It is quite clear from the previous example that a quantitative knowledge of the frictional force and damping functions is essential for any analysis of this type of self-excited vibration. Any variation in either function due to increase in amplitude or velocity of the vibration can have considerable effect on the vibration.

## 11.5.1 The Phase-Plane-$\delta$ Method

Because frictional forces are usually complex functions which require experimentally obtained information, the phase-plane-$\delta$ method of analysis is particularly well suited for studies of self-excited vibration [10].

Assume that in the previous equation the resultant function $(c - \alpha N)$ was not a linear function of the velocity but rather a complicated function $f(\dot{x})$ obtained experimentally. The equation of motion is now:

$$m\ddot{x} + f(\dot{x}) + kx = \mu_a N \qquad (11.6)$$

This equation can be written in the $\delta$ form as:

$$\ddot{x} + \omega_n^2 \left( x + \frac{1}{k} [f(\dot{x}) - \mu_a N] \right) = 0 \qquad (11.7)$$

$$\ddot{x} + \omega_n^2 (s + \delta) = 0 \qquad (11.8)$$

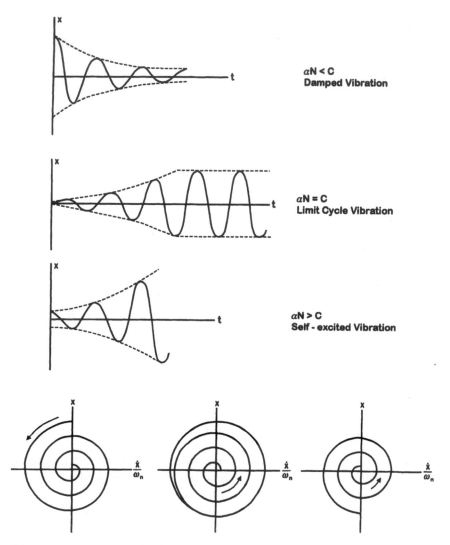

**Figure 11.14**  Dynamic response.

where

$$\delta = \delta_{\dot{x}} + \delta_0$$

$$\delta_{\dot{x}} = \frac{1}{k} f(\dot{x})$$

$$\delta_0 = \frac{-1}{k} \mu_a N = \text{constant}$$

and the problem accordingly will transform to a free vibration with a continuously changing datum. The datum variation $\delta$, which in this hypothetical case, is a function of the velocity $\dot{x}$ can be obtained at any instant of the motion in the phase plane from the $\delta$-curve. By successively plotting small segments of the locus and changing the datum according to the new position, the motion can be graphically represented in the phase plane. This is shown in Fig. 11.14b, which also shows that the motion is stable when the vibration decays or a limit cycle is reached. The motion at the limit cycle will be similar to a free undamped vibration.

Data representing different forms of self-excited vibration of the system shown in Fig. 11.12 are shown in Fig. 11.15. The bearing block of the frictional wheel is moved on the supporting wedge a predetermined amount to produce a particular value of static friction. The motion of the block is indicated by means of a dial gage. The driving motor is then run at different speeds and the resulting vibration of the system is recorded. It should be noted that the frequency of the motion is the same as the natural frequency of the system and is independent of the motor speed.

This type of vibration develops as a result of the negative slope of the friction–velocity function. This is generally the case, with varied degrees in dry friction. The use of grease lubrication causes the frictional resistance to increase with sliding speed, giving a positive slope and consequently avoiding the self-excitation.

## 11.6 PROCEDURE FOR DETERMINATION OF THE FRICTIONAL PROPERTIES UNDER RECIPROCATING SLIDING MOTION

It has been shown in the previous section that friction-induced self-excited vibrations and noise are controlled by the functional dependency of kinetic friction on the relative velocity between the rubbing surfaces. This relationship has to be determined experimentally because it is a function of the materials in contact, surface roughness, and lubrication condition.

The measurement of the coefficient of friction as a function of sliding velocity has been the subject of many studies to determine the influence of such controlling factors as materials, surface roughness, temperature, and lubrication condition. Most of the reported experiments utilized the pin-on-disk tribometer apparatus [16–19] and different types of transducers were used to measure the frictional force directly or indirectly. This type of experiment is generally associated with excessive vibrations due to the rotation of the motor and disk. The irregularities of the disk surface during rotation can also produce variations of the measured coefficient of friction.

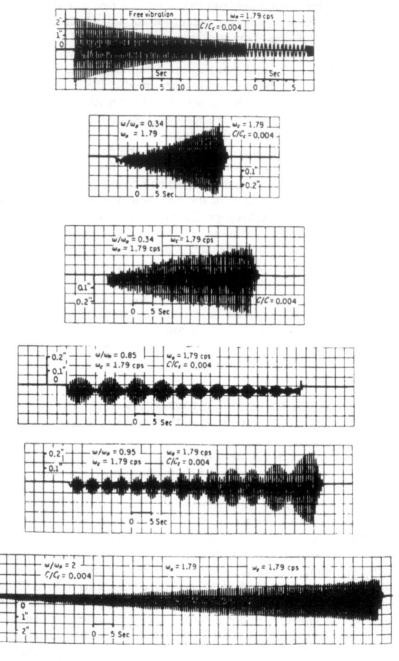

**Figure 11.15**  Examples of self-excited vibration.

This was observed by Godfrey [20] who emphasized the importance of avoiding external vibrations during friction measurements.

Bell and Burdekin [17] utilized acceleration and displacement measurements during one cycle of the friction-induced vibration of slideways to evaluate the frictional force as a function of the instantaneous velocity. The force was calculated from the knowledge of the mass, stiffness, and damping coefficient of the vibrating system by summing the inertia, damping, and restoring forces at each increment of the friction-induced cycle. In 1970, Ko and Brockley [16] developed a technique for determining the friction–velocity characteristic by measuring the friction force versus displacement in one cycle of a quasiharmonic friction-induced vibration using a pin-on-disk apparatus. They reported that their technique proved useful in reducing the effect of changes of the surface and external vibration.

In 1984, Aronov [18] investigated the interaction between friction, wear, and vibration and their dependence on normal load and system stiffness using a pin-on-disk apparatus. The friction-induced vibration, which has been studied by several investigators [16–18, 21, 22], may be classified into three types: stock–slip, vibration induced by random surface irregularities, and quasiharmonic oscillation. These three types of vibration have been observed under certain conditions, which depend on the normal load, sliding speed, and the nature of the surfaces in contact. Tolstoi [23] was one of the early investigators of the stick–slip phenomenon where the vibration in the normal direction to the contact surface is usually of a sawtooth type caused by changes in the coefficient of friction with the relative sliding velocity. The vibration usually occurs when the sliding speed is sufficiently low. At relatively low values of the normal load, normal vibration is produced due to the surface irregularities and waviness. When the normal load and the sliding speed are sufficiently high, quasiharmonic oscillations with nearly sinusoidal waveforms are produced.

Ko and Brockley [16] attempted to minimize the effect of external vibration by using a one-cycle sequence triggering circuit and other electronic devices which permitted the measurement of the kinetic friction force in the presence of friction-induced vibration. However, the problems associated with nonuniformity of the disk surface and the reliability of the measurement continue to present challenges to this approach.

Most analytical and experimental studies which are reported in the literature on friction-induced vibration and friction force measurements are based on a constant relative sliding speed.

Anand and Soom [24] analytically investigated the dynamic effects on frictional contacts during acceleration from rest to a steady state velocity.

The study by Othman and Seireg [25] presents a procedure for evaluating the change of frictional force with relative velocity during reciprocating sliding motion. It utilizes the friction-induced lateral vibration of a rod to evaluate the parameters of the frictional function using a gradient search which minimizes the error between the analytical response and the friction-induced experimental vibrations. The use of sinusoidal sliding motion at the resonant frequencies of the vibrating rod is found to considerably minimize the effect of external vibrations on the experimental results.

The experimental model used for this purpose (Fig. 11.16) consists of a cylindrical steel rod (A), which is pressed on a flexible rod (B), by means of a load $N$. Because the contact area is small, the variations in surface roughness are minimized within the frictional area and a steady-state surface roughness can be rapidly achieved after few reciprocating cycles. The effect of external vibrations on the measurements is also minimized by operating the reciprocating rod at resonant frequencies of the system.

The dynamic motion of the rod $B$ (shown in Fig. 11.16a,b) when subjected to a reciprocating frictional force can be modeled by three degrees of freedom representing the translation $x$ and $y$ at the midspan and the torsional angle $\theta$ about the axis of the rod. By appropriate selection of the rod dimensions each of these movements can be represented by an elastically supported single mass $m$ because the oscillations can be uncoupled for all practical purposes.

For small displacements, the governing equations for the motion of the oscillatory rod $B$ can be represented as:

$$\ddot{y} + 2\zeta\omega_{n_1}(\dot{y} - \dot{y}_i) + \omega_{n_1}^2(y - y_i) = N \tag{11.9}$$

$$\ddot{x} + 2\zeta\omega_{n_2}\dot{x} + \omega_{n_2}^2 x = -\frac{\mu}{m} N \tag{11.10}$$

$$\ddot{\theta} + 2\zeta_t\omega_n^3\dot{\theta} + \omega_{n_2}^2\theta = -\frac{\mu}{I} Nr \tag{11.11}$$

where $\mu$ is the coefficient of friction and $y_i$ is a function describing the disturbance in the $y$ direction resulting from surface waviness. The equations are essentially uncoupled due to the selection of widely separated natural frequencies $\omega_{n_1}$, $\omega_{n_2}$, and $\omega_{n_3}$.

It has been shown [16] that in the case of stick–slip oscillation, the friction force is time dependent during stick and velocity dependent during slip, and in the case of a quasiharmonic oscillation, the motion is govenred by the velocity dependent friction force only.

It is assumed in this study that the friction–velocity relationship is approximated by an exponential function of the following form:

$$\mu = \mu_{\min} + (\mu_{\max} - \mu_{\min})e^{\mu/v_0} \tag{11.12}$$

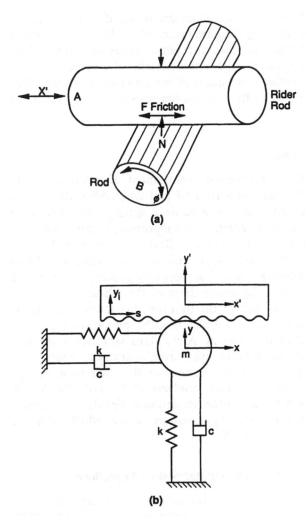

**Figure 11.16** Dynamic model.

where:

$$u = \text{sliding velocity}$$

$$v_0 = \text{exponential constant}$$

$$\mu_{max} \text{ and } \mu_{min} = \text{maximum and minimum bound values of the coefficient of friction}$$

Both the $x$ and $y$ accelerations were monitored during the tests and the latter was found to be consistently negligible in all the performed tests. Because the $x$ motion is designed to be the most dominant mode of oscillation, only Eq. (11.10) is used for evaluating the parameters of the Eq. (11.12), which produce the best fit with the experimental results. This is accomplished by means of a multivariate gradient search to determine the values of $\zeta$, $\nu_0$, $\mu_{min}$, and $\mu_{max}$, which minimize the square of deviation between the calculated and experimental peaks of the acceleration response measured at the center of the beam.

### 11.6.1   Experimental Arrangement

The main features of the experimental arrangement are shown in Fig. 11.17 where the two cylindrical steel rods ($A$ and $B$) (UNS G10100 CD) with mutually perpendicular axes are used as a sliding pair. Rod $B$ is 0.65 m long and 0.0009 m in diameter. It is supported at both ends such that rotation about its axis is constrained. Rod $A$ is connected to an electromagnetic exciter (type B&K 4811 with a force rating of 310 N) and acts as the reciprocating rider. The supporting structure and the rider have natural frequencies far above those of the vibrating rod $B$. A function generator is used to control the reciprocating motion of rod $A$. The $x$ and $y$ oscillations of rod $A$ are measured by piezoelectric accelerometers, which are fixed at its midspan. A standard vibration calibrator (type B & K 4291) is used to check the measuring instrument. The output signals are amplified by a conditioning amplifier then fed to a storage oscilloscope and a real-time spectrum analyzer (type HP 3582A). A chart recorder is also used to record the acceleration signals. The normal load between the rods is controlled by a loading device, which does not affect the natural frequency of rod $B$.

### 11.6.2   Experimental Results and Computer Search Procedure

The natural frequency $\omega_{n_2}$ and the damping ratio of rod $B$ are determined experimentally by impacting the rod in the $x$ direction when the load is 20 N. The system total equivalent vibrating mass is 0.49 kg, which includes the mass of the transducer attached to the rod. The equivalent rod stiffness in the $x$ direction is 7260 N/m. The natural frequency obtained from the peaks of the response is found to be 21.0 Hz. This value is also verified by the real-time spectrum analyzer. The damping ratio, which is calculated using the logarithmic decrement method, is found equal to 0.23.

In order to minimize the effect of the external vibration, the reciprocating frequencies for the test are selected to be equal to 1/3, 2/3, 1, and 4/3 of the natural frequency $\omega_{n_2}$ of the rod $B$. The dotted curves in Figs 11.18a–d are the experimental acceleration responses corresponding to the

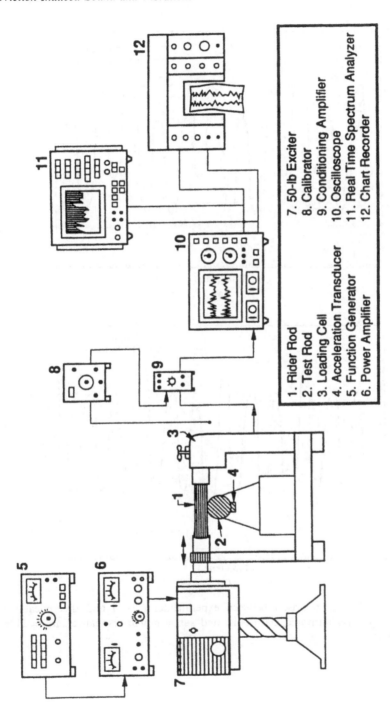

1. Rider Rod
2. Test Rod
3. Loading Cell
4. Acceleration Transducer
5. Function Generator
6. Power Amplifier
7. 50-lb Exciter
8. Calibrator
9. Conditioning Amplifier
10. Oscilloscope
11. Real Time Spectrum Analyzer
12. Chart Recorder

**Figure 11.17** Experimental setup.

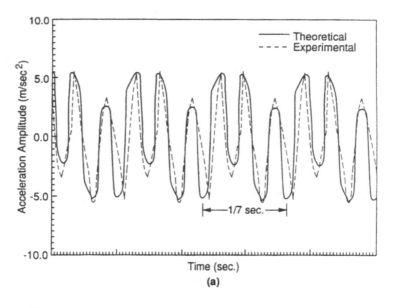

(a)

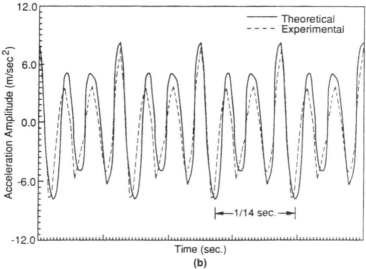

(b)

**Figure 11.18** Comparison between experimental (- - -) and theoretical (—) response at superharmonic, harmonic, and subharmonic resonances. $\alpha_{n_2} = 21 \, \text{Hz}$ in all cases.

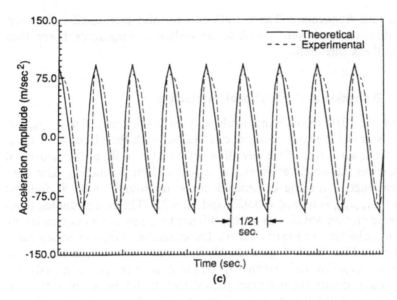

(c)

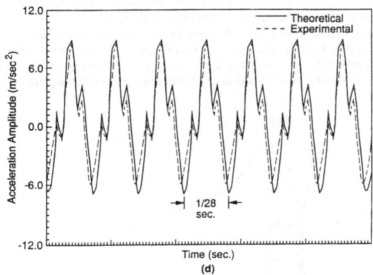

(d)

selected four frequencies. The vibrations were also monitored in the normal $y$ direction and were found to be orders of magnitude lower than those in the $x$ direction.

### 11.6.3   Evaluation of the Frictional Parameters

Equation (11.10) is solved numerically with assumed values of $\mu_{max}$, $\mu_{min}$, $\zeta$, and $v_0$. A gradient search technique is utilized with the objective of minimizing the square error of the deviation between the peaks of the theoretical and experimental acceleration curves as the objective function. The optimum parameters for the considered cases are found to be as follows: $\mu_{max} = 0.11$, $\mu_{min} = 0.06$, $\zeta = 0.023$, and $v_0 = 0.8$. The corresponding theoretical acceleration waveforms for the different reciprocating frequencies are shown by solid lines in Figs 11.18a–d. The evaluated friction–velocity curve is shown in Fig. 11.19. Figure 11.20 illustrates the excellent correlation between the experimental response spectra and the corresponding analytical results obtained with the optimum parameters. It should be noted that the same parameters for the frictional function were obtained for the different frequencies of the reciprocating motion considered in the test.

Although good results were obtained by using an exponential function for the friction–velocity characteristics and by using the peaks of the response curve for computing the error function, the same approach can be used by assuming other functions and minimizing the mean square error

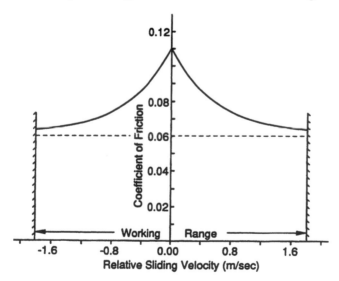

**Figure 11.19**   Evaluated frictional coefficient versus sliding velocity.

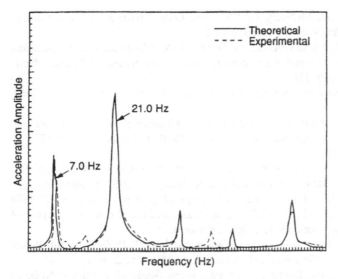

**Figure 11.20** Comparison between the experimental (- - -) and theoretical (—) response spectrum for $\alpha_e = 7\,\text{Hz}$ ($\alpha_e/\alpha_{n_2} = 1/3$).

for the entire response curve. It should be noted that the value of $\zeta$ obtained from the optimization procedure is identical to that obtained from the decay of the experimental free vibration data.

## REFERENCES

1. Jakobsen, J., "On Damping of Railway Break Squeal," Noise Control Eng. J., Sept.–Oct. 1986, pp. 46–51.
2. Matsuhisa, H., and Sato, S., "Noise from Circular Stone-Sawing Blades and Theoretical Analysis of their Flexural Vibration," Noise Control Eng. J., Nov.–Dec. 1986, pp. 96–102.
3. Houjoh, H., and Umezawa, K., "The Sound Radiated from Gears," Trans. JSME, Sept. 1986, Vol. 52(481), pp. 2463–2471.
4. Lyon, R., "Noise Reduction and Machine Diagnostic and Educational Challenge," Noise Vibr. Control Wldwide, Sept. 1985, Vol. 16(8), pp. 221–224.
5. Fielding, B., and Skorecki, J., "Identification of Mechanical Source of Noise in a Diesel Engine; Sound Emitted from the Valve Mechanism," Proc. Inst. Mech. Engrs, 1966–67, Vol. 181, Part I (I), pp. 434–446.
6. Thompson, J., "Acoustic Intensity Measurements for Small Engines," Noise Control Eng. J., Sept.–Oct. 1982, pp. 56–63.
7. Yokoi, M., and Nakai, M., "A Fundamental Study on Frictional Noise," Bull. JSME, Nov. 1979, Vol. 22(173), pp. 1665–1671.

8. Symmons, G., and McNulty, G., "Acoustic Output from Stick-Slip Friction," Wear, Dec. 1986, Vol. 113(1), pp. 79–82.

9. Othman, M. O., Elkholy, A. H., and Seireg, A. A., "Experimental Investigation of Frictional Noise and Surface-Roughness Characteristics," Exper. Mech., Dec. 1990, pp. 328–331.

10. Seireg, A., Mechanical Systems Analysis, International Textbook Co., 1969, p. 412.

11. McCool, J. I., "Relating Profile Instrument Measurements to the Functional Performance of Rough Surfaces," Trans. ASME, J. Tribol., April 1987, Vol. 109, pp. 264–270.

12. Greenwood, J. A., and Williamson, J. B. P., "Contact of Nominally Flat Surfaces," Proc. Roy. Soc. Lond. Series A, 1966, Vol. 295, pp. 300–319.

13. Welbourn, D. B., "Fundamental knowledge of gear noise: a survey," 1979 Conf. Noise and Vibration of Engine Transmissions, Cranfield Institute of Technology, Institute of Mechanical Engineers.

14. Attia, A. Y., "Noise of Gears: a Comparative Study," 1989, Proc. Int. Power Transmission and Gearing Conf., Chicago, ASME, Vol. 2, p. 773.

15. Aziz, S. M. A., and Seireg, A., "A Parametric Study of Frictional Noise in Gears," Wear, 1994, Vol. 176, pp. 25–28.

16. Ko, P. L., and Brockley, C. A., "The Measurement of Friction and Friction Induced Vibration," ASME J. Lubr. Technol. Trans., Oct. 1970, pp. 543–549.

17. Bell, R., and Burdekin, M., "Dynamic Behavior of Plain Slideways," Proc. Inst. Mech. Engrs, 1966–1967, Vol. 181, Part 1, No. 8, pp. 169–184.

18. Aronov, V., D'Souza, A. F., Kalpakjian, S., and Shareef, I., "Interactions Among Friction Wear and Systems Stiffness Part 1: Effect of Normal Load and System Stiffness," ASME J. Lubr. Tribol., Jan. 1984, Vol. 106, pp. 54–58.

19. Brockley, C. A., and Ko. P. L., "Quasi-Harmonic Friction-Induced Vibration," ASME J. Lubr. Technol. Trans., Oct. 1970, pp. 550–556.

20. Godfrey, D., "Vibration Reduces Metal to Metal Contact and Causes an Apparent Reduction in Friction," Trans. ASLE, Apr. 1967, Vol. 10(2), pp. 183–192.

21. Earles, S. W. E., and Lee, C. K., "Instabilities Arising from the Frictional Interaction of a Pin-Disk System Resulting in Noise Generation," J. Eng. Indust., Feb. 1976, pp. 81–86.

22. Aronov, V., D'Souza, A. F., Kalpakjian, S., and Shareef, I., "Interactions Among Friction, Wear, and System Stiffness – Part 2: Vibrations Induced by Dry Friction," ASME J. Tribol. Trans., Jan. 1984, Vol. 106, pp. 59–64.

23. Tolstoi, D. M., "Significance of the Normal Degree of Freedom and Natural Vibrations in Contact Friction," Wear, 1967, Vol. 10, pp. 193–213.

24. Anand, A., and Soom, A., "Roughness-Induced Transient Loading at a Sliding Contact During Start-Up," ASME J. Tribol., Jan. 1984, Vol. 106, pp. 49–53.

25. Othman, M. O., and Seireg, A., "A Procedure for Evaluating the Frictional Properties of Hertzian Contacts under Reciprocating Sliding Motion," Trans. ASME, J. Tribol., 1990, Vol. 112, pp. 361–364.

# 12

# Surface Coating

## 12.1 INTRODUCTION

In tribological systems, the load transfer, relative movement, wear, corrosion, and fatigue damage initiation occur at the surface. Advances in surface coating technology can therefore have considerable impact on improving the performance of such systems and extending their useful life.

The coated layer can be an adsorbed film of the lubricant or a chemical layer formed by the reaction of the materials to the environment. It can also be induced by surface treatment processes which have been known for centuries such as cold working, carburizing, and induction hardening.

The other basic type of surface modification is surface coating. Until recently, coatings were used exclusively as a corrosion inhibitor. Today, engineered coatings can resist abrasive wear, change the friction coefficient, act as a thermal barrier or conductor, and resist corrosion.

## 12.2 COATING PROCESSES

The methods by which surfaces are coated with thin films are often divided into two groups:

1. Hot processes or chemical vapor deposition (CVD)
2. Cold processes or physical vapor deposition (PVD)

Coating properties such as microstructure, substrate adhesion, and wear or abrasion resistance depend on the coating process.

### 12.2.1 Chemical Vapor Deposition

Commercially available CVD hard coatings for tooling include titanium carbide, hafnium carbide and nitride, and aluminum oxide. All of the coatings are applied to a thickness in the range 5–9 μm, generally dictated by the operational requirements for the coated surfaces.

The driving force of the process is the high temperature, typically in the range 1750–1950°F, to which the work pieces are heated, which causes the reactive gases to dissociate and the desired coating compound to form on the work piece surfaces. For example, titanium tetahchloride ($TiCl_4$) would be the reactive gas introduced to provide the titanium and pure nitrogen gas ($N_2$) or ammonia ($NH_3$) would supply the nitrogen to form a TiN coating. Hydrogen chloride gas (HCl) is also formed in this reaction and must be neutralized for safe removal.

Chemical reactions that take place are given below:

$$2TiCl_4 + 4H_2 + N_2 \rightarrow 2TiN + 8HCl$$

Or in the case of TiC coating:

$$TiCl_r + CH_4 \rightarrow TiC + 4HCl$$

Similarly if $Al_2O_3$ is deposited, the gas mixture would consist of $AlCl_3$ (aluminum chloride), $H_2$ and $CO_2$:

$$2AlCl_3 + 3CO_2 + 3H_2 \rightarrow Al_2O_3 + 3CO + 6HCl$$

Important parameters influencing the deposition rate, composition, and structure of the coatings are:

Temperature
Composition of the gas atmosphere
Flow rate of the gas in the coating chamber
Coating time

Due to high coating temperatures, steel parts must be heat treated after coating. It is important that only components which have sufficient dimensional tolerances be coated. In general, parts with tolerance of 0.001 in. make excellent candidates for CVD coatings.

The CVD process is most commonly used for the coating of very large quantities of cemented carbide tools. With respect to equipment for production processes, some additional requirements must be fulfilled, such as large number of components coated in one run with a uniform coating thickness, a minimum rejection rate as a result of the high degree of reproducibility from the process, a high degree of equipment reliability, and low production and maintenance costs.

Production chambers can handle a working diameter of 360 mm and a working height of up to 900 mm. These units have microprocessor-based automated control systems. This enables composition and a sequence of layers. For example, a coating which consists of a sequence of 10 layers can be produced fully automatically in one cycle.

At the present, CVD is primarily used to coat machine tools with TiN. The process starts by placing parts in a chamber and heating to 1000°C. In a few hours, the parts reach a uniform temperature. Gaseous chemicals are introduced into the chamber at atmospheric pressure. Chemical reactions of gaseous material produce the coating material and gaseous byproducts. Coating material crystallizes on the substrate surface. This process takes several hours and is very sensitive to process parameters. However, thick coatings can be applied by this method.

### 12.2.2   Physical Vapor Deposition

This process relies on ion bombardment as the driving force. Temperatures are typically in the range 500–900°F for the deposition of tool coatings. This lower temperature is generally given as the major distinction between CVD and PVD processes. The following are the major PVD coating processes:

1.   Sputter ion plating
2.   Electron gun beam evaporation (ion plating)
3.   Arc evaporation (ion bond)

*Sputter Ion Plating*

Sputter ion plating (SIP) takes place in a vacuum chamber containing argon at a certain known pressure. Parts to be coated are loaded into a standard fixture. The inside surface of the SIP unit as well as all of the exposed surfaces are lined with a sheet of titanium. The titanium acts as a source material. The parts are held at a positive voltage ( + 900 V) with respect to titanium, resulting in a glow discharge (plasma) generated between the workload and the titanium. The ionized argon bombards the titanium, sputtering titanium atoms. These highly energized titanium atoms, through a series of random collisions, migrate to the part and are deposited on the exposed surfaces, forming a thin uniform coating. Nitrogen gas is then bled into the chamber which reacts with the deposited titanium, forming TiN.

To form a fine impurity-free coating, small anodes, biased slightly higher in potential than the work load, are inserted into the chamber in close proximity to the part. The effect is to produce further low-energy sputtering, which produces a microcrystalline structure. This is due to the deposited coating itself being bombarded by high-energy argon atoms.

The first step in the tool coating process is to ensure that the surfaces of the components to be coated are free of oxides, rust preventatives, dust, grease, and burrs, all of which can affect adherence. Cleaning of tools consists of series of mechanical/chemical treatments followed by utrasonic degreasing. It is important to clean tools as carefully as possible to reduce the risk of damaging the cutting edges. Cleaned tools are loaded onto similarly cleaned fixtures. The fixture is then placed into the coating chamber, and argon gas is then flowed into the chamber. The argon is purified before entering the chamber by passing over a heated titanium. The tools to be coated and titanium source material are heated using external radiant heaters to 300°C. The pure argon sweeps away any volatile contaminants which may be in the system or which remains on the parts.

Once the chamber and the work load is at temperature, ion cleaning of the parts takes place. Ion cleaning is accomplished by applying a negative voltage (−500 V) to the parts. This establishes a glow discharge in the chamber from which ions are attracted to the part sputtering the surface. The sputtering action provides a surface which is free of oxides or any barrier to the coating in the chamber. Ion cleaning is essential for good coating adhesion and subsequent surface performance. After the ion cleaning is completed, the bias is reversed, and coating is initiated, as described above.

SIP is a process having excellent throwing power. Because of the large titanium source, the small mean free path of sputtered titanium, and the operation of the system at less than 500°C (927°F), large and small tools of different geometries may be coated in the same cycle. These characteristics of SIP set it off from other PVD processes, and clearly provide greater process flexibility.

### Electron Beam Gun Evaporation

This ion plating process uses a crucible of molten titanium, which is evaporated at low pressure by an electron beam gun to produce titanium vapor, which is attracted by an electric bias to the workpiece. The process is inherently slow, but can be speeded up by ionization enhancement techniques. The main drawback is the constraint that the workload must be suspended above the melting crucible, using water-cooled jigging, and uniformity is difficult to achieve.

### Arc Evaporation

With the arc evaporation (ion bond) method ARE, blocks of solid titanium are arranged around the chamber walls and an arc is struck and maintained between the titainum and the chamber. Titanium is evaporated by extreme local heat from the arc into the nitrogen atmosphere and attracted to the workpiece. The main advantage is that evaporation occurs from the solid rather than the liquid phase. Arc sources therefore may be placed at any angle around the workpiece, which is simply placed on a turntable in the base of the chamber. Thus uniformity is achieved without complex jigging. The kinetic energy of deposition is great enough to give rich plasma of ionized titanium, resulting in good adhesion at a high coating rate and low substrate temperature. Parameters such as coating thickness, coating composition and substrate temperature are easily controlled.

### 12.2.3 Comparison between the CVD and PVD Processes

The principal difference between CVD and PVD processes is temperature. This is of major significance in the coating of high-speed steels as the 1750–1950°F of CVD exceeds the tempering temperature of HSS steel, therefore, the parts must be restored to the proper condition by vacuum heat treatment following the coating process. With properly executed heat treatment, this generally causes no problems. However, in some cases involving extremely fine tolerances, post-coating heat treatment does produce unacceptable distortion.

The high temperature of the CVD process makes it somewhat less demanding than PVD in terms of cleanliness of the workpiece going into the reactor: some types of dirt simply burn off. Additionally, high temperatures tend to ensure a tightly adhering coating, and PVD temperatures are often pushed to the HSS tempering range to enhance coating adhesion.

CVD coatings tend to be somewhat thicker (typically 0.0003 in.) than those deposited by different PVD processes (often less than 0.0001 in. thick). This may be advantageous in some cases, disadvantageous in others. Another characteristic of the CVD coating is that they yield a matte surface somewhat rougher than the substrate to which they are applied. If the application requires it, this can be polished to a high luster, but this is an extra step. PVD coating on the other hand faithfully reflect the underlying surface.

Because the CVD reactions take place within the gaseous cloud, everything within that cloud will be coated. This permits workpieces to be closely packed within a CVD reactor with complete coating of all surfaces except those points on which the parts rest. Even deep cavities and inside diameters will become coated in a CVD reactor.

Except for SIP, all PVD processes have limited throwing power. PVD reactors are therefore less densely packed, and the jigging and fixtures are more complicated.

A cost comparison between CVD and PVD is complex. The initial investment in the equipment is as much as three to four times as great for PVD as for CVD. The PVD process cycle time can be one tenth that of CVD. Mixed components can be coated in one CVD cycle, whereas PVD is much more constrained.

The main advantage of the PVD process is that most metallic and ceramic coatings can be deposited on almost any substrate. The process is very flexible. Several process parameters can be controlled directly and the process is insensitive to slight variation of process parameters. The process is fast and relatively inexpensive because vaporized coating material is carried directly to the substrate where particles condense to form a film.

## 12.3  TYPES OF COATINGS

Surface coatings can be divided into two subgroups, hard and soft coatings. Hard coatings are recommended for heavy load or high-speed applications. Beneficial characteristics are low wear and long operating periods without deterioration of performance. Hard coatings include iron alloys, ceramics like carbides and nitrides, and nonferrous alloys.

Soft coatings are recommended for low-load, low-speed applications. Advantages of soft coatings are low friction, low wear, and a wide range of operating temperatures. Soft metals have received much attention because of their low-load, friction-reducing properties. Many soft coatings are actually solid lubricants (like graphite) that require a resin binder to adhere to the surface. These coatings are typically applied to protect parts during a running in period.

### 12.3.1  Soft Coatings

Soft coatings can be grouped into four main categories: layered lattice compounds such as graphite, graphite fluorides, and $MoS_2$; nonlayered lattice compounds such as $PbO–SiO_2$, $CaF_2$, $BaF_2$, and $CaF_2–BaF_2$ eutectics; polymers; and soft metallic coatings [1].

#### Layered Lattice Coatings
Most layered lattice coatings are hexagonal compounds with slip planes that are oriented parallel with the surface. These compounds are like plates stacked up on top of each other. The plates slip easily when subjected to a shear force. However, they resist movement normal to the surface. Burnishing is an important step that aligns the plates parallel with the surface.

The most common layered lattice compounds are graphite, graphite fluorides, and $MoS_2$. Graphite and graphite fluoride compounds tend to perform better at room temperature in humid environments. Current understanding is that adsorbed moisture helps the plates slip. Higher temperatures drive off moisture and explain a rapid increase in the friction coefficient. Above 430°C, the friction coefficient drops. It is believed that graphites interact with metal oxides that form on the mating surface to reduce friction. Graphite fluorides generally perform better than pure graphite but have a life about ten times longer at room temperature. They are not sensitive to humidity and operate well in a vacuum. However, unlike graphite, wear life decreases proportionally with temperature rise. The compound decomposes around 350°C.

## Nonlayered Lattice Coatings

These coatings are based on inorganic salts. The main characteristic is a phase change caused by frictional heating. The coating is solid at the bulk temperature but becomes a high viscosity melt at the friction interface. Advantages are chemical inertness and effectiveness at high temperatures. Some fluoride salts remain effective at temperatures approaching 900°C. Disadvantages are high friction at low temperatures and manufacturing difficulties. These coatings are very difficult to apply to substrates.

## Polymer Coatings

Polymer coatings are applied to metal and nonmetal surfaces by several different techniques. Traditionally, polymers are used to repel water and resist corrosion. They can resist erosive, abrasive wear caused by impacting particles because the coating is elastic. The coating deforms to absorb particle impact, then returns to its origial shape. Friction is typically very low, especially when polymers are applied to hard substrates.

Polymers are used by industry for bearings, automotive components, pumps, and seals. The coatings are inexpensive and easy to apply. Wide use by industry has helped build a large base of empirical knowledge.

## Soft Metal Coatings

Soft metals are compatible with liquid lubricants, effective at low temperatures and at elevated temperatures (silver and gold are effective near their melting points), can operate in a vast range of normal pressures from vacuum to high pressure, and perform well at high speeds. Soft metals can be applied by several different processes. However, high material costs limit widespread use of soft metals. Bhushan [1] compiles results of several studies performed with ion-plated soft metal coatings. Tests were performed with a pin-and-disk apparatus. The disk was coated; the pin was not. A common characteristic is dependence of friction coefficient on coating thickness where friction reaches a minimum at a critical coating thickness. In the ultrathin region, surface asperities of the mating surface break through the coating and interact with the substrate. Thus in the limit, the friction coefficient reaches that of the substrate material. In the thin region, the real and apparent areas of contact are equal, leading to an increase in friction with increasing coating thickness and reaches an asymptote for thickness above 10 μm.

Studies on the effect of sliding velocity on the friction coefficient and wear life of silver, indium and lead suggest that velocity has little effect on the coefficient of friction. A slight decrease in friction at higher speeds may occur due to thermal softening of the coated material. On the other hand, sliding velocity has a large effect on wear life. Sherbiney [2] reported that wear life is inversely proportional to speed. More recent studies reported by Bhushan indicate that soft metal coatings alloyed with copper or platinum tend to improve wear life and reduce friction [1].

### 12.3.2  Hard Coatings

Hard coatings are recommended for heavy-load, high-speed applications. These coatings exhibit low wear, can be used for long periods without deterioration in performance, and protect against wear and corrosion in extreme conditions. The main types of hard coatings are ferrous alloys, nonferrous alloys, and ceramics. Table 12.1 lists common hard coatings and general properties.

Iron alloys are generally hard and brittle. Steel alloy coatings are more ductile and better able to resist mechanical shock. Nonferrous alloys are primarily used for corrosion resistance at high temperatures. Ceramic coatings are hard, brittle, chemically inert, and against corrosion. Titanium-based ceramic coatings are revolutionizing the machine tool industry.

### Iron-Based Alloys

Iron based alloys are usually applied by weld deposition or thermal spraying. Alloying with cobalt improves oxidation resistance and hardness at elevated temperatures. High chrome and martensitic irons are hard, not as tough as steel coatings. Martensitic, pearlitic, and austenitic steels are recommended for heavy wear and conditions where mechanical or thermal shock are expected. The irons resist abrasion better, but are not recommended for applications involving mechanical and thermal shock.

**Table 12.1**  Hard Coating Reference Chart

| Alloy coating | Properties |
| --- | --- |
| Tungsten carbides | Maximum abrasion resistance, worn surfaces become rough |
| High-chromium irons | Excellent erosion resistance, oxidation resistance |
| Martensitic irons | Excellent abrasion resistance, high compressive strength |
| Cobalt-based alloys | Oxidation resistance, corrosion resistance, hot strength and creep resistance, composition control, several options for coating deposition, good galling resistance |
| Nickel-based alloys | Corrosion resistance, may have oxidation and creep resistance, compositional control, several options for coating processes, relatively inexpensive, poor galling resistance |
| Martensitic steels | Good combinations of abrasion and impact resistance, good compressive strength |
| Pearlitic steels | Inexpensive, fair abrasion and impact resistance |
| Austenitic steels, stainless steels, manganese steels | Work hardening, corrosion resistance, maximum toughness with fair abrasion resistance, good metal-to-metal wear resistance under impact |
| Chromium-based alloys | Good thermal conductivity, resists abrasive wear, low friction coefficient, good corrosion resistance |
| Nickel | High hardness, good abrasion resistance, brittle, low friction coefficient, low wear, corrosion resistant, can be applied to some plastics, weakly ferromagnetic |

## Chrome-Based Coatings

Chrome alloys are usually applied by electrochemical deposition and PVD. CVD less commonly used. Electrochemically deposited chrome has a hardness around 1000 HV that is stable up to 400°C. Th electrochemical process is very slow, thus more costly.

The preferred method of applying chrome coatings is PVD. Hardness of pure chromium coatings can reach 600 HV. Doping with carbon or nitrogen produces hardness of 2400 HV and 3000 HV respectively.

### Nickel Coatings

Nickel applied by electrochemical deposition is one of the oldest known coating methods. This coating is primarily used for corrosion protection and decorative artifacts.

Electrolysis-deposited nickel coatings are better for wear and abrasion resistance. With the addition of phosphorus or boron, hardness can reach 700 HV. The tradeoff is slightly less corrosion resistance. Heat treatment after the deposition process promotes the formation of nickel borides or nickel phosphides, which increases hardness. Adding particles of solid lubricant helps reduce the coefficient of friction.

### Cobalt-Based Alloy Coatings

Cobalt-based alloys are hard and ductile. Uses include high-temperature wear, mild abrasion resistance, and corrosion resistance. With the addition of ceramic carbides such as tungsten carbide, chromium carbide, and coblat carbide, the coating can be used to temperatures of 800°C. Common deposition techniques are welding and plasma spray.

### Nickel-Based Alloy Coatings

Nickel-based alloys were developed as a substitute for cobalt-based alloys. Nickel is much cheaper than cobalt coatings, yet has similar characteristics.

### Ceramic Coatings

Common techniques for depositing ceramic coatings are thermal spray, PVD, and CVD. Ceramic coatings can be applied to metals, ceramics, and cermets. The most common ceramic coatings are oxides, carbides, and nitrides (Table 12.2). Another form of ceramics, hard carbon coatings (graphite based and diamond based), began receiving much attention in the last ten years. At present, titanium nitride (TiN) is the most studied and the most used ceramic coating. It gained acceptance in the machine tool industry because of its high hardness, low friction, chemical inertness in the presence of acids, and extremely long wear life.

Titanium carbide (TiC), Aluminum oxide ($Al_2O_3$) and Hafnium nitride (HfN) are also in use. The use of multiple layer coatings such as TiN over TiC, and $Al_2O_3$ over TiC is also been made. Triple coatings of TiC/$Al_2O_3$/TiN have been proved beneficial, exploiting the characteristics of all three

**Table 12.2** Common Ceramic Coatings

| Class | Type | Depostion process | Properties |
|---|---|---|---|
| Oxides | Alumina | Plasma spray | Good wear resistance at low and high temperatures, low friction coefficient |
| | | PVD | Soft coating used for corrosion resistance |
| | | CVD | Deposited on other alumina coatings listed to improve corrosion resistance |
| | Chromia | Plasma spray, radio frequency sputtering | Excellent wear resistance at ambient and elevated temperatures, thick coatings tend to spall, thin coatings show good substrate adhesion, low friction coefficient at high temperatures |
| Carbides | Titanium | ARE, sputtering, ion plating | HSS tools, very low friction and wear in dry and lubricated conditions, hardness and adhesion is a function of substrate temperature during deposition |
| | | CVD | Hard, brittle, excellent wear resistance, excellent adhesion with substrate, very low friction, couple with TiN and SiC to improve friction and wear |
| | Tungsten | Thermal spray, sputtering | Maintains hardness at elevated temperatures, extremely hard |
| | | CVD | Low deposition temperature, hard, brittle, sensitive to thermal and heavy load cycling |

**Table 12.2** Continued

| Class | Type | Depostion process | Properties |
|-------|------|-------------------|------------|
| | Chromium | Plasma spray, sputtering, PVD | Pure chromium has high friction. Nichrome coatings moderate ductility, good adhesion with substrate, high dependency of friction on sliding speed, hardness approaches 500 HV |
| | | CVD | Limited studies performed |
| | Silicon | CVD, ion plating | High hardness (up to 6000 HV), good oxidation resistance at high temperatures, chemically inert in contact with acids, thermal stability increases with carbon content, requires diffusion barrier between coating and substrate if used on steels, high friction |
| Nitrides | Titanium | CVD | Low friction and wear, chemically inert, excellent adhesion with substrate, high-temperature deposition process removes temper from high-strength tool steels |
| | | PECVD, sputing | Smae as CVD but can be deposited at much lower temperatures |
| | | PVD | Low friction and wear with lubrication, high friction and wear when dry, friction and wear not affected by humidity, chemically inert, excellent adhesion to substrate |

| Hafnium | CVD | High hardness of nitrides at temperatures above 800°C, hardness decreases with temperature, low thermal conductivity (thermal barrier coating for HSS tools) |
|---|---|---|
| Silicon | Sputtering, CVD, PECVD | Good oxidation resistance, good erosion resistance, low thermal expansion, poor adhesion with substrate at elevated temperatures |
| Borides | | Not studied as extensively oxides, carbides, and nitrides. High hardness, high melting point, corrosion resistant, abrasion resistant |

coatings. Generally, these coatings applied to cutting tools can extend useful tool life of cutting tools and wear parts by as much as 300% or more. But the search for even more productive, cost-efficient tooling continues. New coatings, such as titanium diboride, silicon carbide, and silicon nitride are being explored and more effective coating processes are being developed.

One disadvantage of ceramic coatings is the deposition process. Depositing ceramic coatings on substrates is difficult because several variables affect quality and some of them have shown that ceramics are extremely sensitive to the deposition process and substrate temperature. Researchers are investigating solutions to these problems. Meanwhile, coating manufacturers rely on trial and error to find process variables that work.

## 12.4  DIAMOND SURFACE COATINGS

Diamond is the ultimate substance for wear resistance. Desirable properties are low friction, extremely high thermal conductivity, low electrical conductance, high wear resistance, and high abrasion resistance. One disadvantage is that diamond coatings do not adhere well to substrates, and the finished coated surface can be extremely rough. Technical problems associated  with producing these coatings stem from the extreme temperatures and pressures necessary to produce tetrahedral carbon bonds. In the early 1970s, researchers discovered deposition techniques that produced diamond-like coatings at relatively low pressure and temperature (around 800°C). Recently, a 400°C process was discovered, sacrificing deposition rate.

These techniques produced diamond-like films from hydrocarbon gases. It is easy to form graphite-like, three-bond structures from hydrocarbons, but difficult to go the next step from layered hexagonal bonds to the tetrahedral bonds found in diamonds. Thus, the final structure of the coating is a hybrid of graphite- and dimaond-like bonds. The resulting physical properties are between those of graphite and diamond films. For example, diamond-like coatings decompose around 1000°C, between the decomposition temperature of graphite and diamond. This holds true for other physical properties such as thermal conductivity, electrical conductance, and friction coefficient. Current research is focusing on coating techniques and substrate preparations that can improve the process of diamond and diamond-like coatings.

### 12.4.1  Properties of Diamond

Diamond is an exceptional material. Most of its important properties can be labeled as extreme. It has the highest hardness, the highest thermal conductivity, highest molar density and highest sound velocity of any material known. It also possesses the lowest compressibility and bulk modulus of any known material. The thermal expansion coefficient is also very low and ranks among the lowest of known materials. Diamond is also extremely inert chemically, affected only by certain acids and chemicals that act as oxidizing agents at high temperatures.

Table 12.3 summarizes some important properties of diamond.

## 12.4.2 Precoating Surface Treatment

The laser-baked surface treatment under development at the University of Florida represents one of the recent advances in precoating surface treatment [6]. The process involves using a laser to modify a metallic substrate surface. This surface modification produces a uniform roughness which provides nucleation sites for the diamond coating growth and serves to increase the surface area of contact between coating and substrate. The increase in contact surface area improves the adhesion of the diamond layer and allows the interface to grade the surface stresses, effectively reducing the chances of premature debonding.

The surface treatment shows great promise for promoting dimaond film growth on steel and other metallic surfaces. The dramatic thermal expansion mismatch between materials such as steel and diamond makes this endeavour extremely difficult. Silicon has been used successfully as a substrate for this process and results from this work are guiding efforts on other metallic

**Table 12.3**  Properties of Diamond

| Property | Value | Units |
|---|---|---|
| Hardness | $1.0 \times 10^4$ | kg/mm$^2$ |
| Strength, tensile | >1.2 | GPa |
| Strength, compressive | >110 | GPa |
| Coefficient of friction (dynamic) | 0.03 | Dimensionless |
| Sound velocity | $1.8 \times 10^4$ | m/s |
| Density | 3.52 | g/cm$^3$ |
| Young's modulus | 1.22 | GPa |
| Poisson's ratio | 0.2 | Dimensionless |
| Thermal expansion coefficient | $1.1 \times 10^{-6}$ | K$^{-1}$ |
| Thermal conductivity | 20 | W/(cm-K) |
| Thermal shock parameter | $3.0 \times 10^8$ | W/m |
| Specific heat | 0.853 | J/(gm-K) |

*Sources*: Refs. 3–5.

substrates. As an intermediate step, metals with low thermal expansion coefficients – such as molybdenum – are being tested with the process.

Figure 12.1 shows conceptually how the surface modification appears in cross section. Note how the diamond "seeds" place themselves inside the roughness "valleys". The seeds act as nucleation sites for the formation of the diamond film. This feature allows film growth to occur at a lower surface temperature than would otherwise be possible.

**Substrate surface following laser modification**

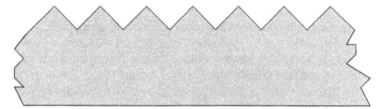

**Substrate surface seeded with 20 $\mu$m diamond particles**

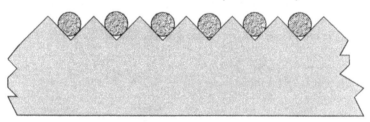

**Substrate after CVD diamond film growth**

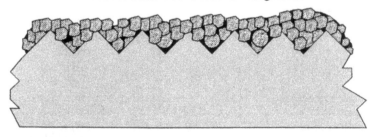

**Figure 12.1** Cross-sectional view of substrate surface following laser modification, seeding, and film growth.

## 12.4.3  Chemical Vapor Deposition of Diamond

Diamond synthesis techniques have been available since the late 1950s [7]. The commercialization of synthetic high-pressure, high-temperature (HPHT) diamond grit occurred in 1959. This grit has been widely used in industrial polishing, cutting, and grinding applications. The HPHT synthesis method essentially emulates nature's way of producing diamond – only at much poorer quality. HPHT methods are basically only capable of producing grit.

The CVD process, a low-pressure synthesis method, was first used to precipitate diamond on diamond seed crystals, using carbon monoxide gas as a source of carbon, in 1952. This method actually predates the HPHT process by several years, but presented more challenges for commercialization.

CVD diamond growth methods use simpler apparatus less subject to mechanical wear, and promise the production of physical forms of diamond other than powder (HPHT) [8]. One of the early drawbacks to CVD methods was the formation of graphite during diamond nucleation. Many variations have been tried in cleaning the graphite structures during diamond growth. Introducing hydrogen to the environment has been effective in "scrubbing" the diamond structures clean from graphite.

Experimental studies found that heating the substrate surface using a plasma source increased the diamond growth rate. This method helps decompose the methane gas into carbon (methane has a high activation energy that causes slow growth rates).

Moustakas reports [8] that microwave-assisted CVD methods are the most prevalent for diamond film growth. The process avoids contamination of the film during growth and produes a higher plasma density over RF (radio frequency) methods. This results in higher concentrations of atomic hydrogen and hydrocarbon radicals necessary for film growth. Typical parameters for deposition using microwave-plasma-assisted CVD, as reported by Moustakas, are shown in Table 12.4.

HPHT diamond is limited in application to planar surfaces. In this respect, HPHT diamond is no better than the natural diamond grit that it replaces. CVD diamond, however, ushers diamond applications to a new

**Table 12.4**  Deposition Parameter Space Used in the Growth of Diamond Films by Microwave-Plasma-Assisted CVD

| Gas mixture | Total pressure (Torr) | Microwave power (W) | Substrate temperature (°C) |
|---|---|---|---|
| $CH_4$ (0.5–2.0%)/$H_2$ | 5–100 | 100–700 | 700–1000 |

level. It offers the potential to deposit large-area, conformable coatings with properties akin to those of natural diamond ([7], p. 592). For the first time, diamond can be used as an engineered material, synthesized to meet specific topological and performance characteristics.

## 12.5   FAILURE MECHANISMS OF SURFACE COATINGS

Several factors may cause poor film adhesion. Some of these are thermal coefficient mismatch, gases adsorbed during the coating process, dirt, oil oxides, substrate defects, solvents, and residual stresses. Sensitivity to these factors depends on coating material, substrate material, coating process, and process parameters. Thus, changing surface preparation techniques or changing coating process parameters may improve adhesion. Many working theories are developed based on research and experience, but none can predict with certainty which surface preparation or what set of coating parameters will produce good adhesion.

Holmberg [9] lists many possible failure modes for surface coatings. He groups coatings into four main areas: hard coating on soft substrate, soft coating on hard substrate, and thin or thick coating.

For thick, soft coating in plastic deformation, wear is primarily mechanical wear caused by overloading. Soft coating debris may break loose and adhere to the mating surface, coating, or be expelled. Microscratches produced by mating surface asperities may initiate flaking, thus accelerating wear. If small, hard particles are present that are smaller than the coating thickness, the particles will embed in the coating during deformation. The addition of thermal fatigue causes flaking and peeling, creating larger particles of soft coating debris. Scratching by mating surfaces asperities and hard particles magnifies the effect.

Adhesive and fatigue wear, or delamination of the coating, may occur where thermal fatigue is not important. Failure is initiated by surface and subsurface stress cracks. Debris consists of either small particles of coating material or large flakes when delamination occurs. Thin, soft coatings and thick, hard coatings share these properties.

Soft coatings are ductile and can resist thermal shock. However, when thermal fatigue becomes more important than mechanical wear, the principal failure mode of thin, soft coatings is delamination. This failure can be caused by a mismatch in thermal expansion coefficient of coating and substrate materials or decomposition of the coating material at elevated temperatures.

Thermal cycling of thick, hard coatings can cause microcracks. Eventually, these cracks grow to form flakes that contribute to abrasive wear. A worse condition occurs when the crack propagate to the coating substrate interface. Depending upon the ductility of the substrate and adhesion strength of the coating, the crack either propagates into the substrate or causes delamination of the coating.

Asperity scratching is to be expected when the mating surface penetrates a thin soft coating and interacts with the substrate. This occurs when the height of asperities is larger than the coating thickness. If the substrate is softer than the mating material, substrate scratching occurs. If not, high rates of mechanical wear usually result. Wear particles contribute to abrasive wear. If hard particles larger than the coating thickness are present, particle scratching occurs. The mating surface pushes particles through the coating, scratching the substrate. However, if soft particles are present, they wedge between the mating surfaces, help support some of the load, and reduce wear.

For thin, hard coatings on a softer substrate, the substrate deformation can cause coating fracture when the coating material has low ductility. Most hard coatings are brittle and cannot withstand internal bending stresses associated with substrate deformation. Thus, the coating fractures. Coatings with better toughness may withstand substrate deformation; however, fatigue life may be significantly reduced. The primary wear mechanisms of tough, hard, thin coatings are surface scratching by the mating surface and abrasive particles.

## 12.6   TYPICAL APPLICATIONS OF SURFACE COATINGS

Most applications of surface coating require the designer to balance several objectives such as abrasion resistance, chemical inertness, high-temperature environment, and low friction. At the present, only experts with extensive experience with a particular coating can make recommendations with respect to that coating.

Matthews [10] listed some difficulties associated with selecting a coating:

There is no systematic analysis and selection procedure which design engineers can use. this is combined with a lack of coating design data in a suitable form for the component designer.

For many of the newly developed coatings, the optimal application areas have not yet been identified.

The quality and capabilities of existing coating processes are continuously improving (i.e., the technology is transient). Accordingly, previous performance data may no longer be applicable, indicating that the knowledge required to make a selection is even greater than previously thought.

There is still a lack of understanding of the mechanisms involved in determining the tribological behavior of coated surfaces.

There is a shortage of coatings experts having a sufficiently broad knowledge of all available techiques to provide a balanced viewpoint.

Matching coatings with applications is a process of trial and error, because no one can fully model yet how the coating and substrate interact.

Another problem is one of matching objectives. Coating engineers concern themselves with process parameters like deposition rate, pressure, and temperature; coating thickness; adhesion; composition; hardness; and wear rate. The design engineer needs to know physical properties like Young's modulus, thermal conductivity, Poisson's ratio, and coefficient of thermal expansion. These quantities are difficult to measure. In most cases, no data exist. Additionally, the knowledge base is not large enough to determine quantitatively which parameters are important in a particular application.

Table 12.5 lists some of the current uses of coating and surface treatment in mechanical applications.

**Table 12.5** Examples of Coatings and Surface Treatments

| Coating/treatment | Applications |
| --- | --- |
| Soft coatings: | |
| Layered solid coatings | Bearings and seals, bearings operating in vacuum and/or high temperatures (aerospace), piston rings, valves |
| Nonlayered coatings | Bearings and seals at high temperatures |
| Polymers | Bearings and seals, pumps, impellers |
| Soft metals | Bearings, electrical contacts |
| Hard coatings: | |
| Metallic | Bearings and seals, piston rings, cylinders liners, and crosshead pins in internal combustion engines and compressors, cannon and gun tubes, metalworking tools, tape-path components of computer tape drives |
| Ceramic | Bearings and seals operative in extreme environments, slurry pump seals, acid pump seals, valves, knife sharpeners, cutting tools, metalworking tools, gas turbine blades, and aircraft engines, gun steels, magnetic heads, tape-path compónents of computer tape drives |
| Surface treatments: | |
| Surface hardening | Camshafts, crank shafts, sprockets and gears, shear blades, and bearing surfaces |
| Diffusion | Cans, camshafts, rockers and rocker shafts, bushes, bearings, transmission gears, drills, milling cutters, extrusion punches, slitting saws, taps, die-casting dies, pump parts, cylinder liners, and cylinder heads |
| Ion implanation | Ball bearings, dies, press tools, injection molds |

## 12.7 SIMPLIFIED METHOD FOR CALCULATING THE MAXIMUM TEMPERATURE RISE IN A COATED SOLID DUE TO A MOVING HEAT SOURCE

### 12.7.1 Single Coated Layer

The model considered in this case is shown in Fig. 12.2. It represents a single-layered substrate with a moving heat source of Hertzian distribution. This heat distribution is equivalent to the heat generated in a contact problem such as a semi-infinite cylinder rolling and sliding over a semi-infinite planar surface. As discussed in Chapter 5, Rashid and Seireg [11]obtained the following relationships for the maximum temperature in the substrate and in the coating, respectively:

$$\frac{(T_s - T_{so})K_o}{q_t} = 1.137\left(\frac{D}{l}\right)^{1.4}\left(\frac{h_o}{l}\right)^{0.47}\left(\frac{l_e}{h_o}\right)^{0.78}$$

$$\frac{(T_o - T_s)K_o}{q_t} = 1.164\left(\frac{l}{D}\right)^{0.026}\left(\frac{K}{K_0}\right)^{0.026}\left(\frac{h_o}{l}\right)e^{-180\times10^{-6}l_e/h_o}$$

$$\frac{T_sK}{q_t} = 1.12\sqrt{\frac{K}{\rho CUl}}$$

$$D = \sqrt{\frac{5Kl}{\rho CU}}$$

$$l_e = \frac{1}{5}\frac{Uh_o^2\rho_oC_o}{K_o}$$

where

$D$ = the temperature penetration depth at the trailing edge
$l_e$ = the required entry distance for temperature penetration across the film
$T_{so}$ = the maximum rise in the solid surface temperature for a layered semi-infinite solid
$T_s$ = the maximum rise in the solid surface temperature for the unlayered semi-infinite solid with the same heat input
$T_o$ = the maximum rise in the layer surface temperature
$K, \rho, C$ = conductivity, density, and specific heat

Two examples are considered to demonstrate the effects of a conductive layer and an insulative layer on a stainless steel substrate. Parameters for these cases are:

$$\text{heat flux } q_t = 100\,\text{W/mm}$$
$$\text{sliding speed } U = 13{,}700\,\text{mm/sec}$$
$$\text{contact length } l = 1.375\,\text{mm}$$
$$\text{coating thickness } h_o = 0 \text{ to } 16\,\mu\text{m}$$

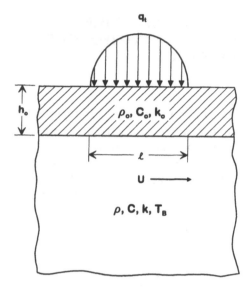

**Figure 12.2** Layered semi-infinite solid moving under a stationary heat source with a Hertzian distribution.

The stainless steel is AISI 304. The temperature rise in the stainless steel is denoted by $T_{ss}$.

### Conductive Layer on Stainless Steel

The conductive layer applied to the AISI 304 substrate is diamond. The temperature rise in the diamond is denoted by $T_d$. $T_{ss}$ and $T_d$ are plotted versus diamond layer thickness in Fig. 12.3. The temperature rise in the diamond shows a slight increase as the layer thickness increases. The temperature rise in the stainless steel decreases approximately 50°C for a 16 μm diamond layer thickness.

### Insulative Layer on Stainless Steel

The insulative layer applied to the AISI 304 substrate is silicon nitride (Si$_3$N$_4$). The temperature rise in the silicon nitride is denoted by $T_{if}$. $T_{ss}$ and $T_{if}$ are plotted versus silicon nitride layer thickness in Fig. 12.4. The temperature rise in the silicon nitride shows a dramatic increase as the layer thickness increases. The temperature rise in the stainless steel decreases approximately 200°C for a 16 μm silicon nitride layer thickness.

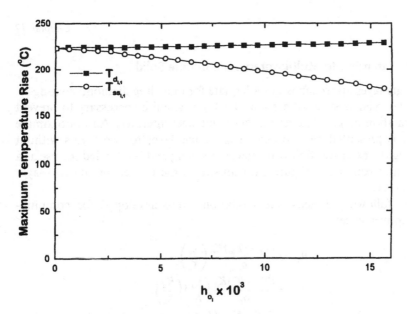

**Figure 12.3** Maximum temperature rise (°C) in AISI 304 substrate ($T_{ss}$) and diamond surface layer ($T_d$) moving under a stationary heat source with a Hertzian distribution. ($U = 12.7\,\text{m/sec}$, $l = 0.25\,\text{mm}$.)

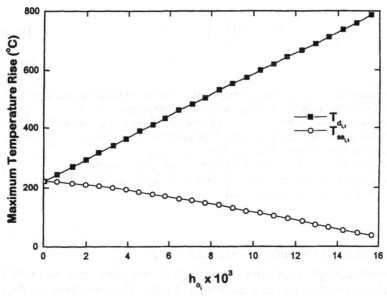

**Figure 12.4** Maximum temperature rise (°C) in AISI 304 substrate ($T_{ss}$) and nitride surface layer ($T_{if}$) moving under a stationary heat source with a Hertzian distribution. ($U = 12.7\,\text{m/sec}$, $l = 0.25\,\text{mm}$.)

## 12.7.2    Extension to Multilayered Semi-Infinite Solid

The single-layer approach is not adequate for modeling diamond coating on
substrates such as steel, where a buffer layer would be necessary to prevent
the formation of graphite at the diamond/steel interface. An approximate
method is presented here to extend the single layer to two layers without
repeating the extensive finite difference modeling and is intended for use as a
design approximation. Figure 12.5 illustrates the model for the two-layer
coating case.

The following dimensionless relations were developed for computational convenience:

$$\theta_{sso} = \frac{(T_s - T_{so})K_o}{q_t}\left(\frac{K_d}{K_o}\right)$$

$$\theta_{os} = \frac{(T_o - T_s)K_o}{q_t}\left(\frac{K_o}{K}\right)\left(\frac{K_d}{K_o}\right)$$

$$\theta_s = \frac{T_s K}{q_t}\left(\frac{K_o}{K}\right)\left(\frac{K_d}{K_o}\right)$$

$$\theta_o = T_o\left(\frac{K_d}{q_t}\right) = \theta_{os} + \theta_s \tag{12.1}$$

$$\theta_{so} = T_{so}\left(\frac{K_d}{q_t}\right) = \theta_s - \theta_{sso} \tag{12.2}$$

where

$\theta_{sso}$ = the dimensionaless maximum temperature rise in the solid surface $T_{so}$
relative to the maximum unlayered surface temperature rise $T_s$
$\theta_{os}$ = the dimensionless maximum temperature rise in the surface layer $T_o$
relative to the maximum unlayered surface temperature rise $T_s$
$\theta_s$ = the maximum unlayered surface temperature rise $T_s$
$\theta_{so}$ = the dimensionless maximum temperature rise in the solid surface
$\theta_o$ = the dimensionless maximum temperature rise in the surface layer
$K_d$ = thermal conductivity of diamond

*Procedure to Predict Multilayer Temperature Rise*

A numerical example illustrates the simplified procedure used to predict
maximum temperature rise in a multilayered solid. The procedure involves
the following steps:

1.  For a given velocity, $U$, and length of contact, $l$, calculate the
    dimensionless temperature rise, $\theta_{o\_d}$ and $\theta_{so\_if}$, for the case of a

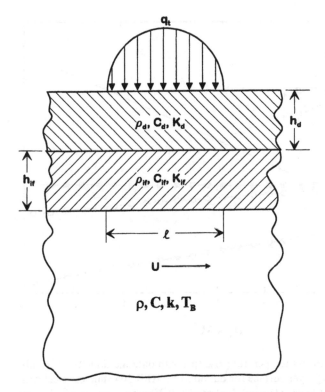

**Figure 12.5** Two-layered semi-infinite solid moving under a stationary heat source with a Hertzian distribution.

diamond layer on the buffer layer substrate using Eqs (12.1) and (12.2). The values for the surface layer thickness, $h_d$, should vary from 0 to a specified upper limit (0.016 mm was used in this study).

2. The ratio between the surface layer and buffer layer maximum temperature rise is now determined. The ratio will remain constant, although the temperature values will be adjusted up or down based on the uncoated steel temperature.

3. For the same $U$ and $l$, calculate the dimensionless temperature rise, $\theta_{o\text{-}if}$ and $\theta_{so\text{-}ss}$, for the case of a buffer layer on the steel substrate using Eqs (12.1) and (2.2). For simplicity, the same values for $h_{if}$ are used for $h_d$ in step 1 above (see Fig. 12.6).

4. The maximum dimensionless temperature rise in the buffer layer, $\theta_{if}$, in the steel substrate, $\theta_{ss}$, and in the diamond layer, $\theta_d$ can

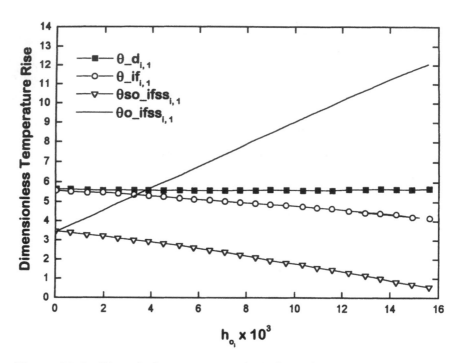

**Figure 12.6**  Dimensionless temperature rise (prior to data manipulation) in substrate surface and coating layer surface for diamond on AISI 304 stainless steel with a silicon nitride interface layer. ($U = 12.7\,\text{m/sec}$, $l = 0.25\,\text{mm}$.)

now be determined. First, the steel substrate temperature rise is scaled by the ratio of the buffer layer temperatures (see Eq. (12.3)). A scaling factor, $\Delta$, is then determined in Eq. (1.24). This factor scales the diamond layer, buffer layer and steel substrate temperatures to the uncoated steel substrate temperature, $\theta_{s,ss}$. This factor insures that the temperature rise predictions all start at the uncoated steel substrate temperature at $h_d = h_f = 0$, and then decrease or increase as the coating thicknesses increase. Equation (12.5) shows how $\Delta$ is used to calculate the dimensionless temperature rise for the diamond and buffer layers ($\theta_d$ and $\theta_{if}$) and the steel substrate ($\theta_{ss}$) (see Fig. 12.7).

5.  The actual temperature rise ($T_{ss}$, $T_{if}$, $T_d$) can now be calculated by Eq. (12.6). Figure 12.8 shows a typical result for $q_t = 1000$ W/ mm$^2$.

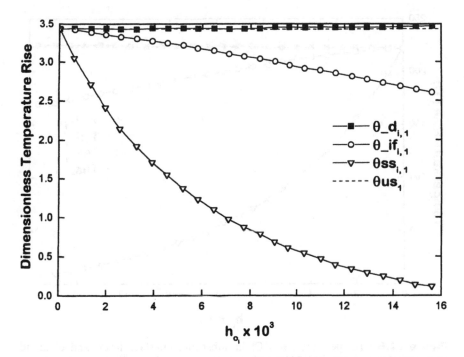

**Figure 12.7** Dimensionless temperature rise (following data manipulation) in substrate surface and coating layer surface for diamond on AISI 304 stainless steel with a silicon nitride interface layer. ($U = 12.7$ m/sec, $l = 0.25$ mm.)

For $i = 1$ to $n$:

$$\theta_{ss}^i = \theta_{so\_ss}^i \left( \frac{\theta_{so\_if}^i}{\theta_{o\_if}^i} \right) \tag{12.3}$$

$$\Delta = \frac{\theta_{s\_ss}}{\theta_{ss}^1} \tag{12.4}$$

For $i = 1$ to $n$:

$$\theta_{ss}^i = \Delta \theta_{ss}^i$$
$$\theta_{if}^i = \Delta \theta_{so\_if}^i$$
$$\theta_d^i = \Delta \theta_{o\_d}^i \tag{12.5}$$
$$T = \theta \frac{q_t}{K_d} \tag{12.6}$$

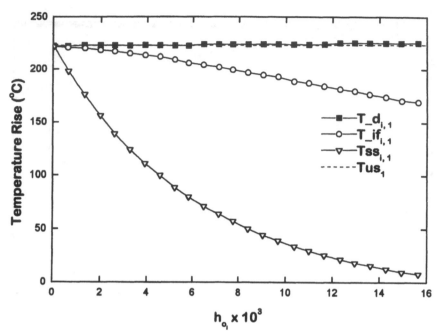

**Figure 12.8**  Temperature rise (°C) in substrate, interface layer, and diamond layer for diamond on AISI 304 stainless steel with a silicon nitride interface layer. ($U = 12.7\,\text{m/sec}$, $l = 0.25\,\text{mm}$.)

### Effects of Varying Parameters

The effects of varying the sliding velocity of the solid, $U$, and the width of contact, $l$, are examined in this section. Because the temperature rise in an uncoated substrate is inversely proportional to the square root of the sliding velocity, $\Delta T \propto 1/\sqrt{U}$, it is expected that the temperature rise for the multi-layered case will follow suit. Figure 12.9 illustrates this case. The parameters, except for $U$, are the same as in Fig. 12.8. As would be expected, the magnitude of the temperature rise for the substrate and layers dropped with the increase in sliding velocity.

   The effect of increasing $l$, the width of contact, is now considered. Because $l$ follows the same inverse relationships as $U$ for the unlayered substrate, $\Delta T \propto 1/\sqrt{l}$, it is expected that the magnitude of the temperature rise will decrease. Figure 12.10 illustrates this case. The parameters, except for $l$, are the same as in Fig. 12.8. As would be expected, the magnitude of the temperature rise for the substrate and layers dropped with the increase in contact width.

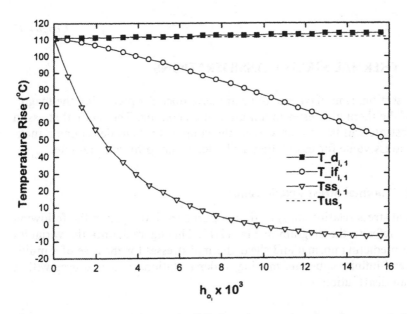

**Figure 12.9** Temperature rsie (°C) in substrate, interface layer, and diamond layer for diamond on AISI 304 stainless steel with a silicon nitride interface layer (velocity increase with length of contact constant). ($U = 50.8$ m/sec, $l = 0.125$ mm.)

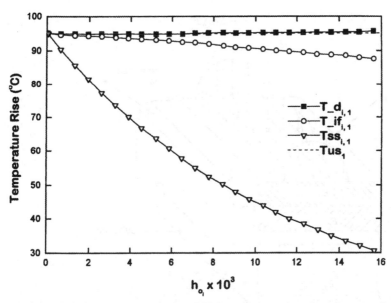

**Figure 12.10** Temperature rise (°C) in substrate, interface layer, and diamond layer for diamond on AISI 304 stainless steel with a silicon nitride interface layer (velocity constant with length of contact increase). ($U = 12.7$ m/sec, $l = 1.375$ mm.)

## 12.8   THERMAL STRESS CONSIDERATIONS

In this section, simplified equations are developed for predicting the magnitude of the thermal stresses in a multilayered coating. The thermal stress in the substrate will be combined with the contact stress to determine a maximum stress value for calculating a life debit due to thermal fatigue.

### 12.8.1   Thermal Stress Relationships

Nominal stress relationships for design purposes developed in the following sections refer to the diagram in Fig. 12.11. This figure defines the variables used in predicting normal and shear thermal stresses for the case of a multilayer semi-infinite substrate moving under a stationary heat source with a Hertzian distribution.

### 12.8.2   Normal Stresses

The "normal" thermal stress in an axial beam built in at both ends is proportional to the increase in temperature and can be expected as:

$$\sigma = E\alpha\Delta T \tag{12.7}$$

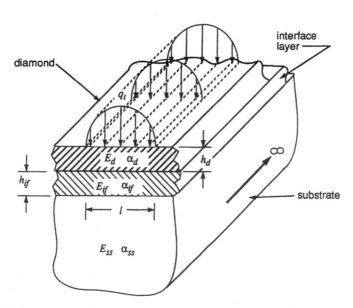

**Figure 12.11**   Model used in calculating thermal stress in multilayer coatings.

If this equation is used for a simplified model, then we have the following equation to describe the normal stress in the diamond, interface coating, and substrate:

$$\sigma_d = E_d \alpha_d T_d$$
$$\sigma_{if} = E_{if} \alpha_{if} T_{if} \qquad (12.8)$$
$$\sigma_{ss} = E_{ss} \alpha_{ss} T_{ss}$$

where $T_d$, $T_{if}$, and $T_{ss}$ are the temperature differentials between each of the layers and its substrate. Note that Eq. (12.8) is a nominal relationship for design approximation only.

### 12.8.3 Shear Stresses

Shear stress can be determined by dividing the shear force, $F_s$, by the shear area, $A_s$. The shear force can be approximated by the difference in normal stresses between two layers times the cross-sectional area of $F_s = (\sigma_2 - \sigma_1)A_c$. The shear stress can now be written as

$$\tau = \frac{F_s}{A_s} = \frac{(\sigma_2 - \sigma_1)A_c}{A_s} \qquad (12.9)$$

Now, referring to Fig. 12.11, if we substitute $A_c = h_\infty$ and $A_s = l_\infty$, we have:

$$\tau = (\sigma_2 - \sigma_1)\frac{h}{l} \qquad (12.10)$$

and can write the following simplified equations for the shear stress between the diamond and interface layers, and the interface layer and the substrate:

$$\tau_{dif} = (\sigma_{if} - \sigma_d)\frac{h_d}{l}$$
$$\tau_{ifss} = (\sigma_{ss} - \sigma_{if})\frac{h_{if}}{l} \qquad (12.11)$$

Figures 12.12 and 12.13 show the calculated nominal thermal stresses and interface shear stresses for the examples considered in the previous sections using equal thickness layers of silicon nitride and diamond on stainless steel.

### 12.8.4 Life Improvement Due to Surface Coating

The effect of thermal stress on the life of a stainless steel substrate is considered with and without protective coating. In this example, coating layers

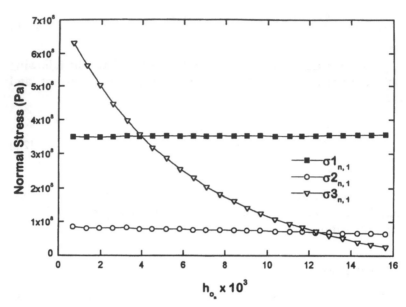

**Figure 12.12** Normal stress (Pa) for various thicknesses (mm) of diamond and interface layer. Substrate is AISI 304 stainless steel and interface layer is silicon nitride. Note: $h_{if} = h_d = h_o$, $\sigma_1 = \sigma_d$, $\sigma_2 = \sigma_{if}$ and $\sigma_3 = \sigma_{ss}$. ($U = 12.7\,\text{m/sec}$, $l = 0.25\,\text{mm}$.)

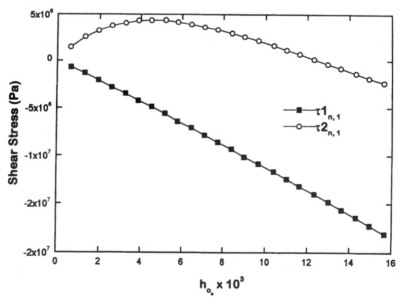

**Figure 12.13** Shear stress (Pa) for various thicknesses (mm) of diamond and interface layer. Substrate is AISI 304 stainless steel and interface layer is silicon nitride. Note: $h_{if} = h_d = h_o$, $\tau_1 = \tau_{dif}$, and $\tau_2 = \tau_{ifss}$. ($U = 12.7\,\text{m/sec}$, $l = 0.25\,\text{mm}$.)

of diamond and silicon nitride of equal thickness are used as in the previous case. The Hertzian contact stress is combined with the thermal stress using the Von Mises distortion energy theory for predicting the relative surface damage. The results for different coating thicknesses are given in Fig. 12.14 as an illustration. It can be seen from the figure that considerable improvement in life can be expected as a result of the coating. The improvement tends towards an asymptotic value for relatively thick layers.

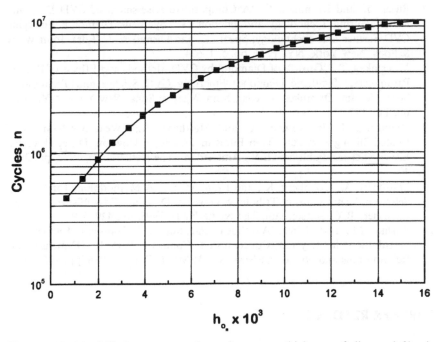

**Figure 12.14** Life improvement in cycles versus thickness of diamond film in mm. Substrate is AISI 304 stainless steel and interface layer is silicon nitride. Note: $h_{if} = h_d = h_o$ and contact stress level is 1000 MPa. ($U = 12.7$ m/sec, $l = 0.25$ mm.)

## REFERENCES

1. Bhushan, B., and Gupta, B. K., Handbook of Tribology, McGraw-Hill, New York, NY, 1991.
2. Sherbiney, M. A., and Halling, J., "Friction and Wear of Ion-Plated Soft Metallic Films," Wear, 1977, Vol. 45, pp. 211–220.

3. Yoder, M., "Diamond Properties and Applications," Diamond Films and Coating: Development, Properties, and Application, Davis, R. (Ed.), Park Ridge, NJ, Noyes Publications, 1993, pp. 1–30.
4. Spear, K., and Dismukes, J. (Eds), Synthetic Diamond: Emerging CVD Science and Technology, John Wiley & Sons, New York, NY, 1994, p. 663.
5. Field, J. (Ed.), The Properties of Natural and Synthetic Diamond, Academic Press/Harcourt Brace Jovanovich, London, England, 1992.
6. Singh, R., Private communications, Dept. of Material Science, Univ. of Florida–Gainesville.
7. Busch, J., and Dismukes, J., "A Comparative Assessment of CVD Diamond Manufacturing Technology and Economics," Synthetic Diamond: Emerging CVD Science and Technology, Spear, K. and Dismukes, J. (Eds), John Wiley & Sons, New York, NY, 1994, pp. 581–624.
8. Moustakas, T., "Growth of Diamond by CVD Methods and Effects of Process Parameters," Synthetic Diamond: Emerging CVD Science and Technology, Spear, K. and Dismukes, J. (eds), John Wiley & Sons, New York, NY, 1994, pp. 145–192.
9. Holmberg, K., Ronkainen, H., and Matthews, A., "Wear Mechanisms of Coated Sliding Surfaces," Thin Films in Tribology, Dowson, D., et al. (Eds), Elsevier Science Publishers B.V., Amsterdam, The Netherlands, 1993, pp. 399–407.
10. Matthews, A., Holmberg, K., and Franklin, S., "A Methodology for Coating Selection," Thin Films in Tribology, Dowson, D., et al. (Eds), Elsevier Science Publishers B.V., Amsterdam, The Netherlands, 1993, pp. 429–439.
11. Rashid, M., and Seireg, A., "Heat Partition and Transient Temperature Distribution in Layered Concentrated Contacts. Part II – Dimensionless Relationships and Numerical Results," ASME J. Tribol., 1986, pp. 102–107.

## FURTHER READING

### Coating

Bell, T., "Towards Designer Surfaces," Met. Mater., August 1991, Vol. 7(8), pp. 478–485.
Gao, R., Bai, C., Xu, K., and He, J., "Bonding Strength of Films Under Cyclic Loading," Surface Engineering Volume II: Engineering Applications, Dotta, P. K. et al. (eds), Royal Society of Chemistry, Cambridge, England, 1992.
Mort, J., "Diamond and Diamond-like Coatings," Mater. Des., June 1990, Vol. 11(3), pp. 115–121.
Rickerby, D. S., and Matthews, A., Advanced Surface Coatings: A Handbook of Surface Engineering, Blackie and Son, New York, NY, 1991.
Sander, H., and Petersohn, D., "Friction and Wear Behavior of PVD-coated Tribosystems," Thin Films in Tribology, Dowson, D. et al. (Eds), Elsevier Science Publishers B.V., Amsterdam, The Netherlands, 1993, pp. 483–493.

Stafford, K. N., Subramanian, C., and Wilkes, T. P., "Characterization and Quality Assurance of Advanced Coatings," Surface Engineering Volume II: Engineering Applications, Dotta, P. K. et al. (Eds), Royal Society of Chemistry, Cambridge, England, 1992.


## Coated Cutting Tools

Anon, "Cutting Tools as Good as gold," Metalwork. Prod., July 1983, Vol. 127(7), pp. 129–144.

Bhat, D. G., and Woerner, P. F., "Coatings for Cutting Tools," J. Metals, Feb. 1986, Vol. 38(2), pp. 68–69.

Bollier, R. D., "Recoating Enhance Resharpening," Mod. Mach. Shop, March 1986, Vol. 58(10), pp. 76–81.

Garside, B. L., "Improvements in Tools and Product Performance Through PVD Titanium Nitride Process," Indust. Heat., Vol. 53(9), pp. 18–20.

Hale, T., and Graham, D., "How Effective Are the Carbide Coatings?" Aust. Mach. Prod. Eng., April 1984, Vol. 37(4), pp. 17–19.

Hatschek, R. L., "Coatings: Revolution in HSS Tools," Am. Machin., March 1983, Vol. 127(3), pp. 129–144.

Hewitt, W. R., and Heminover, D., "TiN Coating Benefits Apply to Solid Carbide Tools Too," Cutting Tool Eng., Jan.–Feb. 1986, Vol. 36(1–2), pp. 17–18.

Jackson, D., "Coatings: Key Factor in Cutting Tool Performance," Mach. Tool Blue Bk., Vol. 81(1), pp. 62–64.

Kane, G. E., "Modern Trends in Cutting Tools," Society of Manufacturing Engineering, 1982, pp. 54–55.

Kane, G. E., "Modern Trends in Cutting Tools," Society of Manufacturing Engineers, 1982, pp. 82–87.

Podop, M., "Sputter Ion Plating of Titanium Nitride Coatings for Tooling Applications," Indust. Heat., Jan. 1986, Vol. 53(1), pp. 20–22.

Schintlmeister, W., Wallgram, W., Kanz, J., and Gigl, K., "Cutting Tools Materials Coated by Chemical Vapor Deposition," Wear, Dec. 1984, Vol. 100(1–3), pp. 153–159.

Walsh, P., and Bell, D. C., "Recoating: A Viable Option of TiN Coating for Special Tooling Applications," Cutting Tool Eng., Feb. 1986, Vol. 38(1), pp. 25–27.

Wick, C., "Coated Carbide Tools Enhance Performance," Manu. Eng., March 1987, Vol. 98(3), pp. 45–50.

Wick, C., "HSS Cutting Tools Gain a Productivity Edge," Manuf. Eng., May 1987, Vol. 98(5), pp. 39–42.

Zichichi, C., "Tool Coatings: Trends and Perspectives," Carbide Tool J., Jan.–Feb. 1986, Vol. 18(1), pp. 18–20.

# 13

# Some Experimental Studies in Friction, Lubrication, Wear, and Thermal Shock

This chapter describes a number of experimental investigations covering different aspects of tribology. The first set of experiments deals with the behavior of Hertzian frictional contacts under different types of tangential loading. In this set, unlubricated spheres pressed against flat surfaces are subjected to oscillatory loads, impulsive loads, and ramp-type loads respectively.

Another experimental procedure is discussed which can be used to investigate the oil film pressure generated by a slider with different geometries undergoing a reciprocating motion at a predetermined distance from a flat surface.

The last two sections describe experimental techniques which can be used to study the effect of the lubricant properties on surface temperature and wear in sliding contacts and the effect of repeated thermal shock on the fatigue life of high-carbon steels.

## 13.1 FRICTIONAL INTERFACE BEHAVIOR UNDER SINUSOIDAL FORCE EXCITATION

This section describes an experimental technique developed by Seireg and Weiter [1] for studying the vibratory behavior of a ball supported between two frictional joints. The setup which is utilized in this investigation for evaluating the "break away" coefficient of friction under sinusoidal tangential forces is also useful in determining the ball response and the energy dissipated per cycle under excitations of different amplitudes and frequencies. Wear and lubrication studies can be readily performed on different contact conditions under sinusoidal tangential forces with frequencies ranging from zero to 2000 Hz and amplitudes from zero to the value necessary to cause gross slip. The main difference between the proposed technique and previous methods is that the tangential force (rather than the displacement) is sinusoidal and remains as such up to the "break away" value.

The effect of an oscillating tangential force on the contact surfaces of elastic bodies has been subject to considerable interest in recent years. Several valuable contributions are available in the literature. Mindlin [2] extended the classical Hertz theory of contact to include the effect of an increasing tangential force with the normal force unchanged. He predicted that slip would occur at the edges of the contact area and progress inwards as the tangential force increases. This slip would occur only on annular ring surfaces. At any point on the contact surface where slip has just taken place, the tangential component of traction has the same sense as that of the slip, and its magnitude is equal to the product of a constant coefficient of friction and the normal component of the pressure at that point. The tractions on and the displacements of the portion of the contact surface where no slip occurs are obtained from the solution of the boundary value problem. Expressions for calculating the relative tangential displacement of distant points on opposite sides of the contact due to a tangential force smaller or equal to that necessary for gross slip are given in Chapter 3. The theory was further extended to calculate the displacement due to an oscillating tangential force within the region of no gross slip. The result is a hysteresis loop and the energy dissipation for the cycle due to friction can readily be calculated. Mindlin et al. [3] found from experiments on polished crown glass lenses that the area of the loop at low loads varied as the square of the displacement, whereas the theory predicts a cube law. The agreement with the theory was good for large displacements. The oscillating force in their test was obtained by utilizing a hollow cylinder of barium titanate for the driving transducer, which is essentially a displacement generator producing sinusoidal tangential displacement. The force was measured by a disk of barium titanate cemented between the driving transducer and the sphere. Johnson [4] utilized a torsional pendulum to apply the tangential force on three unlubricated hard steel balls on hard steel flats under a range of normal loads. Johnson measured the displacements due to static and oscillating tangential forces within the no-gross-slip region. His findings were in general agreement with the previous work. Goodman and Bowie [5] used an apparatus similar to that of Ref. 3 to study the damping effects at the contacts of a 1/2 in. diameter stainless steel sphere pressed between two 1/2 in. square by 1/4 in. thick stainless steel plates. The dynamic hysteresis loops determined in their tests have been shown to conform to the shape predicted by Mindlin's theory. Their results of a dimensionless energy dissipation versus the ratio of peak-to-peak displacement at gross slip were in fair agreement with the theory.

Klint [6] studied the effects of oscillating tangential forces within the region of no gross slip on cylindrical specimens in contact. A horizontal test cylinder of 1/8 in. radius is attached to the piston of a hydraulic cylinder and is forced by the oil pressure against a vertical test cylinder attached to the table of a shaker producing smooth sinusoidal movements. The hydraulic cylinder and consequently the horizontal test cylinder, although spring mounted on the shake table, are essentially fixed in space due to the vibratory characteristic of their support.

A barium titanate force gage was used between the test specimen and the shake table. The tangential compliance and the energy dissipation per cycle were studied for different combinations of materials.

A region within the no-gross-slip region was found where the displacements are primarily elastic and was defined by the "limit of elastic behavior'. The coefficient of friction was calculated from the friction force represented by the flat portion of the force–time relation. Wear and surface damage conditions were also investigated in the test.

In all the previous experimental procedures, the oscillating tangential force was provided by applying sinusoidal relative tangential displacements to the bodies in contact. The force wave forms appear sinusoidal at displacements well below gross slip and then progressively change toward waves with flat tops as the peak displacement is increased.

The investigation described in this section was, therefore, planned to provide a sinusoidally changing tangential force with amplitudes up to the gross slip force, and to study its effects on a sphere pressed between two flat surfaces by a constant normal force. With such a system, it would be possible to study the motion of the ball as a mass supported by a nonconservative hysteretic spring and subjected to sinusoidal excitations (refer to Fig. 13.1).

### 13.1.1  Experimental Setup

The apparatus is illustrated diagrammatically in Fig. 13.2. The main test fixture consists of a $1\frac{1}{4}$ in. ball (a) supported between the flat surfaces of two cylindrical pins 0.572 in. in diameter. One of the pins (b) can be fixed rigidly to the aluminum frame (c) while the other pin (d) acts as a piston in a brass air cylinder (e) attached to the frame to provide the normal force. The air cylinder pressure is controlled by a pressure regulator (f) connected to a 150 psi air supply. When the ball is in place, the pins extend 1/16 in. from the

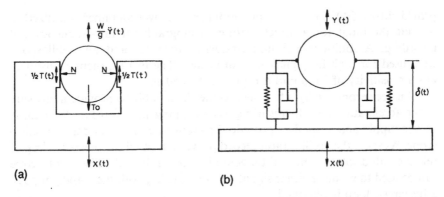

**Figure 13.1** Diagrammatic representation of (a) forces on the ball; (b) the vibratory system.

frame in order to insure maximum rigidity. The test fixture is rigidly fastened to the table (g) of a 50 lbf, 0–2000 Hz electromagnetic shaker.

A differential transformer type displacement transducer (h) is rigidly mounted on the frame with the movable core in contact with the ball and exerting a 12 gf preload. The transducer excitation and amplification is

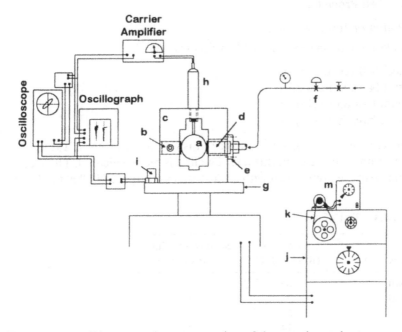

**Figure 13.2** Diagrammatic representation of the experimental setup.

provided by a 2400 Hz carrier preamplifier. A power amplifier is utilized to provide the input for a direct-writing oscillograph for low-frequency test recording. A multichannel high-frequency recorder and an oscilloscope were used for high-frequency measurements. The displacement transducer has a sensitivity of 1 μm and a range of ±0.050 in.

An accelerometer (i) is fastened to the shake table to provide a measure of the acceleration and the resulting dynamic tangential force on the ball. A power supply and amplifiers provide the input to either recorder or oscilloscope. As such, the oscillograph record provides a simultaneous recording of relative ball displacement and tangential force on the ball. The oscilloscope can be used to monitor either signal. Or, by utilizing both the $x$ and $y$ inputs, a hysteresis loop is obtained.

In addition, the output from the calibrated velocity transducer on the shaker can be used to check the accelerometer calibration.

The shaker frequency and amplitude are controlled by the shaker control console (j). A variable speed motor drive and pulleys (k) are connected to the control console to provide a predetermined rate of increase for the shake table amplitude.

### 13.1.2 Test Procedure

*Preparation of Test Specimens*

The material combinations studied in this test are:

> Steel ball on steel pins
> Steel ball on brass pins
> Brass ball on steel pins
> Brass ball on brass pins

The material specifications and surface finish data are given in Table 13.1. Two lubricants were used in this test. The first is methyl alcohol. The second is a mild, extreme pressure oil with additives. Viscosity is SUS 900 to 1000 at

**Table 13.1**  Test Materials

| Test specimen | Hardness (Rockwell C) | Surface roughness rms ($\mu$in.) |
|---|---|---|
| Steel ball | 66 | 2 |
| Brass ball | 28 | 4 |
| Steel pin | 65 | 2 |
| Brass pin | 14 | 1.5 |

100°F. The preparation of the test surfaces is done as follows. The balls are first cleaned with methyl alcohol and paper towels. The oil is applied to the surface by wiping with a clean, predipped cloth. The alcohol is applied by simply dipping the ball (after cleaning) into the alcohol and allowing it to dry in the atmosphere. Throughout this process, the ball is handled using a plastic holder for each contaminating liquid.

The pins are prepared for each test by finishing their test surfaces with 4/9 sandpaper. The cleaning and surface lubrication is done in the same manner as for the balls.

### 13.1.3  Placement of the Specimens

The pins are first placed in the test frame and the ball is then positioned between them by means of a special fixture. The fixture is designed such as when the ball is in location, the contacts are along the central axis of the pins. The stem of the differential transformer will also be touching the ball at the highest point and will be at its null position. The pin (b) is then fixed in place by setscrews and the air pressure is applied on pin (d). The positioning fixture is then withdrawn. The ball is left in position for 3 min before running each test.

### 13.1.4  Static Tests

Static tests were performed to calibrate the normal force and to evaluate a static coefficient of friction. The normal force versus air pressure calibration was done by means of a scale against the movable pin.

For the static friction tests, the specimens are placed in position and the pressure adjusted for a certain normal force. The tangential force is applied at the lowest point of the ball in line with the displacement transducer. The load is increased successively until gross slip is observed by watching the indicator on the transducer bridge unit. The test is repeated four times for each condition. A mercury manometer is also used for the determination of the minimum air pressure required to hold the ball in place when no external tangential forces are applied. In this case, the only acting tangential forces are the ball weight and the spring force from the transducer. In this test, a relatively high pressure is applied after the specimens are in position. The pressure is then allowed to leak out slowly. The manometer reading is taken when gross slip is indicated on the meter of the bridge unit.

### 13.1.5  Dynamic Tests

The frame is fastened rigidly to the shaker table and the test specimens are placed in the frame. The frequency is adjusted to the required value and the switch on both the recorder and the driving motor (m) is turned on. When the ball displacement trace on the recorder (or the displacement signal on the meter of the bridge unit) indicates gross slip, the motor is switched off and then reversed to stop the shaker. The dynamic tests were repeated 10 times for each condition and a calibration for the accelerometer is run on the same chart. The dynamic setup was checked to ensure that no structural resonance or frequency response existed in the frequency range considered in the test and that the movable pin had no relative motion with respect to the frame.

### 13.1.6  Discussion of the Results

Typical oscillograph records of the ball acceleration and displacement are shown in Fig. 13.3. A typical plot of the test results is shown in Fig. 13.4, and the coefficients of friction are given in Table 13.2.

The lines representing the relation between the amplitude $T^+$ of the tangential force versus the air pressure and normal force $N$ were found to fit both the dynamic data and the static data. For all the tests performed in this investigation, there was no evidence of any difference between the statically determined coefficient of friction and that determined by dynamic tests. Gross slip occurred when the tangential force reached the value required to overcome the static friction. This always happened when the ball was at the lowest position of its motion or when the total tangential force was at its maximum value. This can be easily seen in Fig. 13.3b, representing a magnified ball displacement trace.

A series of tests was run for steel balls on steel pins (with oil contamination) for frequencies between 10 and 500 Hz. No frequency effects were observed for this range (refer to Fig. 13.5).

The normal load and consequently the contact stresses and contact area had no obvious effect on the coefficient of friction. The maximum Hertzian contact stress at 40 psi pressure is 105,000 psi for steel on steel, 82,500 psi for steel on brass, and 69,000 psi for brass on brass.

There was also little difference observed in the coefficient of friction between the same specimens when tested with either of the used contaminating liquids.

Ball vibration relative to the pins could be easily measured and recorded. An oscillograph trace is shown in Fig. 13.3. Hysteresis loops

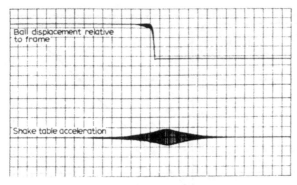

(a)

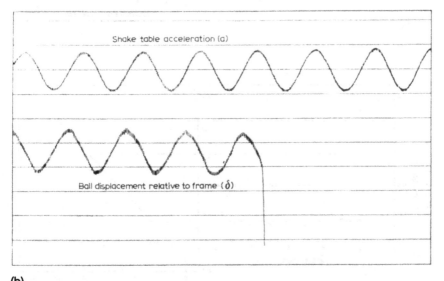

(b)

**Figure 13.3** Recorder traces for relative ball displacement and shake table acceleration: (a) slow speed; (b) high speed.

representing the sinusoidal tangential force versus ball displacement (relative to the pins) were also obtained on the scope screen.

Wear patterns, scars, and weldments (especially for clean surfaces at high frequencies) were observed. In cases when welding occurred very high tangential forces were required to break the frictional joint.

It was also observed that when twisting of the ball occurred during the withdrawal of the positioning fixture, a relatively higher coefficient of fric-

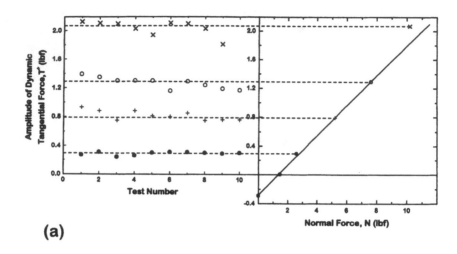

**(a)**

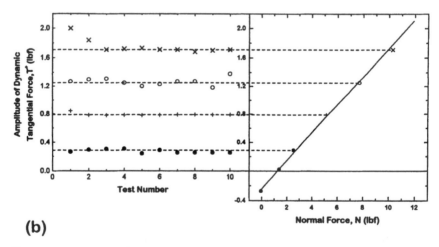

**(b)**

**Figure 13.4**  Brass ball on steel pins: (a) alcohol lubrication; (b) oil lubrication.

tion existed. This is in accordance with the observations reported by Mason [7] and Anderson [8].

Several tests were run to study the effect of the number of cycles at force amplitudes lower than required for gross slip. The shaker frequency was fixed and the amplitude of its vibration was increased in steps. The vibration was allowed to continue for 5 min at each step. Gross slip occurred at the expected value with no obvious effects due to the load history.

**Table 13.2**  Coefficients of Friction

| Materials | | | |
|---|---|---|---|
| Ball | Pin | Lubricant | Coefficient of friction ($\mu$) |
| Steel | Steel | Oil | 0.09 |
| | | Alcohol | 0.09 |
| Steel | Brass | Oil | 0.104 |
| | | Alcohol | 0.110 |
| Brass | Steel | Oil | 0.105 |
| | | Alcohol | 0.105 |
| Brass | Brass | Oil | 0.120 |
| | | Alcohol | 0.121 |

## 13.2  FRICTION UNDER IMPULSIVE LOADING

There is considerable practical interest in determining the frictional resistance under impulsive loading. Gaylord and Shu [9] reported that some materials (steel on steel and titanium on steel) exhibited higher static coefficients of friction under statically applied loads than under shock loads. In

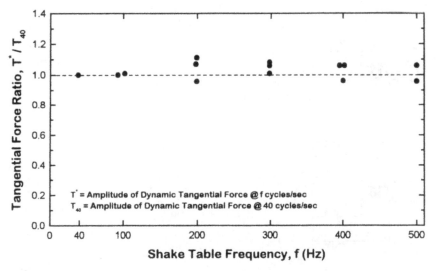

**Figure 13.5**  Sample of results of test on frequency effect. Steel on steel with oil lubrication.

their test, the specimen rested on an inclined plane. The load is applied by dropping a weight on to cushioned stops which are attached to the loading rod. The study described in this section [10] utilizes the same fixture used in the previous section [1] to investigate the frictional behavior of circular contacts under the influence of pulse-type loading. The test arrangement consists of a ball suspended between two flat surfaces (Hertzian contacts) with the impulse load provided by a spherical ball suspended as a ballistic pendulum.

### 13.2.1  Experimental Arrangement

The appratus used in this investigation is represented schematically in Fig. 13.6. The $1\frac{1}{4}$ in. ball (a), is suspended between two cylindrical steel pins with parallel flat ends. One of the pins (b), is rigidly attached to the frame (c), while the other pin (d), acts as an air piston in a precision bore in the fixture to provide the normal force. The magnitude of the normal force is a function of the regulated air pressure from a 1000 psi source (e). The test fixture is rigidly fastened to a steel base (f).

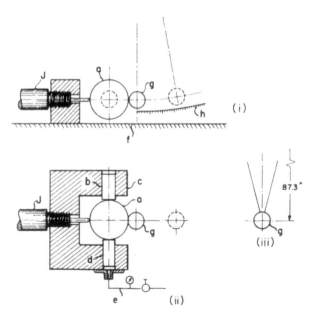

**Figure 13.6**  Diagrammatic representation of the test apparatus.

A steel ball (g), 1/2 in. in diameter is cemented to a fine thread and suspended as shown in Fig. 13.6 to form an 87.3 in. pendulum. This suspension insures the fall and rebound of the ball to be in the same plane. The initial and final position (after rebound) of the pendulum can be read on a graduated arc (h), giving accurate indication on the velocity of the impact ball before and after impact. The velocity ranged between 1 in./sec and 120 in./sec.

The displacement of the ball is measured by means of a differential transformer type displacement transducer (j), rigidly mounted on the frame with the movable core in contact with the ball (a), with a 12 gf preload. The transducer excitation (2400 Hz) and signal amplification is provided by a carrier-type preamplifier coupled to a power amplifier which provides the input to a direct writing oscillograph. The record is calibrated to give the final ball displacement a 1 μin. sensitivity.

Accurate alignment is provided so that the impact between the two spheres is on the same axis as the displacement transducer.

### 13.2.2  Test Procedure

The surface preparation consisted of washing all contacting surfaces of the balls and pins with methyl alcohol. The pins are placed in the frame and the test ball is positioned between the pins by means of the special fixture used in the previous section to insure proper axial location. The air pressure is then applied to provide the desired normal force.

The striking ball is held at the required position on the graduated arc by means of a clean steel bar. The ball is released by rapidly withdrawing the bar forward and away in such a manner as to not interfere with the free fall of the ball. The rebound of the ball is then measured on the graduated arc by observing the maximum position of the ball after impact. The displacement of the ball for each drop (impact) is detected by the transducer and recorded.

### 13.2.3  Theory

The pulse characteristics in this investigation are calculated by means of the Hertz theory of impact [11, 12]. The theory treats the impact of two spheres as a statical problem and takes no consideration to the dissipation of energy during the impact. The result, although both static and elastic in nature, has been widely applied to impact situations where permanent deformations were produced. The application of Hertz theory beyond the limits of its validity has been justified on the basis that it appears to predict accurately most of the impact parameters that can be experimentally verified. In the case of central impact of two stainless steel spheres with 1/2 in. and $1\frac{1}{4}$ in. diameter, respectively, the Hertz theory gives the following expressions for the maximum force and duration of impact:

$$F_{\text{max}} = 1.665v^{1.2} \text{ lbf}$$

$$\tau = \frac{98 \times 10^{-6}}{v^{1/5}} \text{ sec}$$

where

$v =$ velocity of approach (in./sec)

The conservation of momentum and restitution equations for the system under consideration can be written as:

$$m_1 v = m_1 v_f + m_2 V_f$$

and

$$e = -\frac{v_f - V_f}{v - 0}$$

from which

$$\frac{v_f}{v} = \frac{e - \dfrac{m_1}{m_2}}{1 + \dfrac{m_1}{m_2}}$$

and

$$e = \left(1 + \frac{m_1}{m_2}\right)\frac{v_f}{v} + \frac{m_1}{m_2} \tag{13.1}$$

The velocity of the struck ball, $V_f$, after impact can also be expressed in terms of the impact velocity, $v_f$, as:

$$V_f = \frac{v(1 + e)}{1 + \dfrac{m_2}{m_1}} \tag{13.2}$$

The ratio $v_f/v$ of the striking ball under consideration for the different conditions of impact was determined experimentally by reading the angles of drop and rebound of the pendulum. It was found to be $v_f/v = 0.83$ throughout the test.

Substitution in Eq. (13.1) gives $e = 0.949$. Equation (13.2) therefore gives:

$$V_f = (0.1171)v \tag{13.3}$$

The kinetic energy of the struck ball due to the impact is:

$$\tfrac{1}{2}m_2 v_f^2 = \tfrac{1}{2}(7.4 \times 10^{-4}(0.1171)^2 v^2 = 5.075 \times 10^{-6} v^2 \tag{13.4}$$

This energy represents the area under the frictional force–displacement curve up to the peak displacement. The minimum impact energy necessary for gross slip is given by

$$\tfrac{1}{2}m_2 V_f^2 = A_{GS}$$

where

$$A_{GS} = \tfrac{1}{2} C k_e \delta_0^2$$

is the area under the friction force displacement curve up to gross slip and $C =$ factor compensating for the nonlinearity of the force–displacement relationship.

A dimensionless friction force–displacement curve is plotted in Fig. 13.7 according to Mindlin's theory. The graph shows $F/F_0$ versus $\delta/\delta_0$ as represented by

$$\frac{\delta}{\delta_0} = 1 - \left(1 = \frac{F}{F_0}\right)^{2/3} \tag{13.6}$$

$k_e$ is also calculated for the case under consideration:

$$k_e = 2.27 \times 10^5 (N)^{1/2} \tag{13.7}$$

and $C$, which is a dimensionless factor defined by Eq. (13.5), is found from Fig. 13.7 as

$$C = \frac{\text{Area onabo}}{\text{Area oabo}} = 1.228 \tag{13.8}$$

Because there is no way to detect gross slip in this investigation except by means of the permanent displacement of the ball, Mindlin's theory is again utilized to calculate this value. This permanent displacement within the region of no gross slip is given by:

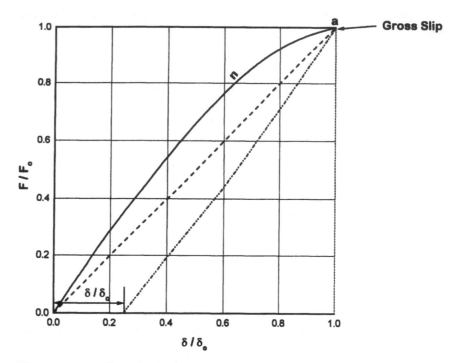

**Figure 13.7** Dimensionless displacement of ball within the region of no gross slip (Mindlin's theory).

$$\frac{d}{\delta_0} = 2\left(1 - \frac{F}{2F_0}\right)^{2/3} - \left(1 - \frac{F}{F_0}\right)^{2/3} - 1 \qquad (13.9)$$

For gross slip $F/F_0 = 1$ and substitution in Eq. (13.9) gives:

$$\frac{d_0}{\delta_0} = 0.26 \qquad (13.10)$$

In this case:

$$\tfrac{1}{2}m_2 V_f^2 = 5.075 \times 10^{-6} v^2 = \tfrac{1}{2} C k_e \delta_0^2 + (2N\mu_k)(d - 0.26\delta_0)$$

which can be written as:

$$\mu_k = \frac{5.075 v^2 - 0.1392(N)^{1/3}\delta_0^2}{2N(d - 0.26\delta_0)} \qquad (13.12)$$

or in the form:

$$d = \frac{5.075v^2 - 0.1392(N)^{1/3}\delta_0^2 + 0.52\mu_k N\delta_0}{2\mu_k N} \tag{13.13}$$

Equations (13.11), (13.12), and (13.13) are utilized for evaluation of the frictional characteristics of the joint from the experimental data as explained in the following.

### 13.2.4 Discussion of Results

Figure 13.8 shows a plot of the permanent ball displacement $d(\mu\text{in.})$ versus the striker velocity $v\,(\text{in./sec})$ for low values of impact under various normal forces. Equation (13.11) representing the conditions for gross slip is also plotted on the graph. The intersection of these lines for each normal force with the corresponding experimental curve gives permanent displacement $(d_0)$ and the striker velocity $(v_0)$ at gross slip. The peak displacement $\delta_0$ for gross slip can therefore be obtained from the equation:

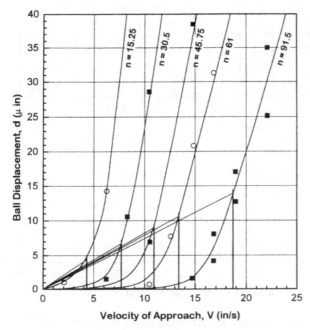

**Figure 13.8** Evaluation of conditions at gross slip.

$$\delta_0 = \frac{d_0}{0.26} \text{ in.}$$

and the tangential force for gross slip is therefore:

$$F_0 = k_e \delta_0 = 2.27 \times 10^5 N^{1/3} \delta_0 \times 10^{-6}$$

from which the coefficient of friction for gross slip is:

$$\mu_s = \frac{F_0}{2N} = \frac{2.27 \times 10^5 N^{1/3}}{2N} \delta_0 = 0.437 \frac{d_0}{N^{2/3}} \tag{13.14}$$

By substituting the value of $d_0$ from Fig. 13.8 in Eq. (13.14) the coefficient of friction for gross slip is readily calculated. They are plotted versus the striker velocity in Fig. 13.9. The average value for this coefficient is found to be $\mu_s = 0.305$.

Equation (13.12) is now used to calculate the kinetic coefficient of friction $\mu_k$ (the average value of the coefficient of friction during slip). For any particular normal force, the value of $\delta_0$ at gross slip is substituted in the equation. Values of $v$ in the region beyond gross slip and the corresponding

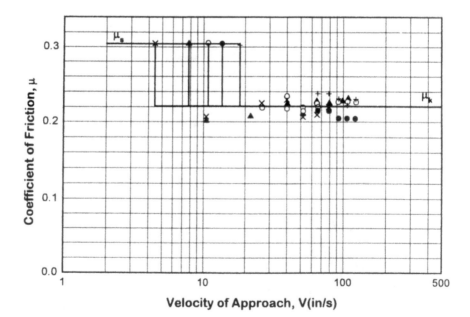

**Figure 13.9**  Values of static and kinetic coefficients of friction. $\times$, $N = 15.25\,\text{lb}$; $\blacktriangle$, $N = 30.5\,\text{lb}$; $\bigcirc$, $N = 45.75\,\text{lb}$; $\bullet$, $N = 61.0\,\text{lb}$; $+$, $N = 91.5\,\text{lb}$.

values of $d$ are then used to calculate $\mu_k$. This kinetic coefficient of friction is also plotted against the striker velocity $v$ in Fig. 13.9. Its value is found to be $\mu_k = 0.22$ and is independent of the normal force.

Equation (13.13) is then used to obtain a theoretical plot of $d$ versus $v$ beyond gross slip. This is shown by the solid lines in Fig. 13.10. The correlation between the calculated curves and the experimental data is evident.

The peak force and the time duration of the impact corresponding to gross slip are listed in Table 13.3 as a function of the normal force. It can be

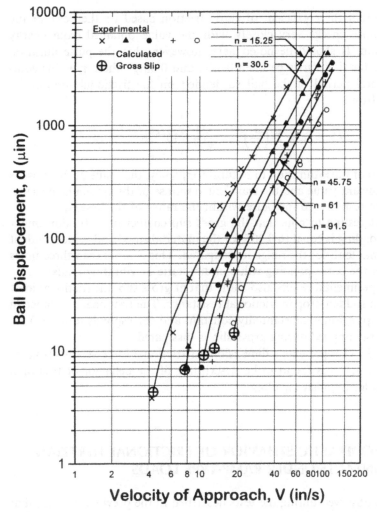

**Figure 13.10** Ball displacement for impacts causing gross slip. (From Ref. 10.)

**Table 13.3**  Pulse Characteristics

| $N$ (lb) | $v_0$ (in./sec) | $v_0^{1.2}$ | $v_0^{0.2}$ | $F_{max}$ (lbf) | $\tau$ ($\mu$sec) |
|---|---|---|---|---|---|
| 15.25 | 4.35 | 5.75 | 1.321 | 9.57 | 74.1 |
| 30.5 | 7.75 | 11.7 | 1.511 | 19.47 | 64.8 |
| 45.75 | 10.95 | 17.5 | 1.600 | 29.10 | 61.2 |
| 61.0 | 13.4 | 22.6 | 1.686 | 37.6 | 58.8 |
| 91.5 | 18.7 | 34.0 | 1.818 | 55.6 | 53.9 |

easily seen that the coefficient of static friction based on the peak of the Hertzian pulse checks very closely with the value obtained from energy consideration. It should be noted here, however, that the pulse duration in this investigation varies between $0.35\tau_n$ and $0.36\tau_n$, where $\tau_n$ is the equivalent natural period of the ball suspended on the frictional support, as calculated from:

$$\tau_n = 2\pi\left(\frac{m_2}{K}\right)^{1/2} = \frac{324}{N^{1/6}} \times 10^{-6} \text{ sec}$$

At this ratio of pulse duration to natural frequency, the shape of the pulse is not of significant value and the transient response of the system is the same as the static response to the pulse within the range of this test [13].

It is interesting to note that the gross slip coefficient of friction under impulsive loading is equal to 0.305 for the materials used and is independent of the normal load for the range investigated. This is more than three times higher than the corresponding value under static or vibrating loads.

The experimental results also show that at gross slip, the frictional joint undergoes a sudden drop in frictional resistance. The frictional resistance in the gross slip region is substantially constant and corresponds to $\mu_k = 0.22$. All the energy applied during gross slip is dissipated.

Figure 13.11 represents a dimensionless frictional force versus displacement plot which was found to be descriptive of the behavior of frictional contacts under impulsive loading.

## 13.3  VISCOELASTIC BEHAVIOR OF FRICTIONAL HERTZIAN CONTACTS UNDER RAMP-TYPE LOADS

It has been observed during the tests described in the previous section, that considerable slip of a "creep" nature may occur under sustained loads with

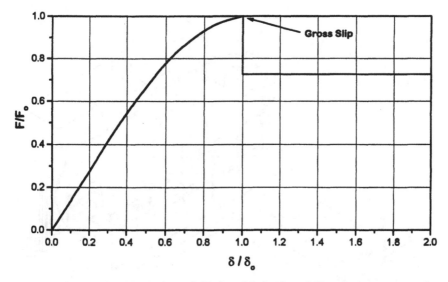

**Figure 13.11** Representation of frictional behavior of Hertzian contacts under impulsive loading.

values below those necessary to produce ball accelerations which characterize gross slip under such conditions.

A phenomenon of "creep" in the frictional contact between a hemisphere of lead and glass flats was detected by Parker and Hatch [14] during their studies on the nature of static friction. Similar observations were also reported by Bristow [15].

The investigation described in this section, was therefore undertaken to study this "creep" phenomenon in the region below gross slip in static friction tests when the tangential loads are applied at relatively low rates and allowed to dwell for relatively short periods. The utilized specimens and surface conditions are the same as those tested previously under vibratory and impulsive loads [16].

### 13.3.1 Experimental Arrangement

The apparatus used in this investigation is schematically represented in Fig. 13.12. A $1\frac{1}{4}$ in. diameter steel ball (a) is suspended between the two parallel flat surfaces of two identical hear steel inserts (b), (c). Insert (b) is fastened to a rigid steel frame (d), whereas (c) is fastened to a solid steel block (e) which can slide tightly with minimum friction in the frame under the influence of air pressure acting on a flexible diaphragm (f). The air pressure is controlled

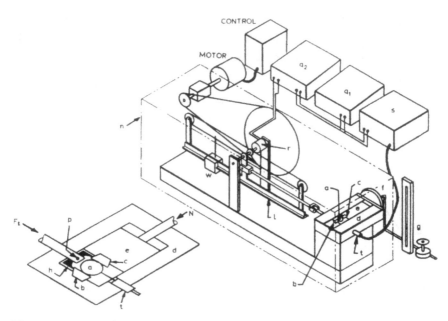

**Figure 13.12**  Diagrammatic representation of the test arrangement.

by a regulator and measured by means of a mercury manometer (g). Calibration of the normal force applied on the ball versus manometer reading is checked periodically by replacing the ball by a ring-type strain gage force meter of $1\frac{1}{4}$ in. outer diameter. The tangential load on the ball is applied by a special loading device capable of different rates of load application ranging from 0.1225 lb/sec to more than ten times this value. The device is composed of a lever (l) carrying a known weight (w) which can be moved on the lever by means of a string on a rotating drum. The rotation of the drum is controlled by pulleys driven by a variable-speed motor. The position of the weight from the center of the lever is indicative of the tangential force on the ball. This can be easily detected by the rotation of the drum. A variable resistance (r) connected to the drum was used in conjunction with a 6 V DC battery to produce a volage which is calibrated to indicate the tangential force on the ball. The calibration was done by utilizing the ring-type strain gage force meter.

The tangential force can be either applied to the ball or to the frame at the inserts by removing or inserting a pin (p) which disengages a special fork (h) to apply the load to the ball or to the frame, respectively. This arrangement makes it convenient to evaluate the apparatus deformations and hysteresis under any particular test condition before applying the tangential load. The ball displacement is measured by a differential transformer-type displacement transducer (t) rigidly fastened to the frame with the movable core in contact with the ball under a 12 g preload. The transducer excitation and signal amplification is provided by a carrier-type preamplifier (s) coupled to a power amplifier which provides the input to the $x$–$y$ plotters $(q_1, q_2)$.

Accurate alignment is provided, ensuring that the load on the sphere is on the same axis as the displacement transducer. The whole apparatus is enclosed in a plastic box (n) which is thermostatically controlled to within $\pm 1°$F.

Two $x$–$y$ plotters were used simultaneously to record the load–displacement and the displacement–time behavior of the ball under different load regimes.

As a general procedure, the apparatus was subjected to two consecutive hysteresis loops corresponding to the highest level of tangential load in all tests (approximately 12 lb). A third load cycle, similar to the expected frictional cycle, was applied to the apparatus and recorded. This cycle was found to give a reproducible hysteresis loop for the apparatus itself. The pin (p) was withdrawn from the loading fork (h) and the particular load regime is then applied to the ball.

The following are samples of the tests performed in this investigation. In the test illustrated in Fig. 13.13a, the ball was subjected to a 2 min dwell at a load level of approximately 0.87 the gross slip value. The load was then released and reapplied until gross slip occurred. The loading, unloading, and reloading up to gross slip were performed at the same rate. The ball was repositioned and subjected to six successive hysteresis loops after a reproducible apparatus loop was obtained. In the seventh loop the load was again sustained for 2 min at 0.87 the gross slip value, after which the load was released and reapplied until gross slip occurred. The figure shows strain-hardening effects in the successive loops with no significant change in the "creep" displacement during the 2 min dwell.

As shown in Fig. 13.13b, the 2 min dwell tests were also performed at different load levels. In each of the these tests, the load was sustained after two successive load cycles. The load was released at the same rate, after which the ball was loaded to gross slip. The data show an exponential increase in the creep displacement as the dwell load approaches the gross slip value.

Figure 13.13c shows typical results from tests where the ball was subjected to several successive 2 min dwell cycles. It can be seen that the "creep" displacement diminished with successive cycles. The decay rate was found to be more pronounced in the early cycles.

Figure 13.14 shows typical time–displacement tests in which the load was applied at the particular rate and allowed to dwell at different points

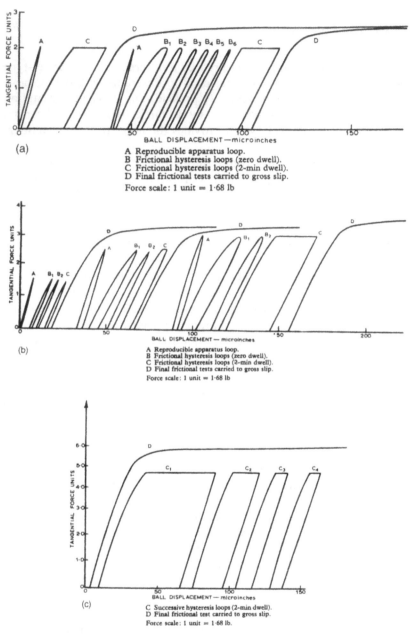

(a)
A Reproducible apparatus loop.
B Frictional hysteresis loops (zero dwell).
C Frictional hysteresis loops (2-min dwell).
D Final frictional tests carried to gross slip.

Force scale: 1 unit = 1·68 lb

(b)
A Reproducible apparatus loop.
B Frictional hysteresis loops (zero dwell).
C Frictional hysteresis loops (2-min dwell).
D Final frictional tests carried to gross slip.

Force scale: 1 unit = 1·68 lb

(c)
C Successive hysteresis loops (2-min dwell).
D Final frictional test carried to gross slip.

Force scale: 1 unit = 1·68 lb.

**Figure 13.13** (a) Load–displacement curves with 18.5 lbf normal force. (b) Frictional loops at different load levels with 30 lb normal force. (c) Successive dwell loops at the same load level with 50 lb normal force. A: reproducible apparatus loop; B: frictional hysteresis loops (zero dwell); C: frictional hysteresis loops (2 min dwell); D: final frictional tests carried to gross slip. (Force scale: 1 unit = 1.68 lbf. Ball displacement: μin.)

*510*

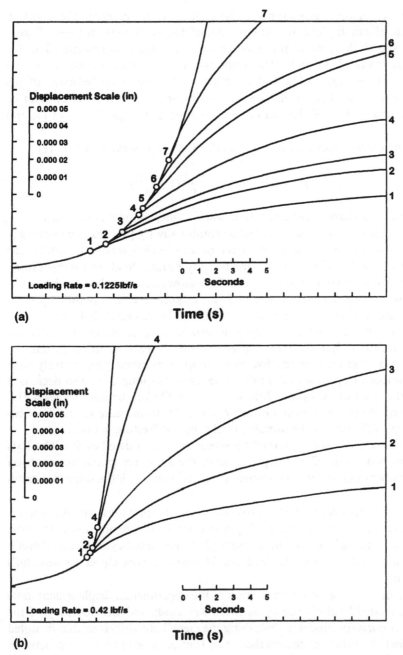

**Figure 13.14** Displacement–time curves for 30 lb normal load. (a) 0.1225 lb/sec loading rate; (b) 0.42 lb/sec loading rate.

within the region of gross slip. The dependency of the creep displacement on the rate of ball displacement at the onset of dwell can be readily seen. Figure 13.15 shows the plot of this relationship from the experimental data for normal loads of 30 and 50 lb, respectively. The experimental results, as illustrated by Figs 13.13a and b, indicate that the creep behavior of the frictional contacts (as in most low-temperature instances of creep) can be approximated by a Boltzmann model as shown in Figs. 13.16, 13.17, and 13.8.

The creep displacement can, therefore, be represented by [17, 18]

$$\chi_c = X_0(1 - e^{-Ct}) = \frac{(\dot{\chi}_c)_i}{C}(1 - e^{-Ct})$$

where $C$ is a characteristic constant when a linear model is assumed.

The factor $C$ has been evaluated empirically by plotting the slope $(\dot{\chi}_c)_i$ of the displacement–time curve at the onset of creep versus the total creep displacement $X_0$. This can be done for any normal load, test temperature, and rate of load application. Figure 13.15 shows an example of such plots at 80°F with normal loads of 30 and 50 lb, respectively. The points represent data from load application rates of 0.1225, 0.2667, and 0.42 lb/sec.

The repeated "no-dwell" cycling tests at load levels below gross slip showed clearly that the largest plastic displacements occurred during the first cycle. The area of the hysteresis loop diminished progressively with the number of cycles and the rate of decay of the loop area also decreased with the number of cycles. Johnson [4] and O'Connor and Johnson [19] observed the phenomenon and attributed it to an increase in the local coefficient of friction in the annulus of slip by the fretting action.

These effects were observed by successive 2 min dwell cycles. The tests also showed no significant dependence of the 2 min creep displacements on the number of no-dwell cycles which preceded them with the same maximum load.

Gross slip curves, when produced without previous cycling history, exhibited considerably higher displacements in the region close to gross slip than expected by Mindlin's theory [2, 3]. Repeated cycling caused strain hardening and brought the load–displacement curves closer to Mindlin's prediction.

Figure 13.17 shows the deviation of the experimental displacement–time curves from Mindlin's theory at the lower loads and the accuracy of the linear viscoelastic model in describing the creep behavior. It should be noted here that the values of the coefficient of friction at gross lip varied between 0.09 and 0.095 in most of the tests. These values are essentially the same as those in the previous tests with vibratory loads.

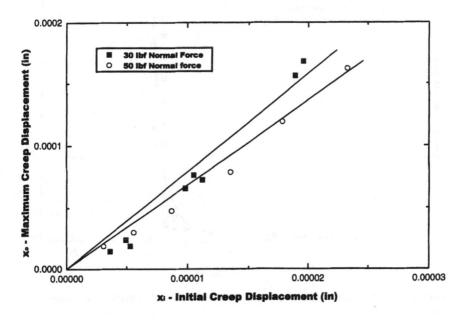

**Figure 13.15** Maximum creep versus initial creep rate.

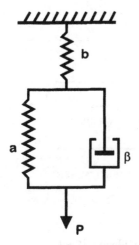

**Figure 13.16** Boltzmann model.

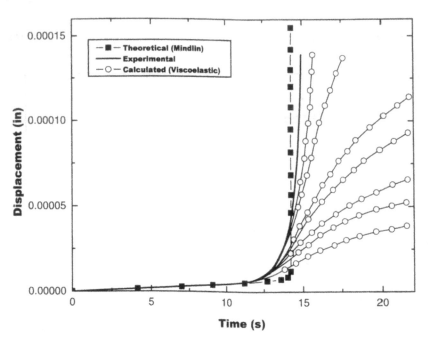

**Figure 13.17** Comparison between experimental and calculated displacement-time curves.

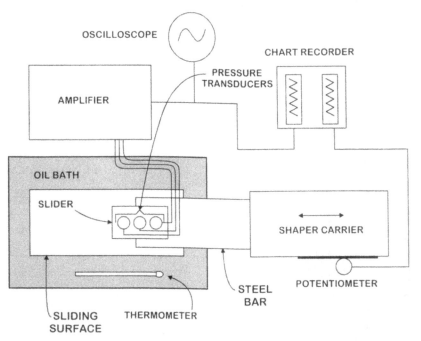

**Figure 13.18** Diagrammatic representation of the experimental setup.

## 13.4 FILM PRESSURE IN RECIPROCATING SLIDER BEARINGS

An experimental procedure for evaluating the oil film pressure in reciprocating slider bearings with arbitrary geometry is presented in this section
[20]. A special test fixture is constructed where the slider is inserted in such a
way as to insure that a specific film geometry is achieved and maintained
throughout the test. The pressure is monitored at three different locations
along the central line of the slider by means of miniature pressure transducers. The setup is capable of producing oscillatory sliding motions with
different strokes and speeds and can be used to simulate a variety of film
geometries and operating conditions. The results indicate the development
of negative pressure and cavitation during the return stroke. The pressure
distribution during the forward stroke follows the same pattern as that
predicted by isoviscous theory. The magnitude of the pressures, however,
can be higher or lower than the isoviscous values depending on the operating conditions. The peak pressure during the forward stroke follows a
square root relationship with the instantaneous sliding velocity, which can
be attributed to the thermohydrodynamic phenomenon discussed in
Chapter 6.

### 13.4.1 Experimental Setup

The experimental setup is diagrammatically represented in Fig. 13.18. The
main components of the setup and a brief explanation of their functions are
listed below:

A shaper carrier is used for providing the reciprocating motion, the
   speed and stroke are adjustable.

A slider block is used for holding the slider in order to maintain the film
   geometry constant relative to the sliding surface. The geometry is
   illustrated in Fig. 13.19.

A steel bar is used for transmitting the reciprocating motion from the
   carrier to the slider. It also serves as a cantilever spring to force
   down the brass surface of the slider block against the sliding surface.

The slider surface is a well-ground steel plate. The slider block is submerged in an oil bath. The temperature of the bath is monitored by
   a thermometer throughout the test.

A potentiometer is used to record the position of the slider.

A two-channel oscilloscope is used for monitoring the signals from the
   pressure transducers.

A multichannel chart recorder is used for recording the signals.

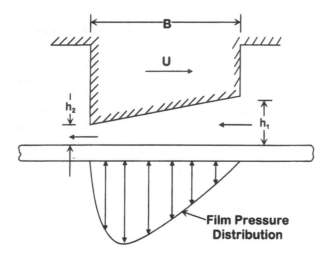

**Figure 13.19**  Geometry of the tapered slider bearing.

Three temperature-compensated strain gauge transducers with a range
from 0 to 100 psi are used for pressure sensing.

The minimum film thickness and slope of the slider is easily adjusted by
changing the metal shims between the slider and the slider block. Three oil
holes of 0.0315 in. diameter, which are located on the slider surface, allow
the transducers to pick up the pressure. The potentiometer is utilized to
provide the information on change of position, and speed as a function of
time. For all the given data, the same film geometry is used with two speeds,
18 and 37 strokes per minute (spm), respectively. The 18 spm produces a
maximum speed of 9.17 in./sec when the slider is moving forward with an
11 in. stroke. The maximum speed corresponding to 37 spm is 18.07 in./sec
with the same stroke. The oil bath temperature was maintained at
$25 \pm 0.5°C$.

### 13.4.2  Experimental Results

A sample of the recorded data is shown in Fig. 13.20. The figure shows the
pressure–position data for the test slider bearing when the SAE 5 oil is used
at a speed of 18 spm. In this case, the minimum film thickness is 0.006 in.
and the slope is 0.0009 in./in.

It can be seen that for the first stroke, at the second and third oil holes,
an oscillatory pressure drop occurs before the peak pressure is reached. A
sample of the pressure data from the three transducers is given in Table 13.1

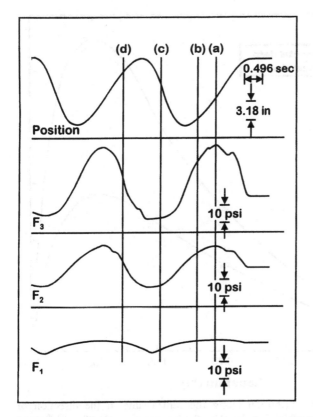

**Figure 13.20** Sample experimental data: $h_2 = 0.0006$ in., $m = 0.0009$ in./in., 18 spm, SAE 5 oil, 25°C.

at four different instances in time, as indicated by (a), (b), (c), and (d) in Fig. 13.20. The corresponding instantaneous velocities of the slider are also tabulated.

It was noticed during the tests that the peak pressure for each stroke dropped gradually as the test continued. This can be attributed to the formation of air bubbles which are generated during the backstroke and accumulate inside the tube leading to the pressure transducer. Before each test, the transducer is filled with oil to ensure that there are no bubbles in it. For the same test conditions, the pressure generated in the first stroke is found to be reproducible, and the peak pressure of that stroke can be used to represent the pressure corresponding to the maximum speed of the slider.

Figures 13.21 and 13.22 show sample results of the pressure distributions along the central line of the slider bearing in the direction of motion.

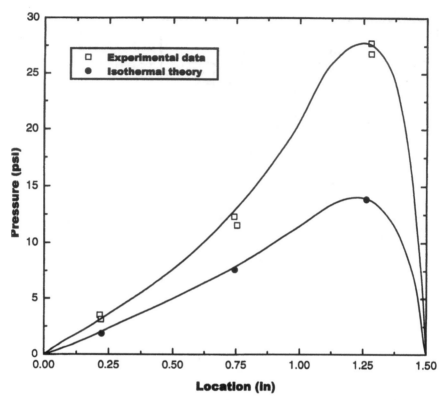

**Figure 13.21** Pressure distribution along the central line in the direction of motion: $h_2 = 0.0006$ in., $m = 0.0009$ in./in., $V = 9.17$ in./sec., $\mu = 31.03 \times 10^{-6}$ reyn.

The experimental results are compared to those predicted by isothermal theory, which takes into consideration the temperature rise in the bearing for the given film geometry, oil, and inlet temperature. It is noticed that the magnitudes of the pressures measured may differ considerably from those predicted by the isothermal hydrodynamic theory. However, the calculated and measured pressures retain the same normalized shape for any given film geometry, oil, and sliding speed.

It is worth noting that the isothermal theory can either underestimate (refer to Fig. 13.21) or overestimate the preak pressure (refer to Fig. 13.22). This has been found to depend on the values of minimum film thickness, viscosity, and speed in a similar maner as discussed in Chapter 6.

Figures (13.23 and 13.24) show the pressure–speed characteristic of test slider bearings. It can be seen that the pressure is proportional to the square root of the speed as was the case for journal bearings.

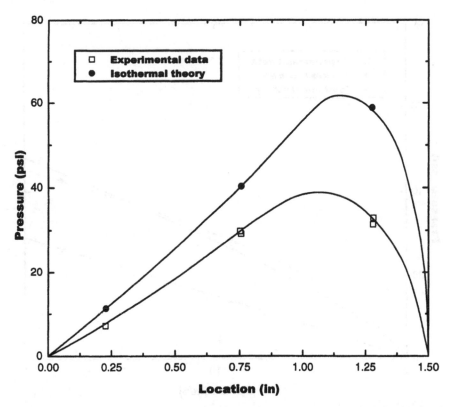

**Figure 13.22** Pressure distribution along the central line in the direction of motion: $h_2 = 0.0010$ in., $m = 0.0011$ in./in., $V = 18.01$ in./sec., $\mu = 1.986 \times 10^{-5}$ reyn.

## 13.5 EFFECT OF LUBRICANT PROPERTIES ON TEMPERATURE AND WEAR IN SLIDING CONCENTRATED CONTACTS

The experimental study discussed in this section deals with investigating the effect of some of the physical properties of lubricants on the contact temperature and wear in heavily loaded Hertzian contacts under sliding conditions [21]. The surface temperature and wear in a rotating mild steel shaft are measured under different loads applied by a tungsten carbide slider. The carbide tip and the shaft are used as part of a dynamic thermocouple system to monitor the contact temperature. Tests are conducted for Hertzian pressures ranging from 1250 to 2140 MPa ($1.81 \times 10^5 - 3.10 \times 10^5$ psi) and sliding speeds from 0.4 to 1.3 m/sec (943–3142 in./min). Temperature and wear

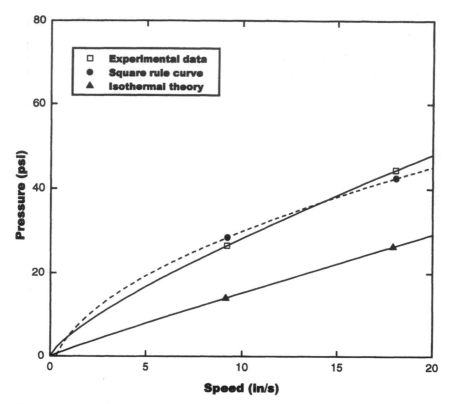

**Figure 13.23** Pressure–speed characteristic of slider bearing: $h_2 = 0.006$ in., $m = 0.009$ in./in., SAE 5 oil, 25°C.

data are given from tests with a heavy duty oil (SAE 80W-90), a high-viscosity residual compound, a vegetable oil, and water-miscible cutting fluid (0.0476% emulsifiable oil by volume). The results show that, for the considered tests, viscosity does not appear to be the significant property of the lubricant temperature rise and wear rate as indicated by the scar depth under similar test conditions.

### 13.5.1   Experimental Setup

The experimental setup used in this study is illustrated diagrammatically in Fig. 13.25. The supporting structure consists of a steel plate (1), pillowblock (2), and loading screws (3). The rotating shaft (4) is supported on bearings and the desired load is applied on it by the loading screws which are mounted on the steel plate and apply the load to the pillowblocks. The

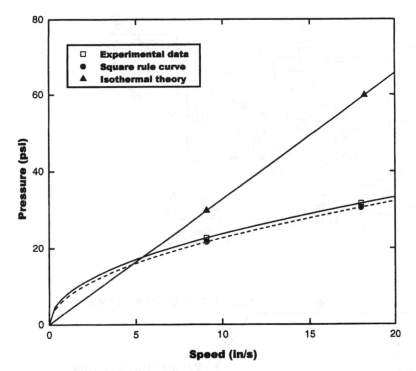

**Figure 13.24** Pressure–speed characteristic of slider bearing: $h_2 = 0.0010$ in., $m = 0.0011$ in./in., SAE 20 oil, 25°C.

steel plate and the loading screws are electrically insolated from the steel shaft to reduce possible noise.

A 5 hp variable speed motor (5) supplies the power necessary to rotate the shaft at the desired rotating speed. It has a speed range from 715 to 5000 rpm. A belt drive (6) with a speed ratio 2.4 is used between the test shaft and motor, making the range of the test from 300 to 2000 rpm.

A strain gage type load cell (7) supports the tungsten carbide tip and provides a continuous record of the load.

A dynamic thermocouple circuit is used to monitor the contact temperature as shown in Fig. 13.25a and illustrated in detail in Fig. 13.25b. The main junction is the low carbon steel shaft (4) (AISI 1020) and the tungsten carbide tip (8) (80% WC, 8%Co). Junction $X$ is kept at constant temperature by running cooling water through it, whereas the other $Y$, which has the least effect on the calibration, is kept at room temperature (20°–24°C). Due to the influence of the variable temeprature of junction $Y$, the system may have an error of approximately 1%.

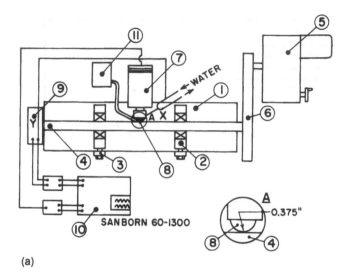

(a)

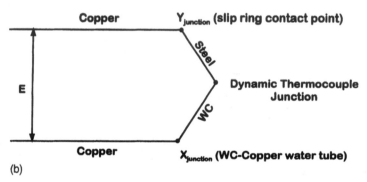

(b)

**Figure 13.25**  (a) Diagrammatic representation of the experimental setup. (b) Thermocouple circuit.

The slip ring (9) is used to transmit the thermoelectric signal from the rotating shaft.

A two-channel recorder (1) is used to simultaneously record the load and the contact temperature. The latter is calibrated in a bath of the lubricant using a mercury thermometer. The rotational speed of the shaft is measured by strobe light.

A lubricant reservoir (11) is mounted at a prescribed height above the test setup. The flow rate of the lubricant is calibrated for the different fluid levels which are maintained within a certain range throughout each test to keep the flow rate approximately constant.

### 13.5.2 Test Specimens

The test specimen is selected as a steel shaft 25 mm (1 in.) diameter supported at a 250 mm (10 in.) span. The rider is tungsten carbide with a cylindrical surface 9.5 mm (0.375in.) radius and a modulus of elasticity of 551.6 MPa ($80 \times 10^6$ psi). The length of the shaft allows for several successive tests to be conveniently run on the same shaft. The flexibility of the shaft also helps in controlling the applied load and minimizing the load fluctuation during the rotation of the shaft or as a result of wear.

### 13.5.3 Test Lubricants

Four different types of lubricants were used in the performed experiments. Two of them represent heavy-duty lubricants, and the others are non-conventional fluids which are not generally used in common lubrication practices. The selected lubricants are as follows:

1. SAE 80W90 oil
2. A very-high-viscosity residual compound
3. Water-miscible cutting fluid (0.0476% emulsifiable oil by volume)
4. Vegetable oil

Lubricants 1, 3, and 4 were applied at the contact location at different flow rates varying from 0.35 ml/sec to 2.1 ml/sec. The residual compound was applied directly on the shaft by a brush because it is too thick to flow. Viscosity data at 38°C (100°F) for the different lubricants used are as follows: 200cSt for SAW 80W90 oil, 583 cSt for the residual compound 0.65 cSt for water, and 40 cSt for the vegetable oil.

### 13.5.4 Wear Scar Measurement

In all the previous tests, wear scar geometry was measured after the tests using a Talysuf model 3 (Taylor–Hobson) surface texture measuring machine.

The shaft is supported on the table by a set of V-blocks in the machine. The chart recorder is calibrated by using a calibration plate with certain scar depth. The wear scar is measured at 90° intervals around the circumference of the shaft by moving the stylus along the shaft surface in the axial direction.

### 13.5.5 Results

Typical data on temperature and wear for different conditions of load and speed are illustrated in the following. Figure 13.26a shows reuslts from experiments run with SAW 80W90 oil lubrication at a particular shaft speed under 45, 89, 133, and 178 N (10, 20, 30, and 40 lbf) in succession without changing the point of load application. The test speeds in this case are 300, 500, 750, and 1000 rpm. The point of load application is changed after each constant speed test to a new location on the shaft.

Summary temperature data from the test with residual compound and water-miscible cutting fluid lubrication are shown in Figs 13.26b and c, respectively.

Some of the wear scars from the previous tests at 300 and 1000 rpm are shown in Fig. 13.27. These scars are the result of six repeated applications of each of the four levels for 15 sec and turning the motor off until the temperature cools back to the ambient temperature. The total test duration corresponds to each scar is approximately 360 sec. The maximum scar depth data are given in Table 13.4. The depth of the wear scar appears to reflect the magnitude of the temperature rise in all the performed tests.

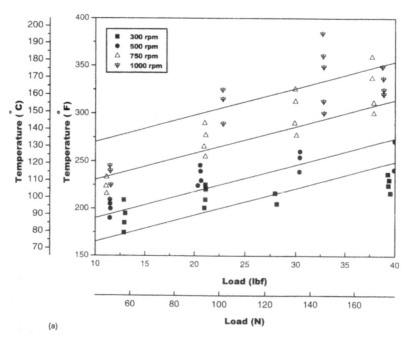

**Figure 13.26** Experimental results of temperature versus load at different speeds: (a) SAE 80W90 oil; (b) residual compound; (c) water-miscible cutting fluid.

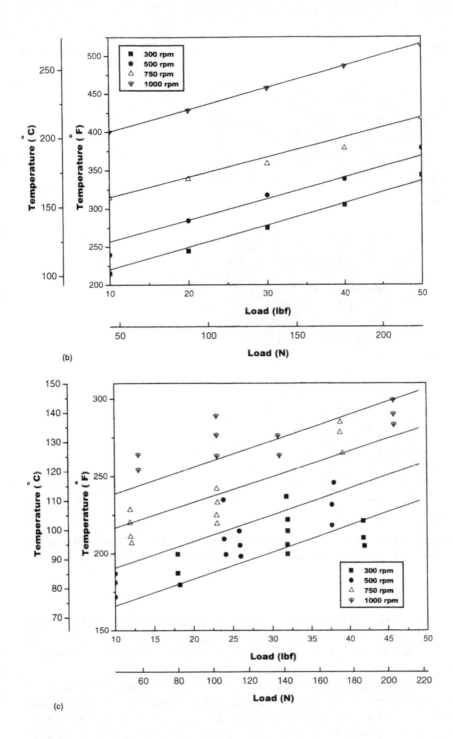

(b)

(c)

525

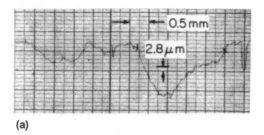

(a)

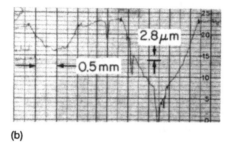

(b)

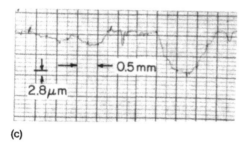

(c)

**Figure 13.27** Wear scar measured with successive load at 300 and 100 rpm with (a) SAE 80W90 oil; (b) residual compound; (c) water-miscible cutting fluid.

The measured temperature rise and wear under the same load and speed for all test conditions are the highest for the residual compound, lower for the SAW 80W90 oil, lower yet for the vegetable oil and lowest for the water-based solution.

This strongly suggests that viscosity is not the primary lubricant property in the performed tests. As can be seen in Table 13.5, it appears that thermal conductivity, specific heat, and $k\rho c$ parameter of the lubricant are significant physical characteristics whose effect on temperature rise and wear in the boundary warrants further careful investigation.

**Table 13.4** Summary Data on Wear Scar Depth

| Fig. no. | Lubricant | Load $N$ | Speed (rpm) | Total test duration | Maximum scar depth ($\mu$m) |
|----------|-----------|----------|-------------|---------------------|------------------------------|
| 13.26a | SAE 80W90 | 178 | 750 | 97 | 19.6 |
| 13.26b | Water-miscible cutting fluid | 178 | 750 | 126 | 11.2 |
| 13.26c | Vegetable oil | 178 | 750 | 96 | 16.8 |
| 13.27a | SAE 80W90 | 45, 89, 133, | 300 | 360 | 14 |
|  |  | 178 | 1000 | 360 | 38.4 |
| 13.27b | Residual compound | 45, 89, 133, | 300 | 360 | 16.8 |
|  |  | 178 | 1000 | 360 | 47.6 |
| 13.27c | Water-miscible cutting fluid | 45, 89, 133, | 300 | 360 | 8.4 |
|  |  | 178 | 1000 | 360 | 28 |

## 13.6 THE EFFECT OF REPEATED THERMAL SHOCK ON BENDING FATIGUE OF STEEL

Many tribological pairs are subjected to high thermal flux rates at the asperity contacts. This can result in transient localized temperature rise on the surface followed by sudden cooling by the lubricating oil. The thermal stress cycles produced by this repeated action can significantly affect the surface fatigue life, especially when high-carbon steel is used.

The study reported in this section [22] investigates the effect of repeated thermal shock on the bending fatigue of two types of high-carbon steels with the same chemical composition but with slightly different carbon content. Two levels of hardness for each type are used. The specimens are subjected to

**Table 13.5** Lubricant Properties

| Lubricant | Viscosity (cSt at 38°C (100°F)) | Thermal conductivity, $k$ (W/(m-°C)) | Density, $\rho$ (kg/m$^3$) | Specific heat $C$ (cal/(g-°C)) | $k\rho c$ |
|-----------|----------------------------------|--------------------------------------|----------------------------|---------------------------------|-----------|
| Residual compound | 583 | — | 1109.3 | 0.35 | — |
| SAE 80W90 | 200 | 0.133 | 913.2 | 0.44 | 53.44 |
| Vegetable oil | 40 | 0.168 | 921.2 | 0.4–0.5 | 134.33 |
| Water-miscible cutting fluid | 0.65 | 0.616 | 1000 | 0.998 | 614.77 |

40 repeated cycles of thermal shock and then subjected to different levels of unidirectional bending stress in order to establish stress–life curves. Identical tests are performed on specimens made of 1020 steel for comparison.

The setup for inducing the thermal cycles is schematically represented in Fig. 13.28. Standard bending fatigue test specimens (a) with 0.125 in. thickness are used throughout the test. Liquid nitrogen is allowed to flow continuously from a container (b) with a controlled rate into a channel (e) and directed to the point where rapid induction heating is applied in an intermittent fashion. This is accomplished by positioning the electrode (c) at an appropriate distance above the test point. The specimen is grounded by means of the clamp (g).

Calibration specimens equipped with four thermocouples placed at different locations through the thickness are shown in Fig. 13.29. They are used for monitoring the transient temperature through the thickness in the process of adjusting the arc voltage and the rate of flow of the liquid nitrogen to achieve repeatable thermal cycles of the desired range. The extrapolated temperature at the surface can be readily determined. A sample recording of such cycles is shown in Fig. 13.30.

The calibration specimen is then replaced by the test specimens and 40 cycles of thermal shock are applied. Aftewards, the specimen with induced thermal shock cycles is subjected to a set level of unidirectional bending stress in a standard bending fatigue tester until fracture occurs and the

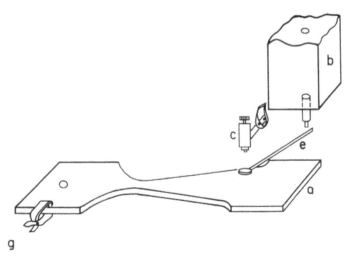

**Figure 13.28**  Setup for inducing thermal cycles.

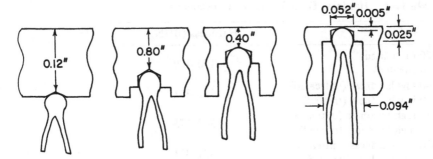

**Figure 13.29** Details of the thermocouples used for calibration.

number of cycles are recorded for the set level of bending stress. The specimens are checked for the location of crack initiation which invariably occurs at the point where the thermal cycles are applied.

Identical control specimens in each case were also tested without being subjected to thermal shock in order to determine the effect of the thermal cycles on the fatigue life.

The high-carbon steel materials tested are 4350 and 4340 steels which are heat treated to produce two levels of hardness in each case (refer to Table 13.6). The heat treatment is performed in such a way that the material hardness of the specimens is not changed for the temperature range considered in the tests.

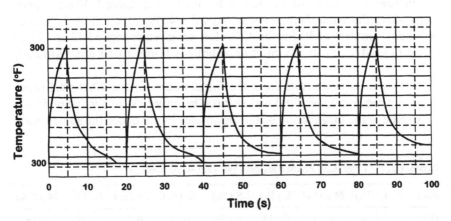

**Figure 13.30** Time history of thermal cycles induced at the selected point on the surface.

**Table 13.6**  Hardness Data for the Test Specimens

| Material | Hardness (BHN) |
|----------|----------------|
| 1020 (untreated) | |
| 1020 (treated) | 170–180 |
| 4340 low hardness (untreated) | |
| 4340 low hardness (treated) | 285–311 |
| 4340 high hardness (untreated) | |
| 4340 high hardness (treated) | 363–375 |
| 4350 low hardness (untreated) | |
| 4350 low hardness (treated) | 321–331 |
| 4350 high hardness (untreated) | |
| 4350 high hardness (treated) | 375–388 |

Similar tests were also performed on 1020 steep specimens for comparison.

The chemical composition of the 4340 and 4350 steel used in the test is given in Table 13.7.

Sample results of the unidirectional bending fatigue test data for the different materials with and without the 40 thermal cycles are given in Figs 13.31–13.35.

It can be seen in Fig. 13.31 that the high-hardness 4350 steel had an endurance limit of approximately 108 ksi, as would be expected from the published data on the material. The 40 thermal cycles in this case caused a reduction of approximately 45% in the value of the endurance limit.

Figure 13.32 shows similar results for the low-hardness 4350 steel specimens where the same thermal cycles caused a 42% reduction in the endurance limit. Similar results for the 4340 steel are given in Figs 13.33 and 13.34 where the reduction in the endurance limit is found to be approximately 20%.

It is interesting to note from the fatigue data for the 1020 steel specimens that the endurance limit was reduced by only 12% to a value of 55 ksi. This value is the mean value of the endurance limits under the same condi-

**Table 13.7**  Chemical Properties of 4340 and 4350 Steel

| Steel | C (%) | Mn (%) | P (%) | S (%) | Si (%) | Cr (%) | Ni (%) | Mo (%) |
|-------|-------|--------|-------|-------|--------|--------|--------|--------|
| 4340 | 0.38 | 0.77 | 0.014 | 0.026 | 0.27 | 0.78 | 1.59 | 0.23 |
| 4350 | 0.49 | 0.66 | 0.009 | 0.025 | 0.25 | 0.77 | 1.61 | 0.23 |

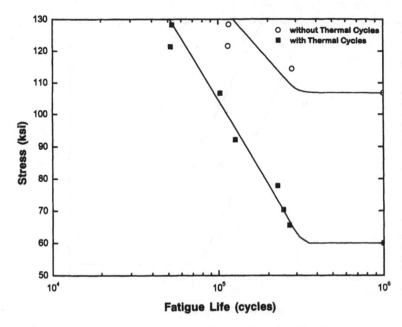

**Figure 13.31** Fatigue data for 4350 steel (high hardness): ○ = without thermal cycles; ■ = with thermal cycles.

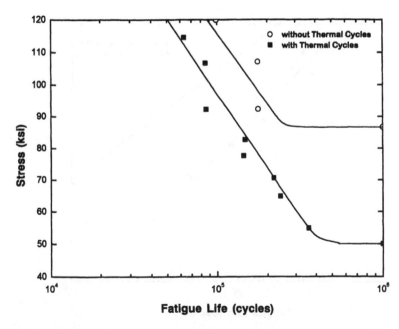

**Figure 13.32** Fatigue data for 4350 steel (low hardness): ○ = without thermal cycles; ■ = with thermal cycles.

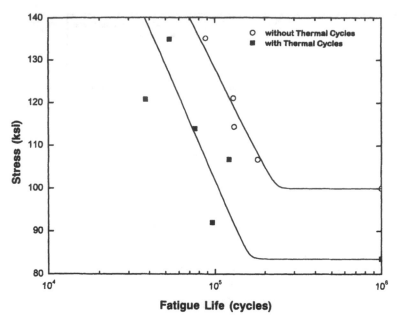

**Figure 13.33** Fatigue data for 4340 steel (high hardness): ○ = without thermal cycles; ■ = with thermal cycles.

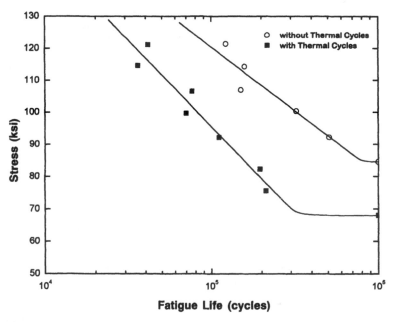

**Figure 13.34** Fatigue data for 4340 steel (low hardness): ○ = without thermal cycles; ■ = with thermal cycles.

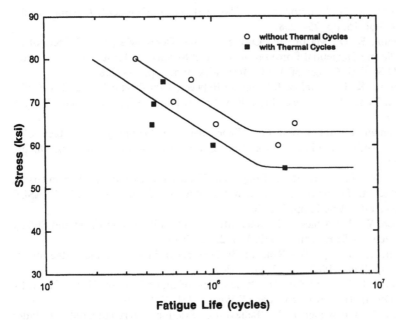

**Figure 13.35** Fatigue data for 1020 steel: ○ = without thermal cycles; ■ = with thermal cycles.

tions for the 4350 steel with high and low hardness. The results demonstrate the considerable deterioration of the endurance limit of high carbon steel due to thermal cycles.

It can be seen that the steel with higher carbon content exhibited considerably more reduction in the bending fatigue strength. Microhardness tests of the specimens showed no appreciable change in the hardness distribution due to the thermal or mechanical stress cycles. Microstructure investigations using the scanning electron microscope showed that microscopic thermal cracks as well as intergranular cracks occurred in the steel with higher carbon content, which may explain the reduction in the bending fatigue life.

## REFERENCES

1.  Seireg, A., and Weiter, E. J., "Frictional Interference Behavior Under Dynamic Excitation," Wear, 1963, Vol. 6, pp. 66–77.

2. Mindlin, R. D., "Compliance of Elastic Bodies in Contact," J. Appl. Mech., 1949, Vol. 16, pp. 259–268.

3. Mindlin, R. D., Mason, W., Osmer, T., and Deresiewiez, K., "Effects of an Oscillating Tangential Force on the Contact Surfaces of Elastic Spheres," Proc. 1st U.S. Natl. Congr. of Appl. Mechanics, 1951, pp. 203–208.

4. Johnson, K. L., "Surface Interaction Between Two Elastically Loaded Bodies Under Tangential Forces," Proc. Roy. Soc. (Lond.), 1955, Ser. A, Vol. 230, pp. 531–548.

5. Goodman, L., and Bowie, G., "Experiments on Damping at Contacts of a Sphere with Flat Plates," Proc. Soc. Exptl. Stress Anal., 1961, Vol. 18, pp. 48–54.

6. Klint, R. V., "Oscillating Tangential Forces on Cylindrical Specimens in Contact at Displacements Within the Region of No Gross Slip," ASLE Trans., 1960, Vol. 3, pp. 255–264.

7. Mason, W. P., "Adhesion Between Metals and Its Effects on Fixed and Sliding Contacts," ASLE Trans., 1959, Vol. 2, pp. 39–49.

8. Anderson, O. L., "The Role of Surface Shear Strains in the Adhesion of Metals," Wear, 1960, Vol. 3, pp. 253–273.

9. Gaylord, E. W., and Shu, H., "Coefficient of Static Friction Under Statically and Dynamically Applied Loads," Wear, 1961, Vol. 4, p. 401.

10. Seireg, A., and Weiter, E. J., "Behavior of Frictional Hertzian Contacts Under Impulsive Loading," Wear, 1965, Vol. 8, pp. 208–219.

11. Love, A. E. H., A Treatise on the Mathematical Theory of Elasticity, Cambridge University Press, London, England, 4th Edn, 1929, pp. 198–200.

12. Goldsmith, W., Impact, Edward Arnold., London, England, 1960.

13. Jacobsen, L. S., and Ayre, R. S., Engineering Vibrations, McGraw-Hill Book Co., New York, NY, 1958, p. 173.

14. Parker, R. C., and Hatch, D., "The Static Coefficient of Friction and the Area of Contact," Proc. Phys. Soc. London., 1950, Vol. 63 (B), p. 185.

15. Bristow, J. R., "Mechanism of Kinetic Friction," *Nature*, Lond., 1942, Vol. 149, p. 169.

16. Seireg, A., and Weiter, E. J., "Viscoelastic Behavior of Frictional Hertzian Contacts under Ramp-Type Loads," Proc. Inst. Mech. Engrs., 1966–67, Vol. 181, Pt 30, pp. 200–206.

17. Lubahn, J. D., and Felgar, R. P., Plasticity and Creep of Metals, Wiley, New York, NY and London, England, 1961.

18. Crussard, B. C., "Transient Creep of Materials," Paper 75, Ft. Int. Conf. on Creep, Inst. Mech. Engrs, London, 1963, Vol. 2, p. 123.

19. O'Connor, J. J., and Johnson, K. L., "The Role of Surface Asperities in Transmitting Tangential Forces Between Metals," Am. Soc. Mech. Engrs, Paper No. 62-Lub-14, 1962.

20. Wang, N. Z., and Seireg, A., "Thermohydrodynamic Performance of Reciprocating Slider Bearings," ASLE, Paper No. 87-AM-3A-3, 1987.

21. Seireg, A., and Hsue, E., "An Experimental Investigation of the Effect of Lubricant Properties on Temperature and Wear in Sliding Concentrated Contacts," ASME Trans., J. Lubr. Technol., April 1981, Vol. 103, pp. 261–265.
22. Seireg, A., and Wang, C. F., "The Effect of Repeated Thermal Shock on Bending Fatigue of High Carbon Steels," M. E. Report, University of Wisconsin–Madison, 1980.

# Author Index

# Subject Index